Methods in Molecular Biology™

Series Editor

John M. Walker
School of Life Sciences
University of Hertfordshire
Hatfield, Hertfordshire, AL10 9AB, UK

T0076445

For other titles published in this series, go to
www.springer.com/series/7651

Transgenesis Techniques

Principles and Protocols

Third Edition

Edited by

Elizabeth J. Cartwright

Cardiovascular Medicine, University of Manchester, Manchester, UK

Humana Press

Editor
Elizabeth J. Cartwright
Cardiovascular Medicine
University of Manchester
Manchester
UK

ISSN: 1064-3745 e-ISSN: 1940-6029
ISBN: 978-1-60327-018-2 e-ISBN: 978-1-60327-019-9
DOI: 10.1007/978-1-60327-019-9
Springer Dordrecht Heidelberg London New York

Library of Congress Control Number: 2009929336

© Humana Press, a part of Springer Science+Business Media, LLC 2009
All rights reserved. This work may not be translated or copied in whole or in part without the written permission of the publisher (Humana Press, c/o Springer Science+Business Media, LLC, 233 Spring Street, New York, NY 10013, USA), except for brief excerpts in connection with reviews or scholarly analysis. Use in connection with any form of information storage and retrieval, electronic adaptation, computer software, or by similar or dissimilar methodology now known or hereafter developed is forbidden.
The use in this publication of trade names, trademarks, service marks, and similar terms, even if they are not identified as such, is not to be taken as an expression of opinion as to whether or not they are subject to proprietary rights.

Printed on acid-free paper

Springer is part of Springer Science+Business Media (www.springer.com)

To Dan, Edward and William with love.

Preface

One of the major challenges currently facing the scientific community is to understand the function of the 20,000–25,000 protein-coding genes that were revealed when the human genome was fully sequenced. This book details the transgenic techniques that are currently used to modify the genome in order to extend our understanding of the in vivo function of these genes.

Since the advent of transgenic technologies, the mouse has become by far the most popular model in which to study mammalian gene function. This is due to not only its genetic similarity to humans but also its physiological and, to a certain extent, its anatomical similarities. Whilst a large proportion of this book is dedicated to the use of the mouse in transgenesis, the mouse is certainly not the only model to provide essential information regarding gene function. A number of other valuable models are used in transgenic studies including *Drosophila*, *C. elegans*, *Xenopus*, zebrafish, and rat. For each of these species, a chapter in this book is dedicated to highlighting how each is particularly suited, for example, to the study of embryonic development, physiological function of genes and to study orthologs of human disease genes. These chapters give detailed practical descriptions of animal production, construct design, and gene transfer techniques; recently developed methods will be described along with highly established classical techniques.

A number of chapters in this book are dedicated to the generation of genetically modified mice by the present classic techniques of injection of exogenous DNA into the pronuclei of fertilised eggs and by gene targeting using homologous recombination in embryonic stem cells. These chapters, as with all the others in the book, have been specifically written for this edition of *Transgenesis* and so contain up-to-date details of the practices in the field. Chapters are included describing optimal transgene and construct design, in-depth technical details for pronuclear microinjection of transgenes and associated surgical techniques, details for the optimal conditions in which to culture embryonic stem cells in order to maintain their pluripotent state, and methods for targeting these cells. A combination of chapters (Chaps. 13–15) describe how to generate chimaeras by microinjection of targeted ES cells into blastocysts or by morula aggregation, and the surgical techniques required to transfer the resulting embryos. For a number of years, the use of Cre/loxP and flp/frt recombination systems has gained in popularity; Chap. 16 describes their use and introduces other state-of-the-art site-specific recombination systems that can be used to manipulate the mouse genome. The generation and use of Cre-expressing transgenic lines are described in Chap. 17. One chapter of the book highlights the large-scale international efforts that are being made to systematically knockout every gene in the genome. The remaining chapters detail the breeding and husbandry skills required to successfully propagate a transgenic line and the increasingly essential methods for cryopreserving a mouse line and recovering lines from frozen stocks.

This book is a comprehensive practical guide to the generation of transgenic animals and is packed full of handy hints and tips from the experts who use these techniques on a day-to-day basis. It is designed to become an invaluable source of information in any lab currently involved in transgenic techniques, as well as for researchers who are newcomers to the field. This book also provides essential background information for scientists who work with these models but have not been involved in their generation.

On a personal note, it has been a great pleasure to edit this latest edition of *Transgenesis*. Firstly, I learnt many of my skills from reading earlier editions of the book and I hope that this edition will help and inspire many others. Secondly, I have been privileged to work with the exceptionally talented researchers in the transgenesis field who have contributed to this book.

Manchester, UK ***Elizabeth J. Cartwright***

Contents

Contributors

IGNACIO ANEGON • *INSERM – Institut National de la Santé et de la Recherche Médicale, Nantes, France*
J. SIMON C. ARTHUR • *MRC Protein Phosphorylation Unit, College of Life Sciences, University of Dundee, Dundee, UK*
KAZUHIDE ASAKAWA • *Division of Molecular and Developmental Genetics, National Institute of Genetics, Mishima, Shizuoka, Japan*
IVANA BARBARIC • *Department of Biomedical Science, University of Sheffield, Sheffield, UK*
MARIE-CHRISTINE BIRLING • *Institut Clinique de la Souris – Mouse Clinical Institute (ICS-MCI), Illkirch, France*
VALÉRIE BRAUN • *genOway SA, Lyon, France*
ELIZABETH J. CARTWRIGHT • *Cardiovascular Medicine, University of Manchester, Manchester, UK*
YACINE CHERIFI • *genOway SA, Lyon, France*
JEAN COZZI • *genOway SA, Lyon, France*
T. NEIL DEAR • *Leeds Institute of Molecular Medicine, St. James's University Hospital, Leeds, UK*
MARTIN FRAY • *Frozen Embryo & Sperm Archive (FESA), Medical Research Council, Mammalian Genetics Unit, Harwell, UK*
ROLAND H. FRIEDEL • *Institute of Developmental Genetics, Helmholtz Center Munich, Neuherberg, Germany*
WENDY J.K. GARDINER • *Mary Lyon Centre, Medical Research Council, Harwell, UK*
FRANÇOISE GOFFLOT • *Institut Clinique de la Souris – Mouse Clinical Institute (ICS-MCI), Illkirch, France*
ANNE-CATHERINE GROSS • *genOway SA, Lyon, France*
JAMES GULICK • *Molecular Cardiovascular Biology, Cincinnati Children's Hospital, University of Cincinnati, Cincinnati, OH, USA*
KOICHI KAWAKAMI • *Division of Molecular and Developmental Genetics, National Institute of Genetics, Mishima, Shizuoka, Japan*
HIROSHI KIKUTA • *Division of Molecular and Developmental Genetics, National Institute of Genetics, Mishima, Shizuoka, Japan*
CHRISTIAN KLASEN • *Transgenic Service, European Molecular Biology Laboratory, Heidelberg, Germany*
NIKOS KOURTIS • *Foundation for Research and Technology, Institute of Molecular Biology and Biotechnology, Heraklion, Crete, Greece*
JANA LOEBER • *Department of Developmental Biochemistry, University of Goettingen, Goettingen, Germany*

VICTORIA A. MCGUIRE • *MRC Protein Phosphorylation Unit, College of Life Sciences, University of Dundee, Dundee, UK*

CHRISTEL MERROUCHE • *genOway SA, Lyon, France*

FONG CHENG PAN • *Vanderbilt University Program in Developmental Biology and Department of Cell and Biology, Vanderbilt University Medical Center, Nashville, TN, USA*

ANGELA PASPARAKI • *Foundation for Research and Technology, Institute of Molecular Biology and Biotechnology, Heraklion, Crete, Greece*

TOMAS PIELER • *Department of Developmental Biochemistry, University of Goettingen, Goettingen, Germany*

ANNE PLÜCK • *Centre for Mouse Genetics, Institute for Genetics, University of Cologne, Cologne, Germany*

MATTHIAS RIECKHER • *Foundation for Research and Technology, Institute of Molecular Biology and Biotechnology, Heraklion, Crete, Greece*

LEONIE RINGROSE • *IMBA – Institute of Molecular Biotechnology GmbH, Vienna, Austria*

JEFFREY ROBBINS • *Molecular Cardiovascular Biology, Cincinnati Children's Hospital, University of Cincinnati, Cincinnati, OH, USA*

KAI SCHUH • *Institute of Physiology I, University of Wuerzburg, Wuerzburg, Germany*

MAXIMILIANO L. SUSTER • *Division of Molecular and Developmental Genetics, National Institute of Genetics, Mishima, Shizuoka, Japan*

NEKTARIOS TAVERNARAKIS • *Foundation for Research and Technology, Institute of Molecular Biology and Biotechnology, Heraklion, Crete, Greece*

LYDIA TEBOUL • *Mary Lyon Centre, Medical Research Council, Harwell, UK*

MELANIE ULLRICH • *Institute of Physiology I, University of Wuerzburg, Wuerzburg, Germany*

AKIHIRO URASAKI • *Division of Molecular and Developmental Genetics, National Institute of Genetics, Mishima, Shizuoka, Japan*

LUCIE VIZOR • *Medical Research Council, Harwell, UK*

XIN WANG • *Faculty of Life Sciences, University of Manchester, Manchester, UK*

XAVIER WAROT • *EPFL FSV – École Polytechnique Fédérale de Lausanne, Lausanne, Switzerland*

SARA WELLS • *Medical Research Council, Harwell, UK*

Part I

Transgenesis in Various Model Systems

Chapter 1

Transgenesis in *Drosophila melanogaster*

Leonie Ringrose

Summary

Transgenesis in *Drosophila melanogaster* relies upon direct microinjection of embryos and subsequent crossing of surviving adults. The necessity of crossing single flies to screen for transgenic events limits the range of useful transgenesis techniques to those that have a very high frequency of integration, so that about 1 in 10 to 1 in 100 surviving adult flies carry a transgene. Until recently, only random P-element transgenesis fulfilled these criteria. However, recent advances have brought homologous recombination and site-directed integration up to and beyond this level of efficiency. For all transgenesis techniques in *Drosophila melanogaster*, microinjection of embryos is the central procedure. This chapter gives a detailed protocol for microinjection, and aims to enable the reader to use it for both site-directed integration and for P-element transgenesis.

Key words: *Drosophila melanogaster*, Embryo, Microinjection, Transgenic, Recombination, Integration, Homologous recombination, phiC31/integrase, Site-directed integration, P-element

1. Introduction

Transgenesis in *Drosophila melanogaster* has undergone something of a revolution in the last few years. The classical technique of random P-element-mediated transgenesis has recently been supplemented by two novel technologies: homologous recombination and ΦC31 integration (for reviews, see *(1)* and *(2)*). In P-element transgenesis *(3)*, a modified transposon vector is used in combination with transient expression of the P transposase enzyme to generate several fly lines with different insertion sites in the genome. These insertions are subsequently mapped and characterised. P-element insertions have been invaluable for mutagenesis screens, but until recently, this was also the only

Elizabeth J. Cartwright (ed.), *Transgenesis Techniques*, Methods in Molecular Biology, vol. 561
DOI 10.1007/978-1-60327-019-9_1, © Humana Press, a part of Springer Science+Business Media, LLC 2009

method available for introducing a transgene of choice into the *Drosophila* genome. The random nature of P-element insertions has several drawbacks for transgene analysis. Mapping of insertion sites is time consuming, and transgene expression levels are subject to genomic position effects, making it difficult to draw comparisons between different constructs.

A recently developed alternative to random insertion is homologous recombination *(4, 5)*. This involves inserting a donor construct at random into the genome by P-element transgenesis, and in subsequent generations, mobilising the donor construct to the correct locus by homologous recombination. This technique had long been lacking to Drosophilists, but has not replaced P-element transgenesis as the method of choice for routine transgene analysis, because both the cloning of donor constructs and the generation of homologous recombinants are more time consuming than for P-element transgenesis.

Recently, ΦC31 integration has been developed *(6)*. This technique allows rapid and efficient generation of site-specific integrants, and relies upon 'docking site' fly lines, which carry a single recognition site (attP) for the phage ΦC31 integrase enzyme, previously introduced into the genome by P-element transgenesis. A donor plasmid carrying a second recognition site (attB) and a source of integrase enzyme is used to generate flies in which the donor plasmid docks to the genomic site. Integration events are highly specific, as the attP site is 39 bp long and does not occur at random in the *Drosophila* genome. Many mapped and characterised docking site lines are now available (*see* **Note 1**), and ΦC31 integration is rapidly becoming widely used for many transgenic applications.

All these transgenic techniques rely upon microinjection of embryos as a first step. In early *Drosophila* embryogenesis, the nuclei share a common cytoplasm for the first nine divisions. Directly after the tenth division, the first cells to become separated are the pole cells, which will later form the adult germ line. Transgenic animals are made by microinjecting DNA and a source of enzyme (P-transposase or ΦC31 integrase, *see* **Note 2**) into the posterior of the embryo where the pole cells will form, at an early stage before they have become separated from the common cytoplasm. DNA can enter the nuclei and is integrated into the genome of some cells. Embryos are allowed to mature and the adults are outcrossed to screen for transgenic flies in the next generation.

This chapter gives a detailed description of microinjection, from preparing DNA to screening for transformants. The main protocol deals with ΦC31 integration as we perform it in our laboratory. Alternatives for both ΦC31 and P element transgenesis are given in the notes.

2. Materials

2.1. Preparation of DNA

1. Donor plasmid containing attB site and transgene of interest (*see* **Note 3**).
2. Helper plasmid expressing ΦC31 integrase (*see* **Note 2**).

Midi- or miniprep kit for preparation of plasmid DNA (Qiagen).

3. Absolute ethanol.
4. 3 M NaOAc, pH 5.2.
5. Sterile distilled water.

2.2. Preparation of Injection Needles (see Note 4)

1. Capillaries: borosilicate glass capillaries, 1.2 mm × 0.94 mm
2. Needle puller: P-97 micropipette puller (Sutter instruments).
3. Needle grinder: Narishige microgrinder EG-400.

2.3. Preparation of Flies for Egg Laying

1. Fly line containing genomic attP site (*see* **Note 1**).
2. Fly bottles.
3. Fresh yeast paste: cubes of fresh baker's yeast cubes are obtainable from large supermarkets. They can be frozen and stored at −20°C for several months. Thaw at room temperature and mix with a little water to give a thick paste.
4. Dried yeast: Mix instant yeast granules with water to give a thick paste. Both fresh and dried yeast paste can be kept at 4°C for up to a week. Do not seal the container tightly, as the paste will expand.
5. Fly cages: PVC plastic tubing of either 50 mm or 90 mm diameter is cut into 100–150 mm sections and sealed at one end with nylon or metal mesh. The other end fits onto to a 50-mm or 90-mm agar plate, which is taped in place for egg collection.
6. Agar plates: Add 18 g agar to 600 mL tap water and bring to boiling point by microwaving. Dissolve 10 g sucrose in 300 mL tap water, heating a little if necessary. Add the sucrose solution to the agar, add 3.5 mL 100% acetic acid and mix well. Pour into petri dishes (90 mm or 50 mm) and allow to cool. Store for 1 day at room temperature to dry before using. Plates can be stored wrapped in plastic at 4°C for several weeks. About 16–20 plates per day of injection are required per cage (*see* **Note 5**).

2.4. Dechorionation and Dessication of Embryos

1. Filtration apparatus consisting of glass funnel, filter support, stopper, sidearm flask, and clamp, suitable for 50-mm membrane filters. Attach the apparatus to water tap as shown in **Fig. 1**.

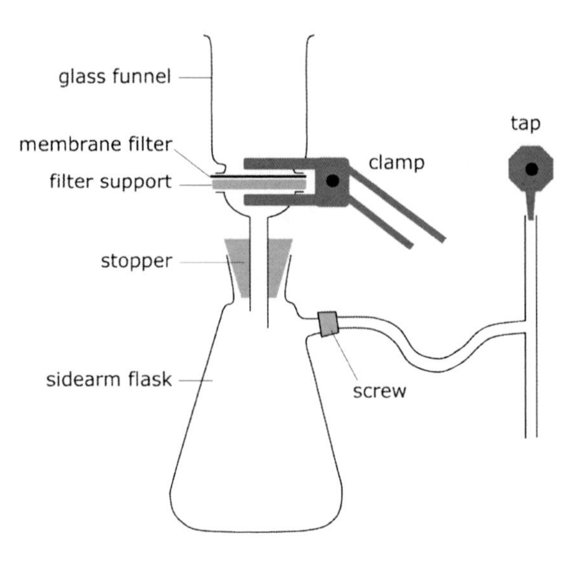

Fig. 1. Filtration apparatus.

2. Bleach solution: mix 50 mL household bleach (2.8% hypochlorite) with 50 mL sterile distilled water. Make fresh every day. Wear a lab coat and gloves when handling bleach, as it bleaches clothes upon contact and is harmful to skin.
3. Membrane filters: mixed cellulose ester membrane filters, black with white grid marking. Circular, 50-mm diameter, 0.6-μm pore size (Schleicher and Schuell, type ME 26/31 ST).
4. Binocular dissection microscope.
5. Fine stiff paintbrush with nylon hairs: cut away hairs until only a few remain, for use in aligning embryos.
6. Dissection needle.
7. Forceps.
8. Microscope slides: use slides with frosted part for labelling, such as Superfrost plus (Fisher).
9. Coverslips: 24 × 24 mm.
10. Embryo glue: Make three balls of 2.5-m Scotch tape Magic 810 (3 M). Add these to 30 mL heptane in a 50-mL falcon tube. Shake vigorously at 28°C for 24 h. Cut a hole in the bottom of the falcon tube and drain solution into a small glass bottle. This glue keeps for several months at room temperature (*see* **Note 6**).
11. Drying chamber: 150-mm petri dish containing orange self-indicating silica gel granules: check that the silica gel granules are orange; if they are not then they are saturated and no longer effective for drying embryos. Change to fresh granules.
12. Halocarbon oil: Voltalef 10S halocarbon oil, or halocarbon 700 oil (Sigma).

2.5. Microinjection of Embryos

1. Microscope: Either a compound or inverted microscope is suitable for injection. We use a Zeiss Axiovert 200 inverted microscope with ×10 objective and ×10 oculars.
2. Micromanipulator and needle holder (Narishige).
3. Microinjection system: Femtojet 5247 programmable microinjector with integrated pressure supply (Eppendorf) (*see* **Note 7**).
4. Microloader pipette tips (Eppendorf).

2.6. Further Handling and Screening for Transgenics

1. Humid box: sealable plastic sandwich box containing damp paper towels.
2. 50-mm Petri dishes.
3. 18 mm × 18 mm cover slips.
4. Flies for crossing to surviving adults: w- or appropriate balancer lines.
5. Fly vials.

3. Methods

3.1. Preparation of DNA

1. Prepare donor and helper plasmids in advance. Use midi- or miniprep (Qiagen quality) DNA. Do not elute the DNA in the buffer provided, as it contains Tris buffer, which is harmful to embryos. Instead, elute in sterile distilled water (*see* **Note 8**).
2. Check the concentration of eluted DNA. If the concentration is sufficient, make an injection mix at 250 ng/µL of donor vector plus 600 ng/µL of helper, in sterile distilled water (*see* **Note 9**).
3. If the DNA concentration is too low, precipitate the DNA: Add 0.1 volume of 3 M NaOAc, pH 5.2, and two volumes of absolute ethanol. Incubate at –20°C overnight. Centrifuge at 4°C for 10 min at 14,000 × *g*. Remove the supernatant, add 70% ethanol to the pellet, and centrifuge at 4°C for 5 min at 14,000 × *g*. Air dry the pellet and resuspend in sterile distilled water.
4. Plasmids and injection mixes can be stored indefinitely at –20°C. For DNA stored in water, however, the absence of a buffering agent may lead to degradation upon repeated freezing and thawing (*see* **Note 8**).

3.2. Preparation of Injection Needles

1. Before beginning to inject, prepare a supply of needles. We use a needle puller (P-97, Sutter instruments) with the following settings: Heat = 595; Pull = 70; Vel = 80). Insert a glass capillary into the needle puller, close the lid, and press 'pull'. This makes two needles from each capillary that are closed at the tapered end (*see* **Note 4**).

2. Open the needles by grinding in the needle grinder. Insert the needle into the holder at an angle of 40° to the grindstone. Keeping a constant flow of water over the grindstone, lower the needle onto the grindstone till the tip bends very slightly and water rises up into the needle. Immediately the water enters; stop moving the needle and allow to grind for 20 s (*see* **Note 9**).

3.3. Preparation of Flies for Egg Laying

1. Expand the fly line that is to be injected to give six bottles. Flip all six each week if large-scale injections are planned. Use flies that are 1-week old and well fed for the best egg laying.
2. One week before injection: flip adult flies every 2 days into bottles with fresh yeast paste. This feeds them optimally, so females lay a lot of eggs. Keep these bottles at 18°C.
3. Two days before injection: transfer flies to cages (use 4–6 bottles per 90-mm-diameter cage). Add a little dried yeast paste on a small square of paper (this facilitates later removal) onto the plates and place the cages at 25°C. Change the plates every 24 h and discard them. This acclimatises flies to the cage environment.
4. On the day of injection: Ensure that plates are at room temperature. Change the overnight plate, and wipe the inner rim of the cage to remove any first instar larvae. Add a very small spot of yeast paste on a square of paper to the centre of each new plate. Change the first plate after 1 h, and discard it. This is because females may keep fertilised eggs for some time before laying them. Use the subsequent plates for collections.
5. Change the plates every 30 min to ensure that embryos can be collected, prepared, and injected before the germ cells form. For an optimal injection workflow, flies should be laying about 200 eggs every 30 min.

3.4. Dechorionation, Lining Up, and Dessication of Embryos

1. Change the plate after 30 min laying, and remove the yeast and paper square. Set a timer for 2 min. Add bleach solution directly onto the plate and incubate for 2 min. Wear a lab coat to protect clothing from bleach.
2. Assemble the filtration apparatus with a fresh membrane filter as shown in **Fig. 1**. Clamp the apparatus together. Add water and filter through to wet the membrane. Start the tap and open the screw on the sidearm flask. After exactly 2 min, tip the bleach from the plates onto the filter. Close the screw just until the liquid goes through, and then open it again, to avoid damaging embryos.
3. Add water to the plates and filter in the same way. Add water to the filter and filter through. Always be aware that too much suction will damage embryos: open the screw on the sidearm flask as soon as the liquid goes through the membrane. Remove

the filter holder and filter from the apparatus; dry excess liquid with paper from underneath. Wash the glass cup to remove embryos sticking to the sides, so that they do not get collected the next time around (*see* **Note 10**).

4. Line up embryos. Use a fine paintbrush with a few hairs or a dissection needle to line up embryos in rows in the same anterior-posterior orientation. Leave a small space between embryos as shown in **Fig. 2a**. Aim for a regular line. The neater the line, the smoother and faster the subsequent injection. Each row can be up to 20 mm long, to fit on a coverslip. Do not line up for longer than 20 min, to ensure that embryos are not too old. With practice, it should be possible to line up 100–200 embryos in 20 min, making several rows of about 60–80 embryos each.
5. Make a line of embryo glue on the edge of a coverslip with a Pasteur pipette and allow to dry for 30 s. Using forceps, very gently touch the line of embryos with the glued edge to pick them up. Take care not to damage the embryos at this step (*see* **Note 6**).

Turn the cover slip and put it on a glass slide with a drop of water to stick the coverslip to the slide as shown in **Fig. 2b**.

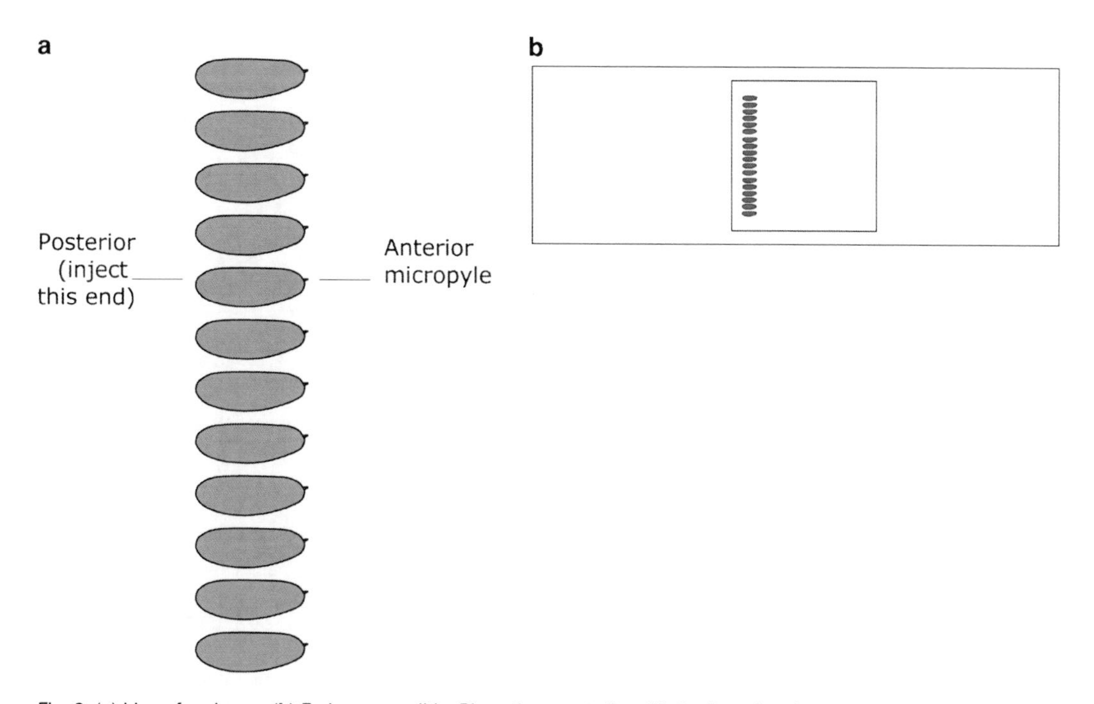

Fig. 2. (**a**) Line of embryos. (**b**) Embryos on slide. Place the cover slip with the line of embryos perpendicular to the long edge of the slide as shown. A drop of water between slide and cover slip is sufficient to prevent movement during injection.

6. Dry the embryos by placing the slide in a large (150 mm) closed petri dish containing self-indicating orange silica gel. The silica gel crystals must be orange. If they are not, replace them with fresh ones. Incubate at 18°C for 15–20 min. The drying time is critical (*see* **Note 11**).

   After drying, cover the line of embryos with Halocarbon oil. This prevents further drying but allows exchange of air. Begin injection. Make sure the needle is mounted and the injection apparatus is ready to start injection immediately after drying (*see* **Subheading 3.5**).

3.5. Microinjection of Embryos

1. Switch on the femtojet and allow to warm up (about 5 min). Set the pressure (pi) to 500 hPa, and injection time to 0.5 s.
2. Mount the needle: Remove the needle holder from the micromanipulator and remove the old needle if necessary. Using a microloader tip, load 2–3 mL of DNA into the new needle, taking care to avoid air bubbles. Mount the new needle into the holder. Mount the holder into the micromanipulator (*see* **Note 12**).
3. Place the slide with embryos onto the microscope stage, and use the micromanipulator to position the needle so it is in the centre of the field. Check that the posterior ends of the embryos are facing the needle. If not, raise the needle and turn the slide around, taking care not to damage the needle.
4. Move the embryos away from the needle. Clean the needle and check that a bubble of liquid of the correct size comes out, as shown in **Fig. 3a**. If the bubble is too small, increase the pressure (pi) but do not exceed 1,000 hPa. If the bubble is still too small, change the needle.

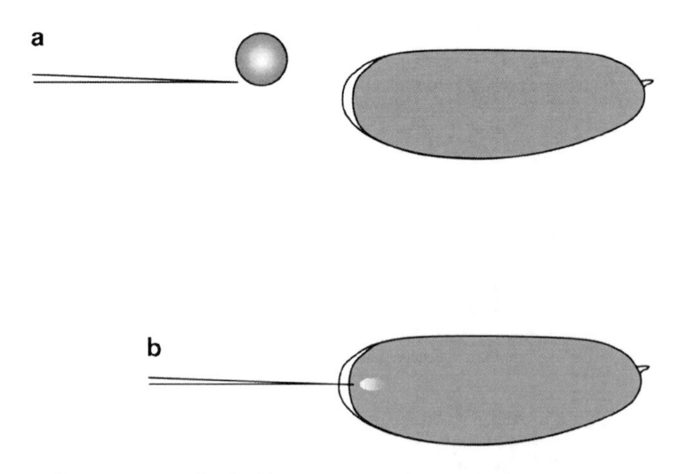

Fig. 3. (**a**) Testing the needle. Position the needle close to, but not touching, the row of embryos. Press the 'clean' button. A bubble of approximately the size shown should emerge. Note, when pressing the 'inject' button, the bubble will be almost undetectable. (**b**) Injecting. Insert the needle into the embryo as shown. Press 'inject'. A small transient movement should be visible.

5. Position embryos and needle. In most setups, the needle is brought to a suitable position using the micromanipulator and is then fixed at that position. The embryos are injected by moving the microscope stage. Use the micromanipulator to position the needle so that it is in the centre of the field of view. Now move the embryos until the needle touches the posterior end of an embryo at its outermost point. This can be tricky and requires some practice. Focus sharply on the outermost posterior point of one embryo and use the micromanipulator to bring the tip of the needle into the same focal plane. The needle should be perpendicular to the point of penetration. From this point on, the needle should no longer be moved. Now by moving the microscope stage, insert the needle into the outer membrane and through, so it just enters the inner membrane, as shown in **Fig. 3b**. A short sharp movement works best.
6. Inject. The drop of injected liquid should be visible as a very small movement, like a small pale cloud transiently appearing in the cytoplasm (*see* **Fig. 3b**). If a large pale spot remains, decrease the injection pressure. Note the desiccation state of the embryos: If they are too dry they will deform under the pressure of the needle. If they are insufficiently dried, they will leak cytoplasm. Adjust the drying time in the next round if necessary (*see* **Note 13**). Inject the row of embryos, cleaning the needle regularly. Inject only embryos that have not yet formed pole cells, as shown in **Fig. 4**. Leave out embryos that are too old (*see* **Note 14**). Inject 50–100 embryos per

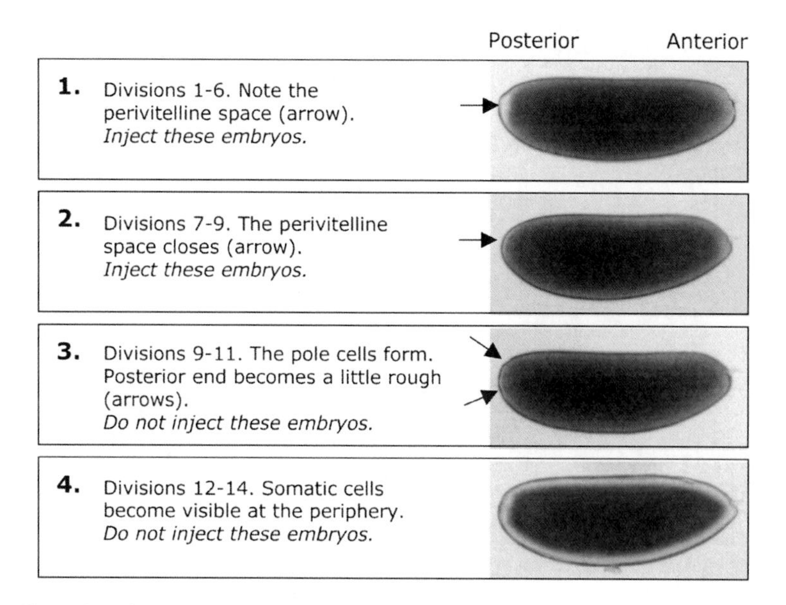

Fig. 4. Age of embryos (*see* **Note 14**).

construct for ΦC31 system, and 150–200 per construct for P-elements.

7. After finishing the row, go back and kill the older embryos either by piercing with the needle, or inserting and cleaning the needle. Some embryos may have a small bubble of liquid from the perivitelline space at the point of injection: remove this bubble to prevent further leakage, but do not kill the embryo unless it has leaked cytoplasm into the bubble. Make a note of how many embryos were injected.
8. With practice, timing can be optimised as shown in **Fig. 5**. One person injecting and lining up alone can inject 100 or more embryos per hour. Only every second plate is used for injection. The procedure can be optimised further with one person lining up and a second person injecting. In this case every plate can be used for injection.

3.6. Further Handling and Screening for Transgenics

1. Remove the slide. Place the coverslip with embryos into a 40-mm petri dish with an 18 × 18 mm coverslip in the bottom. Place the petri dish in a sealed sandwich box at 18°C. 4–6 h later; cover the petri dish with a damp paper towel. Surviving first instar larvae should hatch between 24 and 48 h after injection. After 24 h, start checking for first instar larvae, collect them carefully with a brush or dissection needle, and place into a fresh vial of food. Continue to check for survivors over the next 2 days. Put 40–50 larvae in one vial, and note how many survived to first instar. Make sure this vial does not dry out over the following week. Check every day and add a little water to the food if necessary (*see* **Note 15**).
2. Immediately after injection, set up at least two bottles of w- flies or appropriate balancer lines for collecting males and virgins in 10 days. Keep these bottles at 25°C and flip every few days.
3. Collect pupae: About a week after injection, start checking the vials of injected larvae for pupae. Using a small spatula, transfer each pupa carefully into a single vial. Note how many

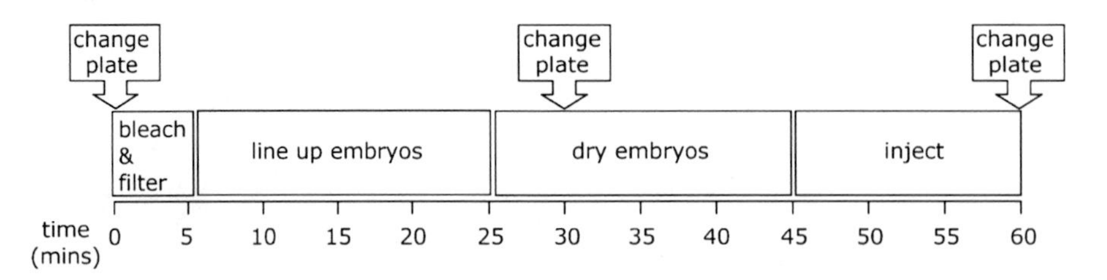

Fig. 5. *Timing*. The workflow for one person injecting alone is shown. It is important to change the plates every 30 min to ensure that embryos do not become too old. Only the second plate (30–60 min) is used for injection, the other is discarded. The time spent lining up and injecting can be adjusted so that both are accommodated within the hour. With practice, several rows can be processed from one plate. The drying time is critical and must be determined (*see* **Note 11**). If two people work together, one should line up and dry, whilst the other injects. In this case, every plate is used.

pupae there are and how many hatch. This gives a measure of survival rate (*see* **Note 16**).

4. When flies hatch, cross each to three flies of the opposite sex (w- or balancer lines) in a fresh vial (the vials containing pupae tend to dry out, and are not optimal for crosses). Some crosses will be sterile: note how many are fertile. Take care that these vials do not dry out. Check regularly and add water if necessary. Screen the progeny of these crosses for red eyes or the appropriate marker.

5. Calculate integration efficiency:

Number of vials with transgenic flies (T) as a percentage of the number of fertile crosses (F): T/F × 100 (*see* **Note 17**).

6. Calculate injection efficiency:

Number of vials with transgenic flies (T) as a percentage of the number of embryos injected (E): T/E × 100 (*see* **Notes 18** and **19**).

4. Notes

1. *Fly lines for injection*. For the ΦC31 system, many docking site lines with attP sites at characterised positions in the genome are available, from http://www.frontiers-in-genetics.org/flyc31/ or from Bloomington Stock Centre, http://flystocks.bio.indiana.edu/. For some lines, integration efficiencies and expression levels have been determined. A 'split white' system has also been described *(7)*, in which reconstitution of the *white* transformation marker requires sequences from both the donor plasmid and the docking site. This ensures that only correct integration events will be recovered. However the frequency of non-specific events is extremely low, and simple attP docking lines are suitable for most applications. Finally, a cassette exchange system has also been developed *(8)*. Docking site flies contain two attP sites flanking a marked cassette, that is exchanged upon recombination with a donor plasmid carrying two attB sites. This has the advantage that only sequences within the cassette are integrated, and that unmarked sequences can be inserted, as integration can be detected by the loss of the original marker. For P-element transgenesis, attP sites are not required. Use a fly line that is mutant for the transformation marker (e.g. w^{-} if the marker is *white* (*see* **Note 3**).

2. *Source of ΦC31 integrase or P-transposase*. For ΦC31 integration, there are three options for providing the integrase enzyme:

(a) Co-injection of RNA encoding the ΦC31 integrase *(6, 9)*. RNA is transcribed from the plasmid pET11 C31polyA, available on request from calos@stanford.edu, as described in *(9)*. For co-injection with RNA, the donor plasmid should be used at the concentration recommended by these authors (1 μg/μL).

(b) We co-inject a DNA plasmid (pKC40, Kuan-chung Su and Barry J Dickson, unpublished) expressing the ΦC31 integrase fused to an NLS, under control of the hsp70 promoter, whose constitutive expression gives sufficient expression for efficient integration without heatshock. pKC40 is available on request from L. Ringrose or B. J. Dickson (dickson@imp.univie.ac.at). We find DNA injection more convenient than RNA injection and obtain integration efficiencies that are equivalent to those reported for RNA injection (S. Bhalerao and L. Ringrose, unpublished; see also **Note 17** on efficiencies).

(c) Expression of ΦC31 integrase in the fly germline. Fly lines with an endogenous source of ΦC31 integrase under the control of vasa or nanos regulatory elements have recently been generated *(7)* and are available from http://www.frontiers-in-genetics.org/flyc31/ or from Bloomington Stock centre, http://flystocks.bio.indiana.edu/. The authors report improved efficiency compared to RNA injection, and that donor plasmid concentrations between 200 and 800 ng/μL gave similar transgenic frequencies. For injection, combined lines carrying a source of integrase and an attP docking site must be used (also available from the aforementioned sites; see also **Note 1**). The integrase must be crossed out in the subsequent generation to avoid unwanted recombination events.

3. *Donor vectors.*

(a) For standard transgenesis, three commonly used donor vectors carrying a multiple cloning site, an attB site, and the *white* gene as a transformation marker are *pUASTattB, pattB, and placZattB*. These constructs are available from http://www.frontiers-in-genetics.org/flyc31/ or the Drosophila Genomics Resource Centre, https://dgrc.cgb.indiana.edu/ *(9)*

For large DNA fragments (Drosophila P1 or BAC clones, up to 133 kb), the attB-P[acman] vectors are suitable (Genbank accession numbers: attB-P[acman]-Cm^R: EF106979 and attB-P[acman]-Ap^R:EF106980) *(10)*. These are conditionally amplifiable BAC vectors carrying an attB site and marked with the *white* transformation marker, and are available from the Drosophila Genomics Resource Centre, https://dgrc.cgb.indiana.edu/

For RNAi hairpin constructs, the Valium vector, containing an attB site, two multiple cloning sites for hairpin construction, and the *vermillion* gene as selectable marker, has recently been described *(11)*.

These attB-containing donor vectors are used in combination with a source of ΦC31 integrase (*see* **Note 2**), and are injected into attP containing 'docking site' fly lines (*see* **Note 1**).

(b) For standard P-element transgenesis, commonly used vector backbones are pCaSpeR and pUAST. These carry recognition sites for the P-transposase, a multiple cloning site, and a *white* transformation marker. These and other P-element vectors are available from the Drosophila Genomics Resource Centre, https://dgrc.cgb.indiana.edu/. For P-element transgenesis with large fragments, the P[acman] series is suitable *(10)*. P-element vectors are used in combination with a source of transposase (*see* **Note 2**) and are injected into fly lines that are mutant for the transformation marker (e.g. w^- if the marker is *white* (*see* **Note 1**).

For P-element transgenesis, the P-transposase is commonly provided by co-injection with the helper plasmid pUChsΔ2–3, available from the Drosophila Genomics Resource Centre, https://dgrc.cgb.indiana.edu/ This plasmid contains a modified transposase gene under control of the hsp70 promoter, whose constitutive expression gives sufficient transposase expression for P element integration without heatshock *(12, 13)*.

4. *Injection needles.* We make our own needles, but an alternative is to buy them readymade: Eppendorf Femtotips (sterile injection capillaries with 0.5 μm inner and 0.7 μm outer diameter) are suitable. These needles are closed and need to be opened by grinding or breaking (*see* **Note 9**) before use.

5. *Apple Juice plates.* There are many alternative recipes for agar plates, traditionally containing apple juice or grape juice: instead of dissolving 10 g sucrose in 300 mL water, dissolve 10 g sucrose in 300 mL apple juice or grape juice. We find the acetic acid recipe to be more economical, and the flies lay equally well as on apple juice.

6. *Double-sided tape.* An alternative to embryo glue is double-sided tape. Use 3 M double-coated tape 415. Note that some other brands contain glue that is toxic to embryos. For other brands of tape, a survival test should be performed. Make a line of dechorionated embryos, pick up on tape, cover with oil, and allow them to develop without injection. More than 80% of the embryos should survive if the tape is not toxic *(13)*.

7. *Injection apparatus.* We use the Eppendorf femtojet for injection as it is easy to use, and allows automated regulation of injection pressure and volume. Some authors prefer to use a manual set up with a syringe. This is described in *(12)* and *(9)*.

8. *Water*. When using water to elute DNA from columns provided in Qiagen kits, check that the pH is between 7.0 and 8.5. Lower pH results in decreased DNA yield. The bottled sterile water provided in many institutes can be rather acidic, typically around pH6. Adjust pH with NaOH. DNA dissolved in sterile water gives good injection efficiencies. Alternatively, use DNA in injection buffer: 5 mM KCl, 0.1 mM phosphate buffer, pH 6.8.
9. *Preparing needles*. Both home-made and commercial needles are closed at the tip and have to be opened before injection. The quality of the needle tip is critical for successful injection. Needles with a slanting tip rather than a flat one are ideal and do the least damage to embryos (for excellent photos of good needles see *(9)*). We find that grinding makes uniform needles with consistent quality and ensures a sloped end. Test a few needles by injecting water into a drop of oil at the injection microscope. A good measure for correct preparation of needles is the size of the bubble they produce upon cleaning (*see* **Fig. 3a**). Establish a grinding protocol that gives reproducibly good needles and make a supply of 20 or more. We keep closed and open needles in a closed petri dish to protect from dust, supported on strips of plasticine. An alternative to grinding is breaking the needle tip. This is done at the injection microscope, by gently pushing the needle tip, immersed in oil, against the edge of the cover slip. Breaking has the disadvantage that the needle is broken with DNA loaded. If breaking is successful, the needle can be used, but if not a new one must be mounted. An excellent guide to breaking needles is given in *(9)*.
10. *Dechorionation*. We prefer the filter method of dechorionation and collection as it is very rapid and requires minimum handling of embryos. Many alternative protocols are available, in which embryos are collected in a small sieve, and aligned on different supports. See for example *(9, 12, 13)*.
11. *Embryo drying time*. The drying time for embryos is absolutely critical. Overdrying of embryos will kill them, whilst embryos that are not dried enough will leak cytoplasm. Be aware that the embryos begin to dry whilst they are being lined up. If you are working in a room that is not climate controlled, then the prevailing weather conditions can have a profound effect on drying during this stage, and the silica gel drying time may have to be adjusted accordingly: for Northern European climates, drying times on silica gel of between 10 and 20 min are typical. In Mediterranean climates, drying times of as little as 2 min are sometimes necessary (L. Ringrose, unpublished observation). It is advisable to monitor humidity in the injection room and to install humidifiers if drying out during lining up becomes a problem.

12. *Blocked needle.* The needle can easily become blocked if there is dust or other particles in the DNA. Immediately before loading, centrifuge the DNA injection mix at 14,000 × *g* for 5 min and take DNA from the top one-third of the solution. Take care when loading the needle not to scrape the microloader on the glass edge of the capillary: this can file off tiny slivers of plastic that drop into the needle and cause blockage. Take care that the total DNA concentration in the injection mix does not exceed 1 μg/μL, as DNA viscosity can also block the needle.
13. *Correctly dried embryos* should resist deformation by the injection needle (See *(9)* for pictures of correct and over-dried embryos).
14. *Age of embryos.* **Figure 4** shows early embryos of increasing age. Up to nuclear division 9, all nuclei share a common cytoplasm. Injected DNA has access to the pole cell nuclei. Directly after division 10, the pole cells are made, separating the germ cell nuclei from the rest of the embryo. Do not inject embryos after this point. See http://flymove.uni-muenster.de/ under 'stages' for an excellent movie of early embryogenesis.
15. *Larval survival.* Fish et al. *(9)* recommend removing excess oil from the embryo line after injection to enhance survival. First instar larvae can be collected by hand and placed on food to optimise survival. The number of larvae per vial is also important. About 50 per vial is optimal. Too few larvae can result in drying out of the food. This can be prevented by regular checking and addition of a little water.
16. *Expected survival rates.* For P-element injections, approximately 30% of injected embryos typically survive to give fertile adults. For ΦC31 integration using co-injected RNA as a source of integrase, this number is reported to be slightly lower, around 20% *(9)*. We also observe similar rates of about 20% injected embryos giving fertile adults using co-injected DNA as a source of integrase (L. Ringrose and S. Farina Lopez, unpublished observation).
17. *Expected integration efficiencies.* For P-element injections, 10–50% of surviving fertile flies typically produce transgenics. For ΦC31 integration, efficiencies of between 20 and 60% are typical *(7, 9)*. The use of integrase expressed from the germline has been reported to increase efficiency over RNA co-injection *(7)*. Using DNA as a source of integrase we achieve equivalent efficiencies to those reported for both RNA injection and germline expression (typically between 20 and 60%; S. Bhalerao, S. Farina Lopez, L. Ringrose, unpublished). ΦC31 integration efficiencies have been reported to

vary from one landing site to another (http://www.frontiers-in-genetics.org/flyc31/). We also find that the same landing site can give quite different integration efficiencies with different donor constructs.

18. *Expected injection efficiencies.* Injection efficiency combines survival rate and integration efficiency, and gives an indication of how many embryos need to be injected. For both P-element transgenesis and ΦC31 integration, typical injection efficiencies are between 5 and 10% (i.e. 5–10% of injected embryos give transgenics). For P-element transgenesis, one wishes to recover multiple lines, requiring injection of 150–200 embryos per construct. For ΦC31 integration, a single transgenic line is sufficient; thus, only 50 embryos per construct need to be injected if survival and integration rates fall within the ranges mentioned. The success of transgenesis can be greatly reduced if the transgene has any adverse effects on survival or viability. This can usually be overcome by injecting more embryos.

19. *Self-cross.* With ΦC31 landing sites that are well characterised and have good integration efficiencies, and transgenes that do not reduce viability, single fly crosses are not required and can be replaced by self-crossing. Allow the vial of injected embryos to develop into adults, and screen the progeny of these for transgenes. Note that this method saves an enormous amount of time but does not allow for calculation of integration or injection efficiencies. This is not an option for P element transformation, where each transgenic fly has a different genomic insertion site.

20. *Fatigue.* Injecting or lining up for long periods of time can be very tiring and can cause back, neck, and eye problems. Ensure that the height of chairs and microscopes enables a comfortable working position and that the lenses of dissection and injection microscopes are clean and properly adjusted.

21. *Trouble shooting.* If landing site fly lines are kept expanded for too long, this can result in loss of integration efficiency and viability. Re-expand landing site lines from stock vials every 3 months or so. If efficiencies do not improve, crossing out to balancers and re-establishing homozygote stocks of landing site lines can also help. If embryos survive well, but no transgenes are found, check DNA preps for degradation by agarose gel electrophoresis. Remaking midiprep DNA can also help.

22. *Injection services.* Several commercial companies exist that offer injection of embryos as a service. A comprehensive listing of companies is given at http://flybase.bio.indiana.edu/static_pages/allied-data/external_resources5.html

Acknowledgements

I wish to thank Sara Farina Lopez and Sylvia Oppel for sharing technical expertise, and Sheetal Bhalerao for sharing unpublished data and for helpful discussions during preparation of this chapter.

References

1. Venken, K.J. and Bellen, H.J. (2005) Emerging technologies for gene manipulation in *Drosophila melanogaster*. *Nat Rev Genet*, **6**, 167–178
2. Venken, K.J. and Bellen, H.J. (2007) Transgenesis upgrades for *Drosophila melanogaster*. *Development*, **134**, 3571–3584
3. Rubin, G.M. and Spradling, A.C. (1982) Genetic transformation of Drosophila with transposable element vectors. *Science*, **218**, 348–353
4. Gong, W.J. and Golic, K.G. (2003) Ends-out, or replacement, gene targeting in Drosophila. *Proc Natl Acad Sci U S A*, **100**, 2556–2561
5. Xie, H.B. and Golic, K.G. (2004) Gene deletions by ends-in targeting in *Drosophila melanogaster*. *Genetics*, **168**, 1477–1489
6. Groth, A.C., Fish, M., Nusse, R. and Calos, M.P. (2004) Construction of transgenic Drosophila by using the site-specific integrase from phage phiC31. *Genetics*, **166**, 1775–1782
7. Bischof, J., Maeda, R.K., Hediger, M., Karch, F. and Basler, K. (2007) An optimized transgenesis system for Drosophila using germ-line-specific phiC31 integrases. *Proc Natl Acad Sci U S A*, **104**, 3312–3317
8. Bateman, J.R., Lee, A.M. and Wu, C.T. (2006) Site-specific transformation of Drosophila via phiC31 integrase-mediated cassette exchange. *Genetics*, **173**, 769–777
9. Fish, M.P., Groth, A.C., Calos, M.P. and Nusse, R. (2007) Creating transgenic Drosophila by microinjecting the site-specific phiC31 integrase mRNA and a transgene-containing donor plasmid. *Nat Protoc*, **2**, 2325–2331
10. Venken, K.J., He, Y., Hoskins, R.A. and Bellen, H.J. (2006) P[acman]: a BAC transgenic platform for targeted insertion of large DNA fragments in *D. melanogaster*. *Science*, **314**, 1747–1751
11. Ni, J.Q., Markstein, M., Binari, R., Pfeiffer, B., Liu, L.P., Villalta, C., Booker, M., Perkins, L. and Perrimon, N. (2008) Vector and parameters for targeted transgenic RNA interference in *Drosophila melanogaster*. *Nat Methods*, **5**, 49–51
12. O'Connor, M.J. and Chia, W. (2002) Gene transfer in Drosophila. *Methods Mol Biol*, **180**, 27–36
13. Voie, A.M. and Cohen, S. (1998) Germ-Line Transformation of *Drosophila melanogaster*. *Cell Biology: A Laboratory Handbook*, Academic Press, New York, vol. **3**, pp. 510–518

Chapter 2

Transgenesis in *Caenorhabditis elegans*

Matthias Rieckher*, Nikos Kourtis*, Angela Pasparaki, and Nektarios Tavernarakis

Summary

Two efficient strategies have been developed and are widely used for the genetic transformation of the nematode *Caenorhabditis elegans*, DNA microinjection, and DNA-coated microparticle bombardment. Both methodologies facilitate the delivery of exogenous DNA into the developing oocytes of adult hermaphrodite animals, which then generate transgenic worms among their progeny. Although both approaches share the common underlying principle of introducing foreign DNA into the germline of *C. elegans*, they offer distinct transformation outcomes. In this chapter, we present DNA microinjection and bombardment methods for transgenesis in *C. elegans* and provide time-tested procedures for their implementation. We also discuss their relative advantages as well as their limitations and evaluate their potential for a range of applications.

Key words: Biolistic transformation, Chromosomal integration, DNA-coated microparticle bombardment, Extrachromosomal arrays, Genetic transformation, Germline, GFP, Microinjection Nematode, transgenic animals

1. Introduction

Caenorhabditis elegans is a small (approximately 1 mm) soil-dwelling, free-living nematode worm. In the laboratory, animals feed on an *E. coli* diet and complete a reproductive life cycle in 2.5 days at 25°C, progressing from fertilized embryos through four larval stages to become hermaphroditic adults, which then live for about 2 weeks and lay ~300 eggs. While the dominant sexual form is the hermaphrodite (genotype: XX), males (genotype X0) can also be propagated and used to construct strains carrying multiple mutations *(1)*. Importantly, *C. elegans* strains can be

* Matthias Rieckher and Nikos Kourtis contributed equally.

Elizabeth J. Cartwright (ed.), *Transgenesis Techniques*, Methods in Molecular Biology, vol. 561
DOI 10.1007/978-1-60327-019-9_2, © Humana Press, a part of Springer Science+Business Media, LLC 2009

stored indefinitely in liquid nitrogen, making it feasible for any laboratory to possess an unlimited collection of mutants. Because *C. elegans* can reproduce by self-fertilization it is possible to raise genetically identical populations that do not undergo inbreeding depression. The simple body plan, the transparent egg and cuticle, and the nearly invariant developmental plan of this nematode have facilitated exceptionally detailed developmental and anatomical characterization of the animal (see http://www.wormatlas.org/).

C. elegans has proven instrumental in providing insights into the molecular mechanisms underlying numerous biological processes, such as cell death, cell cycle, sex determination and neuronal function, as well as in the genetic dissection of signalling pathways that were subsequently found to be conserved in mammals. Similarly, *C. elegans* has contributed decisively to our understanding of human neurodegenerative disorders *(2)*, cancer *(3)*, and ageing *(4)*. The availability of a fully sequenced worm genome, in which 60–80% of the genes have a human counterpart, the wealth of information available at Wormbase (http://www.wormbase.org/ *(5)*) relevant to gene structures, mutant and RNAi phenotypes, microarray data and protein-protein interactions, and the availability of numerous mutant strains, have led to the rapid adoption of the worm as a model organism.

DNA manipulated *in vitro* can be introduced into animals for functional assays. Vectors are available for identification of transformants (**Table 1**), cell-specific expression, and generation of fusions to numerous marker genes, such as *E. coli* β-galactosidase, the firefly luciferase, and the jellyfish Green Fluorescent Protein (GFP) so that individual cells can be visualized in stained or living animals *(6–11)*. Transformation is also routinely used to rescue mutant genes, to analyse regulatory elements, to ectopically express altered genes and reporters, and ultimately to investigate gene function *in vivo*. In the following sections we describe the two widely used methods for obtaining transgenic *C. elegans* animals, DNA microinjection, and DNA-coated microparticle bombardment, aiming to provide the essential information required for their successful implementation.

1.1. Microinjection

Microinjection is the most widely used transformation method in *C. elegans*. Transgenic animals are generated by injecting appropriate DNA fragments or plasmids. DNA is commonly injected into the distal gonad syncytium (**Fig. 1**) or directly into the developing oocytes *(12, 13)*. Injected DNA forms large extrachromosomal arrays incorporating between 50 and 300 copies, which are inherited by the progeny of injected animals *(10, 11)*. Identification of transgenic individuals is facilitated by the use of several available transformation markers, which are co-injected along with the DNA of interest (**Table 1**). A varying number of

Table 1
***C. elegans* transformation markers**

Plasmid	Phenotype	Comments	Disadvantages	References
pDPMM0016b	Rescues *unc-119(ed3)*	*unc-119(ed3)* mutants are uncoordinated and dauer-formation defective. Only transformants carrying stable integrated transgenes survive starvation. Useful for enrichment of integrants after bombardment	*unc-119(ed3)* is difficult to grow. Selection process takes a long time	*(17, 26)*
pC1 or pBX	Rescues *pha-1(e2123ts)*	Temperature-sensitive strain. Allows selection of transformed worms at the restrictive temperature. Convenient for obtaining large numbers of transgenics	Requires *pha-1(ts)* mutant background. Selection requires growth at 25°C which can be problematic for some temperature-sensitive transgenes	*(27)*
pGK10	Expresses GFP in muscles of the worm using the SERCA gene promoter	Select for worms with bright stained muscles	Not suitable for the characterization of GFP expression patterns	*(28)*
pRF4	Dominant roller	The *rol-6(su1006)* mutant phenotype is easy to detect	Not suitable for studies of locomotion and other behaviours. Larval lineage analysis is problematic. Mating efficiency reduced	*(11)*
pPD10.46	Dominant twitcher	The *unc-22 antisense* mutant is easy to detect	Similar to *rol-6(su1006.)* More difficult to detect	*(9)*
pPD122.11	GFP expression in pharyngeal muscles	The *myo-2::GFP* mutant is a dominant visible fluorescent marker	Requires fluorescent stereo microscope. Interferes with other GFP-fusion-based expression studies	*(11)*
pTG96_2	GFP expression in somatic cells	*sur5::GFP* is a dominant visible fluorescent marker	Similar to pPD122.11	*(29)*
plin-15EK	Rescues *lin-15(765ts)*	*lin-15(765ts)* develop multiple vulvae (muv) at 22.5°C growth temperature, selection of *lin-15(wt)* transformants, less labour for maintenance	Requires *lin-15(765ts)* mutant background. Overexpression of *lin-15(wt)* might lead to misdevelopment of the nervous system	*(30)*

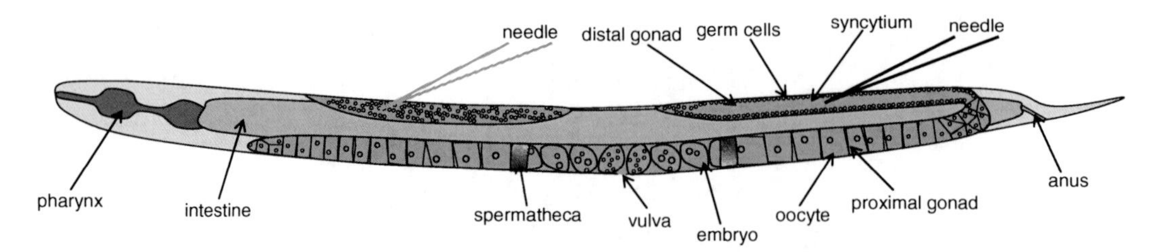

Fig. 1. Schematic drawing of *C. elegans*, indicating the sites of microinjection. The needle injects DNA at the syncytium of the two distal gonad arms, containing mitotic and undifferentiated germ cells. An angle of about 40° between needle and worm facilitates entry through the cuticle. Injected DNA is taken up by developing oocytes which will ultimately get fertilized in the spermatheca to give rise to transgenic embryos.

F1 progeny that carry the desired extrachromosomal array can thereby be identified by characteristic phenotypes associated with each transformation marker. For example, the dominant *rol-6(su1006)* allele induces a distinctive rolling phenotype in transgenic F1 progeny. Extrachromosomal arrays are inherited in a non-Mendelian manner to subsequent generations. Typically, a varying percentage (between 5 and 80%) of next-generation progeny will carry the extrachromosomal array *(11)*. To overcome low-transmission problems and the consequent, gradual loss of extrachromosomal arrays, they can be incorporated into a chromosome by inducing non-homologous, double-strand DNA break repair. We describe a method for integrating extrachromosomal arrays carried by transgenic *C. elegans* lines in the last section of this chapter.

1.2. Bombardment

The relatively recently developed, DNA-coated microparticle bombardment method (biolistic transformation) is also used to transform *C. elegans(14–16)*. Gold microparticles are first coated with the DNA to be introduced to animals and then 'shot' into the worms at high speeds, using a specialized biolistic bombardment instrument or 'gene gun'. The commercially available microparticle bombardment devices most commonly used are the Bio-Rad Biolistic® PDS-1000/He Particle Delivery system, with or without the Hepta adapter and the Bio-Rad Helios Gene system *(14–18)*. A helium gene gun can also be constructed in the laboratory, using off-the-shelf custom parts at a low cost. Instructions for obtaining parts or required accessories and for building the device can be found at Ralf Schnabel's laboratory home page (http://www.ifg.tu-bs.de/Schnabel/ce-home.html).

DNA bombardment offers distinct features and possibilities compared to microinjection **(Table 2)**. Contrary to transgenics obtained by microinjection, which, in their majority, carry large repetitive extrachromosomal arrays, a significant number of transformed individuals generated by bombardment are integrants

Table 2
Features and comparison of transformation methods in *C. elegans*

Microinjection	Bombardment
Transformation of individual hermaphrodites	Transformation of large worm populations
Generally less time consuming	More time consuming
Extrachromosomal arrays carrying ~80–300 copies of the transgene	Low-copy extrachromosomal or integrated lines and homologous recombinants
Frequently extrachromosomal array expression shows extensive mosaicism	Integrated transgenes are expressed in reproducibly consistent manner
Germline expressed genes are silenced	Germline expression of many transgenes remain stable
Transgenic animals are easily identified in a short period of time after injection	Suitable marker for strong selection is necessary
Widely used	Less common
Less expensive: A light DIC (Nomarski) microscope is standard in a *C. elegans* laboratory	Expensive equipment
Inexpensive consumables	Expensive consumables
Requires considerable practice for consistent results	Easier to perform, with less delicate steps

containing a low number of transgene copies *(15)*. Biolistic bombardment has also been used successfully for homologous gene targeting in the worm *(18)* and for expressing transgenes in the *C. elegans* germline, a tissue refractory to expression of genes on conventional extrachromosomal arrays due to germline silencing of highly repetitive DNA *(15)*. In addition, microparticle bombardment has been utilized to facilitate the routine use of dsRNA hairpin constructs (hpRNAi) *(19)* and to target tissues refractory to other RNAi delivery methods *(20)*.

The major advantage of the microparticle bombardment technique over the germline injection method is that stable integrated transgenic strains can be obtained directly. Moreover, germline expression of many transgenes remains stable, increasing in that way the ability of researchers to investigate the regulation of germline development and function *(15)*. In addition, low-copy integrants are likely to circumvent the problems associated with mosaic loss of the extrachromosomal arrays. Integrated transgenic lines created by microparticle bombardment express GFP reporter constructs in reproducible and consistent patterns *(15)*. Finally, the method can be easily scaled up, allowing for isolation of strains with homologous gene replacements *(18)*. This capacity enables the implementation of powerful knock-in and knock-out methodologies. Transformation of *C. elegans* by microparticle bombardment also has limitations that should be taken into consideration before choosing which method to implement **(Table 2)**. The preparation of nematodes for gene bombardment is significantly more laborious compared to microinjection, and transgenic animal generation takes significantly longer. In addition, gene bombardment requires access to specialized equipment and reagents. A suitable transformation marker for strong selection is necessary, since large populations of worms must be bombarded to generate the relatively rare desired integration events **(Table 1)**. Attention should be paid to avoid markers that interfere with the function of the gene of interest. In addition, it is still difficult to express dose-sensitive genes because even in integrated lines, expression of the gene of interest may not reflect that of the endogenous gene. Finally, during the integration events, the transgene may disrupt or interfere with the expression of genes located at the site of integration. Here, we describe the most commonly used microparticle bombardment approach for the transformation of *C. elegans*, using the Bio-Rad Biolistic PDS-1000/He Particle Delivery system with and without the Hepta adapter *(15–18)*.

1.3. Integrating Extrachromosomal Arrays

Transgenic animals obtained by microinjection (and occasionally by bombardment) carry and inherit the exogenous DNA as an extrachromosomal array, which results in mosaic expression and gradual loss between successive generations unless actively selecting for transgenic animals *(11)*. To decrease mosaicism and alleviate loss of extrachromosomal arrays they can be integrated; this also facilitates genetic manipulations and screening experiments with transgenics *(10, 21)*; *see* **Note 5**). Integration is triggered by causing chromosomal breaks through the use of mutagenic chemicals (e.g. Ethylmethane Sulphonate) or irradiating animals with an X-ray, gamma-ray, or UV source. The protocol described here refers to the use of an irradiation source *(22)*. Selection is based on the selection markers used for the initial transformation **(Table 1)**. It is advisable to carry out the integration procedure with 2 or 3 independent transgenic lines carrying the extrachromosomal array.

2. Materials

2.1. Microinjection

1. Wormpick: Cut 3 cm of platinum wire (90% platinum, 10% iridium wire, 0.010 in. diameter; e.g. Tritech Research, Los Angeles, CA) and flatten one end using pincers or a light hammer. Break off the thin part of a glass Pasteur pipette. Melt

the glass at the site of breakage on a Bunsen burner and attach the sharp end of the platinum wire. When using the wormpick always sterilize the tip over a flame.

2. 10× TE (Tris/EDTA) buffer: 100 mM Tris–HCl of pH 7.4–8.0 and 10 mM EDTA of pH 8.0 in distilled H_2O. Autoclave and store at room temperature.
3. 10× microinjection buffer: 2% polyethylene glycol (PEG 6000), 200 mM potassium phosphate, 30 mM potassium citrate (adjust to pH 7.5) in distilled H_2O. Autoclave and store at room temperature.
4. NGM agar plates with bacterial lawn: For 1 L NGM agar, combine 3 g NaCl, 2.5 g bactopeptone, 0.2 g streptomycin, and 17 g agar. Autoclave including stir bar and cool to 55°C while stirring. Add 1 mL 1 M $CaCl_2$ (stock: 14.7 g $CaCl_2$ in 100 mL distilled H_2O, autoclaved), 1 mL 1 M $MgSO_4$ (stock: 12.03 g $MgSO_4$ in 100 mL distilled H_2O, autoclaved), 1 mL cholesterol (5 mg/mL in 100% ethanol; *Caution*: flammable), 1 mL nystatin (Sigma, Cat. No. N1638; 10 mg/mL in 70% ethanol; *Caution*: flammable), and 1 mL KPO_4 (stock: 102.2 g KH_2PO_4 and 57.06 g K_2HPO_4 in 1 L distilled H_2O, autoclaved). Use 60 mm Petri dishes and pour 11.5 mL of NGM agar per plate (adjust volume to size of Petri dishes). Let cool until agar hardens. Streak a drop (~50 μL) of an *Escherichia coli* OP50 liquid culture in the centre of each plate and grow at 37°C for 8 h or at room temperature overnight.
5. M9 buffer: 22 mM KH_2PO_4, 22 mM Na_2HPO_4, 85 mM NaCl, and 1 mM $MgSO_4$. Sterilize by autoclaving for 15 min.
6. Recovery buffer: 0.1% salmon sperm DNA, 4% glucose, 2.4 mM KCl, 66 mM $CaCl_2$, 3 mM Hepes, pH 7.2.

2.2. Bombardment

1. Enriched peptone plates: For 1 L, combine 1.2 g sodium chloride, 20 g peptone, and 25 g agar. Autoclave and cool to 55°C. Add sterile 1 mL cholesterol (5 mg/mL in ethanol; *Caution*: flammable), 1 mL 1 M $MgSO_4$, and 25 mL 1 M potassium phosphate, pH 6.
2. M9 buffer: 22 mM KH_2PO_4, 22 mM Na_2HPO_4, 85 mM NaCl, and 1 mM $MgSO_4$.
3. NGM plates: 3 g NaCl, 2.5 g Bactopeptone, 0.2 g Streptomycin, and 17 g agar. Autoclave. Let cool to 55–60°C and add 1 mL cholesterol (5 mg/mL in ethanol; *Caution*: flammable), 1 mL 1 M $CaCl_2$, 1 mL 1 M $MgSO_4$, 1 mL Nystatin, and 25 mL phosphate buffer, pH 6.
4. Materials, reagents, and solutions required for preparing gold beads include: 1 μm-diameter gold beads (Bio-Rad, Hercules, CA; store in a dry, non-oxidizing environment to minimize agglomeration), 0.1 M spermidine (Sigma; free base, tissue

culture grade; filter sterilized; store at –20°C), sterile 2.5 M $CaCl_2$, dehydrated ethanol, and 50% sterile glycerol.

5. High purity and concentrated (~1 mg/mL) DNA is required for transformation. Purification using a commercial kit is recommended.
6. We describe the procedure of microparticle bombardment using the Bio-Rad Biolistic PDS-1000/He particle delivery system (Bio-Rad). The basic transformation protocol can be scaled up by using the Hepta adapter (Bio-Rad).
7. Information about setting up and operating the bombardment apparatus can be found in the manufacturer's manual. Supplies required for the biolistic transformation include: 1,350 psi rupture discs for the conventional system or 2,000 psi rupture discs (Bio-Rad) if using the Hepta adapter, biolistic microcarriers (Bio-Rad) and Hepta stopping screens (Bio-Rad).

2.3. Integrating Extrachromosomal Arrays

For animal irradiation, an X-ray, a gamma-ray (^{137}Cs; 2,000–4,000 rad), or a UV source can be obtained from commercial sources. A UV crosslinker with the appropriate UV bulbs (254 nm) and energy measurement sensor offers the most convenient and widely used alternative (Bio-Link Crosslinker, Vilber Lourmat, Marne-la-Vallée, FR or Stratalinker, Stratagene, La Jolla, USA).

3. Methods

3.1. Microinjection

3.1.1. Preparing Agarose Pads

Animals are fixed stably on the surface of an agarose pad for injection.

1. Prepare 10–50 mL of 2% agarose solution in distilled H_2O. Boil in a microwave oven until agarose is dissolved and cool down the solution in a water bath (60°C) or at the bench for a short while.
2. Align a number of microscope cover glasses (24 × 50 mm; Paul Marienfeld GmBH, Lauda-Königshofen, Germany) next to each other on the bench and fix at both ends with two layers of adhesive tape.
3. Put a drop of melted agarose (~100 μL) by pipetting in the centre of the cover slide and flatten by dropping a microscope slide on top, immediately (75 × 25 × 1 mm; Paul Marienfeld GmBH, Lauda-Königshofen, Germany). The double-layered tape on both sides of the microscope cover glass serves as a spacer to ensure adequate thickness of the agarose area.

4. Let the agarose harden for at least 5 min and then remove the cover slide from the top of the agarose by carefully slipping it off. Dry the agarose pad before use by leaving on the bench at room temperature for several hours or in a vacuum oven at 100°C for about 1 h. The pads can be stored indefinitely until used.

3.1.2. Preparing DNA Samples

Any DNA including plasmid, phage, or linear DNA as obtained from a PCR reaction can be used for transformation of *C. elegans*. Standard methods are used for purification of DNA for injection *(23)*. We typically prepare plasmid DNA using a standard alkaline lysis protocol combined with a final purification step using a commercial purification column, followed by isopropanol or ethanol precipitation. DNA is dissolved in either distilled H_2O or TE buffer. Note that DNA prepared by boiling lysis or by phenol-chloroform extraction might be ineffective due to remaining impurities *(21)*. A critical factor for successful transformation is the final concentration of the DNA. Samples are routinely injected at a concentration of between 50 and 100 μg/mL *(11)*. Occasionally, lower DNA concentrations may be used to alleviate toxicity problems or lethality due to gene overexpression *(21)*. We inject DNA dissolved in distilled H_2O or 1× TE buffer. A special microinjection buffer can also be used and is particularly advantageous in cases of direct injection into oocytes *(13)*. Care should be taken to remove dust and other microparticles which could clog the injection needle.

1. Set up DNA samples for injection to a final volume of 20–50 μL DNA in distilled H_2O, or 1× TE buffer, or 1× microinjection buffer, at a final concentration of 100 μg/mL.
2. Centrifuge the solution in a microcentrifuge at maximum speed for 10–15 min, preferably at 4°C to remove dust and other microparticles.
3. Carefully pipette 3–5 μL from top of the DNA solution to a clean reaction tube and use it to directly load injection needles. The remaining DNA sample can be frozen at –20°C for later use.

3.1.3. Preparing and Loading Microinjection Needles

Pulling a Microinjection Needle

To pull injection needles use borosilicate glass capillaries (Borosil 1.0 mm OD × 0.5–0.75 mm ID; Capillary tubing FHC Inc., Bowdoin, ME; FLG10, or equivalent) and a needle puller (Microelectrode Puller PN-30, Narishige Group, Tokyo, Japan, or equivalent). Pulling needles with proper tips is essential for successful and efficient microinjection: The aim is to pull needles which are sealed at the tip and are opened later by breaking off the closed end of the tip. The tip of the needle should taper constantly and quickly (5–7 mm in length). Stubby tips may result in openings too wide after milling and tips tapering too long and steady might not be stiff enough. Depending on the type of glass capillaries and needle puller individual settings for making an optimal needle need to be acquired by trial and error. The adjustable parameters of the needle puller are usually heating time, coil temperature, and pulling force which determines pulling speed. Needles can be stored for infinite time in a clean Petri plate modified with wax or double-sided tape as holder. Pre-pulled microinjection needles can also be obtained from several companies.

1. Fix one glass capillary (100 mm) on the needle puller using the two fixing screws. About 4 cm of the capillary should protrude over the sliding unit and about 1 cm at the fixed part of the holder. Close the acrylic hood covering the heating filament.
2. Turn on the machine and choose appropriate settings. For the PN-30 Puller, use magnet (main) 70, magnet (sub) 60 (together these values affect the pulling force and speed of the sliding unit), and heater 80 (determines the temperature of the platinum heater), as the starting values.
3. Start the pulling process of the needle.
4. Open the acrylic hood and the screws. Remove the needle carefully and check the tip under a stereo microscope at high magnification.

Loading the Needle

1. Prepare capillary mouth pipettes: Hold a glass capillary micropipette (Einmal-Mikropipetten intraMARK (100 μL), Brand GMBH, Wertheim, Germany) at the tips and heat its centre over the flame of a Bunsen Brenner until the glass begins to melt. Take the micropipette off the flame and carefully pull it apart, creating a very thin middle part. Break the micropipette at the thin middle part to generate two mouth pipettes.
2. Dip the tip of a mouth pipette into the DNA sample and allow the liquid (estimated amount is about 0.2–0.5 μL) to enter by capillary force.
3. Stick the thin tip of the mouth pipette into the backside opening of a microinjection needle and push it down towards the tip. When close to the tip, blow the liquid out of the mouth pipette and into the injection capillary. Tilt the injection needle to allow the liquid to settle at the tip. More than 50 worms can be injected with one loaded needle.

Breaking the Tip of the Needle

1. Take an agarose pad (*see* **Subheading 3.1.1**) and slightly scratch the edge of the agarose area with a scalpel. Fix the loaded needle to the needle holder of the injection system attached to the inverted DIC (Nomarski) microscope (for microscope details *see* **Note 1**).
2. Focus on the cragged region of the agarose area using the 10× lens and align the needle at a 90° angle to the rough edge of agarose.

3. Change to the 40× lens and bring the tip of the needle and the agarose edge to the same focal plane. Carefully move the edge of the pad towards the needle until the tip bends.
4. Gently tap the sliding sample stage of the microscope towards the needle and break it, creating an opening <1 μm. Apply nitrogen gas pressure to the needle and allow some of the liquid to flow out (*see* **Note 2**). Check for excessive or limited flow which indicates a too-wide or too-narrow opening respectively.
5. Lift the needle along the z-axis into a safe resting position. For an alternative method to break open the needle *see* **Note 3**.

3.1.4. Preparing and Fixing Animals for Microinjection

The number of worms fixed and injected each time on a single agarose pad depends on the experience of the experimenter. It is recommended to start with one animal at a time and progressively raise the number of worms manipulated simultaneously. The limiting factor is the gradual desiccation of animals, which occurs after mounting to the dry agarose pads and depends on the thickness of the agarose layer and the size of the worm *(13)*. Worms shrink and die about 10–15 min after fixation.

1. Prior to the day of microinjection, move worms to be injected, at L4 larval stage, from a freshly grown culture to seeded NGM plates. Let animals grow overnight at 20°C.
2. One hour prior to microinjection, move worms to 15°C, which results in lower locomotion activity making animals easier to handle.
3. Working under a stereo microscope, place a drop of halocarbon oil (Halocarbon oil 700, Sigma-Aldrich Co., St. Louis, MO, Sigma; or Series 700 Halocarbon oil, Halocarbon Products, River Edge, NJ) on a glass slide next to the agarose pad area. Alternatively, Heavy Paraffin Oil can be used (BDH Chemicals, Poole, England; Gallarol-Schlesinger, Carle Place, NY). Transfer the desired amount of worms into the oil (dipping the wormpick into the oil simplifies collecting and transferring the worms). Let animals move in the oil for some minutes to get rid of bacteria attached to their cuticle.
4. Transfer the worms in another drop of oil placed on top of the agarose and gently press them to the agar surface. Alternatively, worms can directly be transferred along with some oil to the clean surface of the agarose. Note that it is impossible to inject moving animals, so be sure that they are attached to the agarose properly.
5. Animals are carefully arranged into position, with their distal gonad arms facing the side of the needle tip (*see* **Fig. 1** and **Note 1**). When injecting several worms at a time align them along their anterior–posterior axis on the pad. These prearrangements simplify the process of injection. Quickly proceed with the final steps of the procedure.

3.1.5. Microinjecting Animals

This step is the most delicate of the protocol: Worm and microinjection needle need to be positioned correctly by using the gliding stage and micromanipulator attached to the inverted DIC (Nomarski) microscope (*see* **Note 1**). For successful microinjection aim at the syncytium of the distal part of one of the gonad arms (*see* **Fig. 1**). Harshly moving the needle into the worm will damage the cuticle or other organs of the animal. Working fast is important to avoid desiccation.

1. After mounting animals on the agarose pad place it on the gliding base of the light microscope and localize the first worm to be injected through the 10× lens. Bring the roughly focused needle tip next to the worm and move the gliding base in order to get a sharp angle of about 20° to 40° between the needle and worm.
2. Focus the worm under the 40× lens and locate the distal gonad arms. Focus on the gonadal syncytium and bring the tip of the needle to the same focal plane next to it (*see* **Fig. 1**).
3. Using the gliding stage gently move the worm towards the needle. Tap the stage gently to move the needle tip into the gonad.
4. Apply nitrogen gas pressure (*see* **Note 2**) to the needle to push the DNA solution into the gonad. Observe the flow of the liquid and the characteristic ballooning of the gonad, which has a maximum capacity estimated to 1–10 pL. Gently remove the needle by moving the gliding stage while constantly applying pressure, which prevents backpressure from the gonad clogging of the needle. For recovering a clogged needle *see* **Note 4**.
5. Repeat for other worms on the agarose pad. Injecting between 10 and 20 animals is usually sufficient to obtain several independent transgenic lines; however, this number can vary considerably for different DNA samples.
6. After injecting all animals on the pad lift the needle along the z-axis into a resting position, remove the agarose pad and quickly proceed with the recovery of the injected worms.

3.1.6. Recovery of Injected Animals

1. To recover the worms, put a drop (about 50 μL) of M9 buffer or recovery buffer (*see* **Note 5**) on top of the halocarbon oil above the agarose area they are attached to. Animals will detach and float away from the agarose and into the recovery buffer.
2. After about 5 min of recovery in the buffer, transfer them with a wormpick into another drop of recovery buffer placed on a NGM agar plate next to the bacterial lawn. Transfer no more than five injected worms to a single plate to allow for easy selection of transgenic individuals among their progeny.

3.1.7. Selecting Transformed Animals and Maintaining Transgenic Nematode Lines

Transformed animals are selected among the F1 progeny of injected worms based on specific phenotypes associated with the appropriate co-injection markers used during injection. The most commonly used co-injection markers for both microinjection and bombardment are presented in **Table 1**.

1. After microinjection let the animals grow at 20°C for about 3–4 days until F1 progeny develop into young adults.
2. Pick transgenic F1 animals, each to one fresh, seeded NGM agar plate. Allow F1 animals to grow and examine their progeny for transformed individuals (*see* also **Note 6**). Keep plates with transgenic F2 animals. These represent independent transgenic lines carrying different extrachromosomal arrays.
3. To maintain transgenic lines, periodically transfer ~10 animals from each plate containing transgenic worms to fresh seeded NGM plates.

3.2. Bombardment

3.2.1. Preparing Animals for Bombardment

Microparticle bombardment transformation requires large numbers of animals *(15–18)*. Growth conditions vary depending on the strain selected to transform **(Table 1)**.

1. Create a master plate by evenly dispersing ~100 worms of the appropriate strain to be transformed on an enriched peptone plate, and completely spread with NA22 bacteria. Let worms grow until they have starved for 2–3 days (~7 days total for *unc-119(ed3)*). At this stage, the plate contains a large population, which consists mainly of L1 larvae. Always maintain master plates for *unc-119* mutant strains, which are slow growing.
   (a) When using the Bio-Rad Biolistic PDS-1000 device without the Hepta adapter, wash off the starved L1 larvae with M9 buffer and transfer approximately one-sixth of the worms onto a 100 mm enriched peptone plate. Grow worms at the appropriate temperature until they become young adults (*see* **Note 7**). Each plate will be used for approximately 1–2 bombardment(s). It is recommended to prepare five or more plates for each transformation.
   (b) For the Hepta adapter protocol, wash off the starved L1 larvae with M9 buffer from the master plate and transfer approximately one-sixth of the worms onto each of six 100 mm enriched peptone plates. Grow worms to starved L1 larvae at the appropriate temperature. Wash off worms from the six plates with M9 buffer. Resuspend worms in 60 mL M9 buffer. Disperse worms evenly on 60 enriched peptone plates. Grow until they become young adults (*see* **Note 7**).

2. Wash off young adults with M9 buffer and transfer to a centrifuge tube. Allow worms to settle by leaving the tube on the bench for 5 min. Remove supernatant and transfer the worm pellet to a dry (1 week post-bacterial spreading) bombardment plate by using a 200 μL micropipette with the end of the tip cut off.
   (a) If using the Bio-Rad Biolistic PDS-1000 device, transfer worm pellet onto a round-shaped bacterial lawn, seeded on dry 60-mm NGM plates. A 0.25-mL worm pellet, harvested from five 100-mm enriched peptone plates, is sufficient for five to six bombardments.
   (b) If using the Bio-Rad Hepta adapter, transfer the worm pellet onto dry 100 mm rich media plates completely seeded with bacteria. A 2–4 mL worm pellet, harvested from sixty 100 mm enriched peptone plates, is sufficient for six to seven bombardments.
3. Leave the covers off the bombardment plates to allow them to dry. To prevent worms from moving off the bacterial lawn, pre-chill plates to 4°C and keep them on ice.

3.2.2. Preparation of Gold Particles

1. Weigh 18 mg of 1 μm gold beads into a siliconized 1.5 mL microcentrifuge tube (*see* **Note 8**).
2. Add 1 mL 70% ethanol. Vortex for 5 min, allow beads to settle for 15 min, short spin in a microcentrifuge, and remove supernatant (*see* **Note 9**).
3. Add 1 mL sterile water. Vortex for 1 min, allow beads to settle for 1 min, short spin in microcentrifuge, and remove the supernatant.
4. Repeat **step 3** three times.
5. Resuspend the gold bead pellet in 300 μL 50% sterile glycerol (*see* **Note 10**).
6. Vortex beads for 5 min and add 10 μL prepared gold beads in a siliconized microcentrifuge tube (*see* **Note 11**).
7. While vortexing on medium speed, add the following reagents per bombardment:
   a. 1 μL DNA (concentration ~1 mg/mL; *see* **Note 12**).
   b. 10 μL of 2.5 M $CaCl_2$.
   c. 4 μL of 0.1 M spermidine.
   Vortex for at least 3 min, allow beads to settle for 1 min, spin briefly in a microcentrifuge, and remove supernatant.
8. Add 30 μL 70% ethanol, flick tube to mix, spin briefly, and remove supernatant (*see* **Note 13**).
9. Add 30 μL 100% ethanol, flick tube to mix, spin briefly, and remove supernatant.

10. Resuspend in 10 μL 100% ethanol. Vortex for at least 3 min to ensure that all large clumps of beads are broken up (*see* **Note 14**).
    a. If not using the Hepta adapter, spread approximately 10 μL of the gold bead/DNA suspension over the central region of a macrocarrier and let dry. Prepare one macrocarrier per bombardment.
    b. If using the Hepta adapter, scale up preparation by preparing three tubes of gold beads/DNA each of which contains a final volume of 100 μL DNA/bead suspension (for each tube increase by tenfold the reagents listed in **steps 6–10**). Transfer 6–7 μL of the beads onto a macrocarrier, spread it around with your pipette tip, and let dry. Only spread on area around the hole in the holder. Repeat for all macrocarriers.

3.2.3. Bombardment Using the Biolistic PDS-100/He Particle Delivery System with the Hepta Adapter

1. Set up the Bio-Rad Biolistic PDS-1000/He particle delivery system with the Hepta adapter following the manufacturer's instructions.
2. Wipe down all the gene gun components with 70% ethanol, then autoclave the Hepta adapter and the macrocarrier holder components. Wipe down the bombardment chamber with 70% ethanol.
3. Place the seven macrocarriers onto the Hepta adapter macrocarrier holder using the special tool.
4. Place a 1,500–2,000 psi rupture disc soaked in isopropanol in the retaining cap of the Hepta adapter, and tighten the adapter onto the chamber.
5. Place Hepta stopping screen and macrocarrier holder in chamber as described in the instrument manual.
6. Secure the uncovered worm plate into lowest shelf of the chamber using a rolled piece of adhesive tape.
7. Evacuate chamber to 26 in. Hg, press fire button, and hold until disc ruptures. Release vacuum and remove plate. Repeat for all six plates.

3.2.4. Bombardment Using the Biolistic PDS-100/He Particle Delivery System Without the Hepta Adapter

1. Set up the Bio-Rad Biolistic PDS-1000/He particle delivery system without the Hepta adapter. The manual accompanying the instrument provides detailed information on the procedure and its requirements.
2. Wipe down all the gene gun components and the bombardment chamber with 70% ethanol.
3. Set up a 0.25 in. gap distance between the rupture disc holder and the macrocarrier launch assembly.
4. Place a 1,350 psi rupture disc into the retaining cap and tighten with the provided torque wrench.

5. Using the macrocarrier insertion tool, place DNA-coated macrocarrier into the macrocarrier holder and screw into the macrocarrier launch assembly according to the manufacturer's instructions.
6. Remove lid and place worm plate on the second target shelf from the bottom.
7. Evacuate chamber to 27 in. Hg, press fire button, and hold until disc ruptures. Release vacuum and remove plate.

3.2.5. Recovery of Animals After Bombardment

1. After bombardment, allow worms to recover for ~1 h.
2. Wash worms off of the bombardment plate with M9 buffer and transfer them to two or more 100 mm enriched peptone plates per bombardment (*see* **Notes 15** and **16**).
3. If using the Hepta adapter, wash off worms from six to seven bombardment plates with M9 buffer, pellet and transfer to 80 (diameter: 100 mm) enriched media plates.
4. Incubate plates at appropriate temperature. Screening for transformed lines depends on the specific transformation marker and selection strategy used.

3.3. Integrating Extra-chromosomal Arrays

1. Move 20–30 freshly grown, transgenic worms (at the L4 larva stage) to a clean NGM agar plate and irradiate. Irradiation conditions need to be adjusted empirically for optimal results (~30–70 mJ total energy, ~50% lethality).
2. Transfer irradiated worms to fresh, seeded NGM agar plates (3–5 on each plate). Let them grow at 20°C for 3–4 days.
3. Pick about 500 transgenic F1 progeny from all plates and transfer them each to an individual seeded NGM plate. Let them grow at 20°C for 3–4 days.
4. Pick 2–4 adult worms from individual F2 lines on a single plate. After another 4–5 days observe the F3 generation for integration: Transmission of the transgenic animal phenotype should be close to 100% (for an alternative growing and selection method *see* **Note 17**).
5. To minimize undesirable mutations due to irradiation treatment, integrated lines should be backcrossed several times *(22, 24)*.

4. Notes

1. Working with and microinjecting *C. elegans* requires a stereoscopic microscope and an inverted DIC (Nomarski) microscope with two lenses (10× and 40×), and a gliding stage,

a micromanipulator with fine controls, and a needle holder attached to a nitrogen tank through a pressure adjusting system and an electromechanical valve for controlling pressure applied to the needle. Microscopes and complete systems can be obtained from companies such as the Carl Zeiss AG, Jena, Germany, or Nikon Corporation, Tokyo, Japan, and others. Several sources for obtaining additional information about microscopy for *C. elegans* work are available both online and in the literature *(25)*.

2. The standard nitrogen gas pressure depends on the system used and the type of needle that has been pulled. For long and thin tips it is necessary to raise the pressure; for short and stubby tips apply a lower pressure.
3. The tip of the needle can also be unsealed by dipping into 15 μL of hydrofluoric acid (HF) for about 1 s. Stop the reaction by immediately rinsing it in a drop of distilled H_2O. The size of the opening can be controlled by the time in the HF *(11)*. Depending on the type of needle puller, needles with an open tip can be produced (some trial and error may be required to obtain the proper settings on the machine).
4. If a needle gets clogged during injection the tip can be carefully broken again. Sometimes it is enough to just apply some higher pressure to recover the needle. It is advisable to prepare more than one needle at a time as a replacement backup.
5. After microinjection highly desiccated animals can be recovered more effectively by using a special recovery buffer for a longer time period (up to some hours).
6. Transgenic animals frequently display mosaic expression of transgenes, resulting in a variable phenotype. Mosaicism and low inheritance of extrachromosomal arrays may improve by maintaining the transgenic lines over several generations. It is recommended to obtain and study at least 3 independent transgenic lines.
7. Young adult hermaphrodites with 5–10 eggs inside transform best.
8. Siliconized microcentrifuge tubes prevent beads from sticking to the tube walls.
9. A vortex mixer with a tube holder is particularly useful when handling multiple samples (available from Fisher Scientific).
10. The bead stock can be stored at 4°C for at least 2 weeks. Vortex prepared gold beads for at least 5 min prior to use to ensure complete suspension.
11. Keep gold beads in suspension during transfers.
12. To optimize transformation efficiency try a range of DNA amounts (0.5–2×). When co-transforming plasmids, use equal amounts of DNA.

13. If a ring of beads forms on the tube wall, scrape it off with pipette tip.
14. Resuspending the beads may be difficult of crosslinking by DNA. Vortex manually instead of using the turbomixer. To test resuspension, put 1–2 μL on a slide and view under a compound microscope. You should see single beads and small clusters.
15. Save original bombardment plates, as these will occasionally also yield transformants.
16. You may examine worms under the microscope; beads are visible at 100×.
17. Instead of transferring a high number (~500) of worms to a single plate, worms from **step 2** can be grown to starvation on the same plates (7–14 days, depending on the size of the plate and bacterial lawn). F2 progeny are generated on the same plates. The agar substrate containing starved animals is then cut into pieces of similar size (about 1 cm^2) and transferred to 9-cm NGM plates each. After 2–3 days, 15–20 adult transgenic worms are selected and moved to individual plates. The progeny of these animals are screened for genetic transmission frequencies approaching 100%, which indicates stable integration events. Once integrants are identified, proceed with **step 5**.

Acknowledgements

We thank Ekkehard Schulze from the BaumeisterLab for valuable information on DNA-coated microparticle bombardment methodologies and members of the Tavernarakis Lab for comments and suggestions. Work in the authors' laboratory is supported by grants from the EU 6th Framework Programme. N. K. is the recipient of a Manasaki doctoral fellowship.

References

1. Brenner, S. (1974) The genetics of *Caenorhabditis elegans*. *Genetics*, 77, 71–94
2. Baumeister, R. and Ge, L. (2002) The worm in us – *Caenorhabditis elegans* as a model of human disease. *Trends Biotechnol*, **20**, 147–148
3. Poulin, G., Nandakumar, R. and Ahringer, J. (2004) Genome-wide RNAi screens in *Caenorhabditis elegans*: impact on cancer research. *Oncogene*, **23**, 8340–8345
4. Kenyon, C. (2005) The plasticity of aging: insights from long-lived mutants. *Cell*, **120**, 449–460
5. Chen, N., Harris, T.W., Antoshechkin, I., Bastiani, C., Bieri, T., Blasiar, D., Bradnam, K., Canaran, P., Chan, J., Chen, C.K. et al. (2005)

WormBase: a comprehensive data resource for Caenorhabditis biology and genomics. *Nucleic Acids Res*, **33**, D383–D389

6. Chalfie, M., Tu, Y., Euskirchen, G., Ward, W.W. and Prasher, D.C. (1994) Green fluorescent protein as a marker for gene expression. *Science*, **263**, 802–805
7. Miller, D.M., III, Desai, N.S., Hardin, D.C., Piston, D.W., Patterson, G.H., Fleenor, J., Xu, S. and Fire, A. (1999) Two-color GFP expression system for C. elegans. *Biotechniques*, **26**, 914–918, 920–911
8. Fire, A., Harrison, S.W. and Dixon, D. (1990) A modular set of lacZ fusion vectors for studying gene expression in *Caenorhabditis elegans*. *Gene*, **93**, 189–198
9. Fire, A., Kondo, K. and Waterston, R. (1990) Vectors for low copy transformation of *C. elegans*. *Nucleic Acids Res*, **18**, 4269–4270
10. Mello, C. and Fire, A. (1995) DNA transformation. *Methods Cell Biol*, **48**, 451–482
11. Mello, C.C., Kramer, J.M., Stinchcomb, D. and Ambros, V. (1991) Efficient gene transfer in C. elegans: extrachromosomal maintenance and integration of transforming sequences. *EMBO J*, **10**, 3959–3970
12. Kimble, J., Hodgkin, J., Smith, T. and Smith, J. (1982) Suppression of an amber mutation by microinjection of suppressor tRNA in *C. elegans*. *Nature*, **299**, 456–458
13. Fire, A. (1986) Integrative transformation of Caenorhabditis elegans. *EMBO J*, **5**, 2673–2680
14. Jackstadt, P., Wilm, T.P., Zahner, H. and Hobom, G. (1999) Transformation of nematodes via ballistic DNA transfer. *Mol Biochem Parasitol*, **103**, 261–266
15. Praitis, V., Casey, E., Collar, D. and Austin, J. (2001) Creation of low-copy integrated transgenic lines in *Caenorhabditis elegans*. *Genetics*, **157**, 1217–1226
16. Wilm, T., Demel, P., Koop, H.U., Schnabel, H. and Schnabel, R. (1999) Ballistic transformation of *Caenorhabditis elegans*. *Gene*, **229**, 31–35
17. Praitis, V. (2006) Creation of transgenic lines using microparticle bombardment methods. *Methods Mol Biol*, **351**, 93–107
18. Berezikov, E., Bargmann, C.I. and Plasterk, R.H. (2004) Homologous gene targeting in *Caenorhabditis elegans* by biolistic transformation. *Nucleic Acids Res*, **32**, e40
19. Tavernarakis, N., Wang, S.L., Dorovkov, M., Ryazanov, A. and Driscoll, M. (2000) Heritable and inducible genetic interference by double-stranded RNA encoded by transgenes. *Nat Genet*, **24**, 180–183
20. Johnson, N.M., Behm, C.A. and Trowell, S.C. (2005) Heritable and inducible gene knockdown in *C. elegans* using Wormgate and the ORFeome. *Gene*, **359**, 26–34
21. Jin, Y. (2005) *C. elegans* – A practical approach. Oxford University Press, Oxford
22. Rosenbluth, R.E., Cuddeford, C. and Baillie, D.L. (1985) Mutagenesis in *Caenorhabditis elegans*. II. A spectrum of mutational events induced with 1500 r of gamma-radiation. *Genetics*, **109**, 493–511
23. Sambrook, J. and Russell, D.W. (2001) *Molecular Cloning: A Laboratory Manual.* 3rd edition. Cold Spring Harbor Laboratory Press, Cold Spring Harbor, NY
24. Hodgkin, J. (2005) *C. elegans* – a practical approach. Oxford University Press, Oxford
25. Schnabel, R. (2005) *C. elegans* – a practical approach. Oxford University Press, Oxford
26. Maduro, M. and Pilgrim, D. (1995) Identification and cloning of unc-119, a gene expressed in the *Caenorhabditis elegans* nervous system. *Genetics*, **141**, 977–988
27. Granato, M., Schnabel, H. and Schnabel, R. (1994) pha-1, a selectable marker for gene transfer in *C. elegans*. *Nucleic Acids Res*, **22**, 1762–1763
28. Zwaal, R.R., Van Baelen, K., Groenen, J.T., van Geel, A., Rottiers, V., Kaletta, T., Dode, L., Raeymaekers, L., Wuytack, F. and Bogaert, T. (2001) The sarco-endoplasmic reticulum Ca2+ ATPase is required for development and muscle function in *Caenorhabditis elegans*. *J Biol Chem*, **276**, 43557–43563
29. Gu, T., Orita, S. and Han, M. (1998) Caenorhabditis elegans SUR-5, a novel but conserved protein, negatively regulates LET-60 Ras activity during vulval induction. *Mol Cell Biol*, **18**, 4556–4564
30. Clark, S.G., Lu, X. and Horvitz, H.R. (1994) The *Caenorhabditis elegans* locus lin-15, a negative regulator of a tyrosine kinase signaling pathway, encodes two different proteins. *Genetics*, **137**, 987–997

Chapter 3

Transgenesis in Zebrafish with the *Tol2* Transposon System

Maximiliano L. Suster, Hiroshi Kikuta, Akihiro Urasaki, Kazuhide Asakawa, and Koichi Kawakami

Summary

The zebrafish (*Danio rerio*) is a useful model for genetic studies of vertebrate development. Its embryos are transparent and develop rapidly outside the mother, making it feasible to visualize and manipulate specific cell types in the living animal. Zebrafish is well suited for transgenic manipulation since it is relatively easy to collect large numbers of embryos from adult fish. Several approaches have been developed for introducing transgenes into the zebrafish germline, from the injection of naked DNA to transposon-mediated integration. In particular, the *Tol2* transposable element has been shown to create insertions in the zebrafish genome very efficiently. By using *Tol2*, gene trap and enhancer trap vectors containing the GFP reporter gene or yeast transcription activator Gal4 gene have been developed. Here we outline methodology for creating transgenic zebrafish using *Tol2* vectors, and their applications to visualization and manipulation of specific tissues or cells in vivo and for functional studies of vertebrate neural circuits.

Key words: Zebrafish, Transgenesis, *Tol2*, Transposon, Microinjection, Gal4-UAS, Neuronal function, Tetanus toxin, GFP

1. Introduction

Transgenesis is used in a wide variety of research applications ranging from the study of gene expression to the creation of animal models of human diseases. The zebrafish (*Danio rerio*) is particularly well suited for these applications because adults are easy to breed, and embryonic development proceeds rapidly from the fertilized egg to an embryo containing most major organs within the first 2 days of life *(1)*. The speed of this process and the transparency of the zebrafish embryo are two major advantages for analysis of vertebrate gene function in this model organism.

Elizabeth J. Cartwright (ed.), *Transgenesis Techniques*, Methods in Molecular Biology, vol. 561
DOI 10.1007/978-1-60327-019-9_3, © Humana Press, a part of Springer Science+Business Media, LLC 2009

This can be accomplished using various strategies, including chemical mutagenesis *(2, 3)*, retroviral insertional mutagenesis *(4)*, gene knockdown via antisense RNA oligonucleotides *(5)*, as well as ectopic expression of dominant-negative constructs *(6–8)*. Transgenic tools enabled in vivo labeling and detailed observation of specific cell types using fluorescent reporters and thereby complement mutagenesis by facilitating targeted genetic screens *(9, 10)*.

Transgenic zebrafish were first generated by microinjection of naked DNA *(11)*. In this technique, a plasmid DNA is linearized with a restriction enzyme, purified, and then microinjected into the cytoplasm of one-cell stage embryos. On average, a few percent of the injected fish transmit the transgene to the next generation. These transgenes tend to form tandem arrays or concatamers at the integration site, which in some cases may lead to variegated expression or silencing in the following generation. Thus, while numerous transgenic fish have been generated with this technique *(12–14)*, the germline transmission frequency and the reliability of transgene expression have been low. Pseudotyped retroviral vectors have also been successfully used for transgenesis in zebrafish, particularly for genome-wide insertional mutagenesis *(15, 16)* and enhancer trapping *(17)*. However, the retroviral vectors can only carry inserts of small sizes and their application in the laboratory is labor intensive.

To further improve the rate and ease of transgenesis in zebrafish and to create vectors that are useful for genetic analyses in this model vertebrate, several researchers turned to transposable elements *(18–21)*. Among them, the *Tol2* transposable element from the medaka fish appears to have the highest rate of genomic integration in the germ lineage and is now widely used as a genetic tool (reviewed in *(22, 23)*). In addition, a vector containing the minimal DNA sequences required for *Tol2* transposition has been described, making preparation of new transgene constructs relatively easy *(24, 25)*. Many *Tol2* vectors have been reported to date, including vectors that facilitate the rapid construction of promoter- or protein-GFP fusions through the Gateway technology *(26–28)*. These vectors can be used for expression of any foreign genes in zebrafish embryos by transient and stable transgenic approaches. Furthermore, *Tol2* gene and enhancer trap vectors have been developed for gene expression studies and mutagenesis *(29–31)*. More recently, *Tol2* was used to create transgenic zebrafish for targeted gene expression in specific tissues and cells using the binary Gal4-UAS system *(32–34)*. The aim of this chapter is to provide an overview of zebrafish transgenesis using the *Tol2* transposon system, including methods for transient gene expression, generation of stable transgenic lines, isolation of gene and enhancer trap fish expressing fluorescent proteins, and targeted gene expression studies with the Gal4-UAS system.

2. Materials

2.1. Making Transgene Constructs with Tol2 Vectors

2.1.1. Construction Using a Minimal Tol2 Vector

1. pT2AL200R150G *(24)* **(Fig. 1a)**.
2. Restriction enzymes and T4 DNA ligase (e.g., TaKaRa, Japan).

2.1.2. Construction Using the Gateway Technology

1. pT2H-GW-hspGFP **(Fig. 1b)**.
2. pCR8/GW/TOPO TA cloning kit (Invitrogen).
3. Gateway LR clonase enzyme mix (Invitrogen).
4. QIAprep spin miniprep kit (QIAGEN).

2.2. Transient and Stable Transgenesis

2.2.1. Transcription of Tol2 Transposase mRNA In Vitro

1. pCS-TP (a plasmid to synthesize transposase mRNA) *(29)*.
2. *Not*I.
3. mMESSAGE mMACHINE SP6 kit (Ambion).
4. Quick Spin Columns for RNA purification (Roche).

2.2.2. Microinjection

1. QIAfilter plasmid maxi kit (QIAGEN).
2. Mating boxes (AQUA SCHWARZ, Germany).
3. Puller (e.g., cat. no. PC-10, Narishige).
4. Glass capillary (Narishige).
5. Micropipette holder (Leica).
6. Teflon tube (inside diameter: 0.9 mm, outside diameter: 2.0 mm).
7. DNA/RNA mixture for microinjection: 10 μL of 0.4 M KCl, 2 μL of Phenol red solution (Gibco), 2 μL of 250 ng/μL transposase mRNA, 2 μL of 250 ng/μL plasmid DNA, and 4 μL of water to reach a final volume of 20 μL.
8. Petri dish, 60 mm in diameter.
9. Dissecting microscope (e.g., Nikon SMZ645).

2.2.3. Excision Assay

1. PCR 8-strip tube (e.g., BMBio, Japan).
2. Embryo lysis buffer: 10 mM Tris-HCl of pH 8.0, 10 mM EDTA, 200 μg/mL proteinase K.
3. Reagents for PCR: Expand High-Fidelity PCR system (Roche); 10× reaction buffer IV and 25 mM $MgCl_2$ (ABgene); dNTP mixture (2.5 mM each).

2.2.4. Transient Transgenesis: Application to Analysis of a Genomic cis-Regulatory Element

1. Fluorescent stereoscope (e.g., MZ16 FA, Leica).

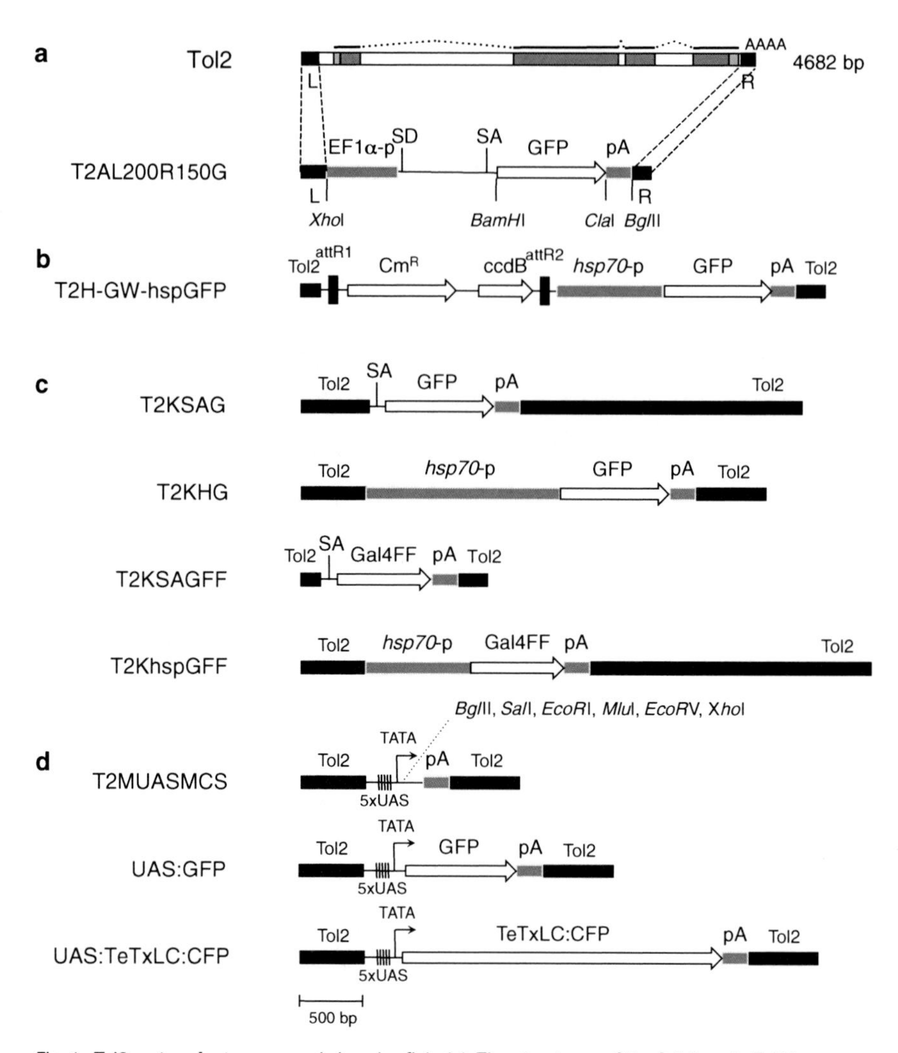

Fig. 1. *Tol2* vectors for transgenesis in zebrafish. (**a**). The structures of the full-length *Tol2* transposable element and the minimal *Tol2* vector T2AL200R150G. *Tol2* is 4682 bp in length and encodes mRNA for the transposase (*dotted lines* indicate introns). T2AL200R150G contains 200-bp and 150-bp DNA from the left (L) and the right (R) terminals of *Tol2*, the *Xenopus* EF1α promoter (EF1α-p), the rabbit-β-globin intron (*thin line* between SD and SA) the GFP gene and the SV40 polyA signal (pA). Unique restriction enzyme sites are indicated. (**b**). The structure of the enhancer assay vector T2H-GW-hspGFP. A DNA fragment cloned on an entry clone can be transferred between attR1 and attR2 of T2H-GW-hspGFP by the Gateway technology. The Gateway cassette is composed of Cm^R (the chloramphenicol resistance gene), *ccdB*, attR1, and attR2. *hsp70*-p: the zebrafish *hsp70* promoter. (**c**). The structures of the gene and enhancer trap constructs. The T2KSAG gene trap construct contains the splice acceptor (SA) from the rabbit β-globin gene and the EGFP gene. The T2KHG enhancer trap construct contains the *hsp70* promoter and the EGFP gene. The T2KSAGFF gene trap construct contains the splice acceptor and the Gal4FF gene. The pT2KhspGFF enhancer trap construct contains the *hsp70* promoter and the Gal4FF gene. (**d**). The structures of the UAS reporter and effector constructs. T2MUASMCS is a cloning vector that contains five tandem repeats of the Gal4-target sequence (5x UAS), followed by a minimal TATA sequence and multiple cloning site (unique cloning sites are indicated) upstream of polyA. UAS:GFP contains the GFP gene downstream of 5x UAS. UAS:TeTxLC:CFP contains a fusion of the tetanus toxin light-chain gene (TeTxLC) and blue fluorescent protein (CFP) downstream of 5x UAS.

2.2.5. Stable Transgenesis

1. DNA extraction buffer: 10 mM Tris–HCl of pH 8.2, 10 mM EDTA, 200 mM NaCl, 0.5% SDS, 200 μg/mL proteinase K.
2. Fish food: paramecia, brine shrimp, Tetramin flakes.
3. Fish tanks and systems for maintenance of fish (e.g., XR-1, Marine Biotech).

2.3. The Tol2-Mediated Gal4 Gene Trap and Enhancer Trap Methods in Zebrafish

2.3.1. The Gal4 Gene Trap and Enhancer Trap Screens

1. pT2KSAGFF *(34)*(**Fig. 1c**).
2. pT2KhspGFF *(34)*(**Fig. 1c**).
3. 9-well glass depression plate (Corning).

2.3.2. Construction of UAS Reporter and Effector Fish

1. pT2MUASMCS *(34)* (**Fig. 1d**).

2.3.3. Targeted Gene Expression with the Gal4-UAS System and Its Application to Genetic Studies of Vertebrate Neural Circuits

1. UAS:TeTxLC:CFP effector fish *(34)* (**Fig.** 1d).
2. 4% PFA/PBS: 1× PBS, pH7.4 (Gibco) containing 4% paraformaldehyde.
3. 1% BSA/PBS: 1×PBS containing 1% bovine serum albumin (Sigma).
4. PBS-TX: 1× PBS containing 0.5% Triton X-100.
5. Primary antibodies: rabbit polyclonal anti-GFP antibody (Invitrogen), mouse monoclonal anti-Hb9 antibody (Developmental Studies Hybridoma Bank), mouse monoclonal zn-12 antibody (the Zebrafish International Resource).
6. Secondary antibodies: goat antirabbit Alexa Fluor 488 and goat antimouse Alexa Fluor 633 (Molecular Probes).
7. Confocal microscope (e.g., Zeiss LSM510META).
8. High-speed digital video camera (e.g., FASTCAM-512PCI, Photoron).

2.4. Genomic Analysis of Tol2 Insertions

2.4.1. Analysis of Tol2 Insertions by Southern Blot Hybridization

1. Tricaine: 0.025% ethyl 3-aminobenzoate methanesulfonate salt.
2. DNA extraction buffer: 10 mM Tris–HCl of pH 8.2, 10 mM EDTA, 200 mM NaCl, 0.5% SDS, 200 ng/μL proteinase K.
3. Primers for amplification of Gal4FF DNA: Gal4FF-f2 (ATG AAG CTA CTG TCT TCT); Gal4FF-r2 (TCT AGA TTA GTT ACC CGG).
4. QIAquick gel extraction kit (QIAGEN, Germany).
5. Prime It-II, random primer labeling kit (Stratagene).
6. Mini Quick Spin DNA columns (Roche).
7. 10× SSC: 1.5 M NaCl, 0.15 M sodium citrate.

8. Hybridization buffer: 0.25 M Na_2HPO_4 (pH adjusted to 7.4 with H_3PO_4), 1 mM EDTA, 10 g/L bovine serum albumin, 7% SDS.
9. Wash solutions: 0.1×SSC; 0.1% SDS and 0.1× SSC.
10. Spectrometer.
11. Vacuum transfer apparatus (e.g., BIO CRAFT, Japan).
12. Hybond-N+ (Amersham).
13. Hybridization oven and tubes (e.g., Robins Scientific).
14. X-ray cassette.

2.4.2. Identification of Tol2 Integration Sites by Adaptor-Ligation PCR

1. The GATC adaptor: hybridize AL (CTA ATA CGA CTC ACT ATA GGG CTC GAG CGG CCG CGG GGG CAG GT) with GATC-AS (GAT CAC CTG CCC CCG CTT).
2. Primers for the first PCR: Ap1 (GGA TCC TAA TAC GAC TCA CTA TAG GG); 175L-out (TTT TTG ACT GTA AAT AAA ATT G) for the left end; 150R-out (AAT ACT CAA GTA CAA TTT TA) for the right end.
3. Primers for the second PCR: Ap2 (CAC TAT AGG GCT CGA GCG G); 150L-out (GAG TAA AAA GTA CTT TTT TTT CT) for the left end; 100R-out (AGA TTC TAG CCA GAT ACT) for the right end.
4. Sequencing primers: 100L-out (AGT ATT GAT TTT TAA TTG TA) for the left end; 100R-out for the right end.
5. *Mbo*I and other appropriate restriction enzymes; T4 DNA ligase.
6. Reagents for DNA sequencing: BigDye terminator v3.1 cycle sequencing kit; 5× sequencing buffer for diluting enzyme; HiDi formamide (Applied Biosystems).
7. ABI PRISM 3130x1 DNA sequence analyzer (Applied Biosystems).

2.4.3. Discovery of the Trapped Gene by 5′ RACE

1. Trizol Reagent (Invitrogen).
2. A primer for RT: Gal4-r3 (TTC AGA CAC TTG GCG CAC TTC GG).
3. Nested primers for PCR: Gal4-r2 (TTT AAG TCG GCA AAT ATC GCA TG); Gal4-r1 (TTC GAT AGA AGA CAG TAG CTT CA).
4. 5′RACE system for rapid amplification of cDNA ends, version 2.0 (Invitrogen).
5. TOPO TA cloning (Invitrogen).
6. Teflon tissue grinder (Wheaton).

3. Methods

3.1. Making Transgene Constructs with Tol2 Vectors

3.1.1. Construction Using a Minimal Tol2 Vector

A minimal *Tol2* vector T2AL200R150G contains 200-bp and 150-bp DNA on the left and the right terminal regions of the full-length *Tol2 (24)* (**Fig. 1a**). Any DNA fragments can be cloned between these sequences, namely between the *Bgl*II and *Xho*I sites (**Fig. 1a**). The *Tol2* vector has a large cargo capacity. At present, it has been shown that a DNA insert of ~10 kb can be cloned into the vector without reducing the transposition efficiency *(24, 25)*.

1. Digest pT2AL200R150G with *Xho*I and *Bgl*II, and replace the DNA fragment containing the GFP expression cassette with a construct of your interest. Also, a promoter and cDNA can be placed upstream of the polyA signal, namely between *Xho*I and *Cla*I, and a promoter can be cloned between *Xho*I and *Bam*HI to express GFP.

3.1.2. Construction Using the Gateway Technology

To ease the cloning step, vectors compatible with the Gateway technology have been described *(26–28)*. pT2H-GW-hspGFP, which contains the zebrafish *hsp70* promoter and the EGFP gene, is designed to detect an enhancer activity (**Fig. 1b**). When the *hsp70* promoter is activated by an enhancer cloned between attR1 and attR2 through the Gateway technology, GFP is expressed (**Figs. 1b** and **2**).

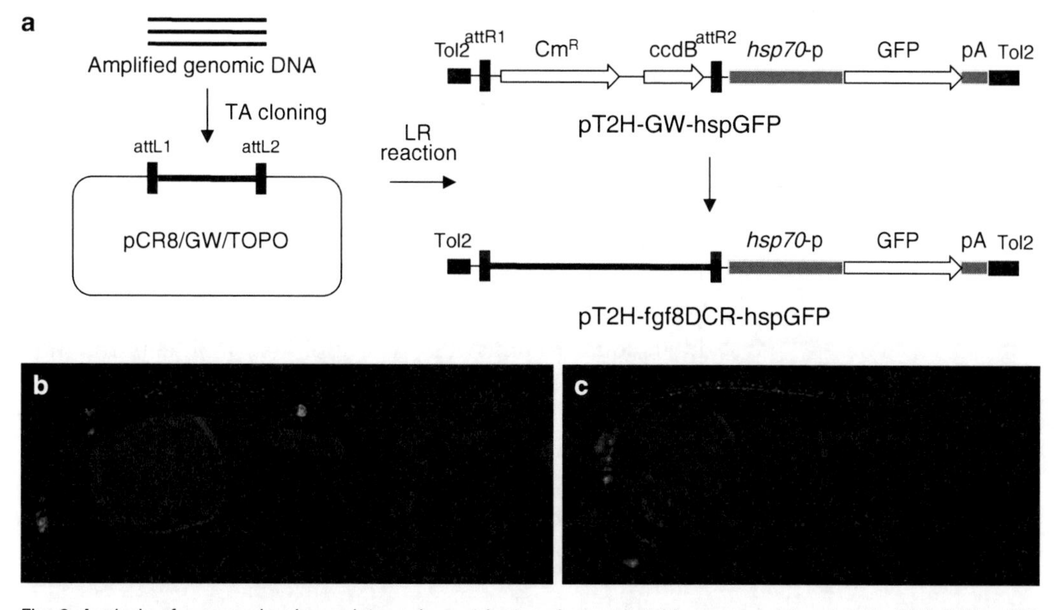

Fig. 2. Analysis of a genomic *cis*-regulatory element by transient and stable transgenesis using the T2H-GW-hspGFP enhancer assay vector. (**a**). A scheme for construction. A putative enhancer from the *fgf8* locus (fgf8DCR) is amplified by PCR and cloned into pCR8/GW/TOPO between the attL1 and attL2 sites. The fgf8DCR fragment is transferred to pT2H-GW-hspGFP between the attR1 and the attR2 through LR recombinase-mediated reaction, resulting in pT2H-fgf8DCR-hspGFP (**b**) Transient transgenesis. Mosaic GFP expression is observed at 24 hpf in the spinal cord and other regions in an embryo injected with pT2H-fgf8DCR-hspGFP and transposase mRNA. (**c**). Stable transgenesis. Nonmosaic GFP expression is observed at 24 hpf in the spinal cord in an F1 transgenic embryo.

1. Amplify a genomic DNA of interest by PCR (*see* **Note 2**) and clone the PCR product into pCR8/GW/TOPO by TA cloning according to the manufacturer's instructions. Confirm the structure of the cloned DNA by sequencing.
2. Prepare plasmid DNA with QIAprep spin miniprep kit. Mix the plasmid DNA with pT2H-GW-hspGFP and LR clonase enzyme mix according to the manufacturer's instructions. Perform LR reaction and transfer the genomic DNA fragment (in this case, fgf8DCR, an enhancer in the *fgf8* locus) from the pCR8/GW to pT2H-GW-hspGFP, giving rise to pT2H-fgf8DCR-hspGFP (**Fig. 2a**).

3.2. Transient and Stable Transgenesis

A scheme for transient and stable transgenesis in zebrafish is shown in **Fig. 3**.

3.2.1. Transcription of Tol2 Transposase mRNA In Vitro

A transcript of the *Tol2* element was identified and its cDNA was cloned *(35)*. pCS-TP contain the cDNA of the transposase gene (**Fig. 1a**) downstream of the SP6 promoter *(29)* and is used to synthesize transposase mRNA in vitro.

1. Digest pCS-TP with *Not*I, purify once by phenol–chloroform extraction, precipitate with ethanol, rinse once with 70% ethanol, and dissolve in nuclease-free water at the concentration of ~1 μg/1μL.
2. By using 1 μg of the linearized DNA as a template and mMESSAGE mMACHINE SP6 kit, synthesize mRNA according to the manufacturer's instructions.
3. After mRNA synthesis and treatment with DNase I, add nuclease-free water up to 100 μL, and purify the sample with a minispin column according to the manufacturer's instructions. Adjust the volume of the eluted solution to 135 μL. Add 15 μL ammonium acetate stop solution and extract RNA once with an equal volume of phenol/chloroform and then with an equal volume of chloroform. Transfer the aqueous phase to a new tube.
4. Add an equal volume of isopropanol, mix well, chill the mixture at least for 15 min at –20°C, and centrifuge at 4°C for 15 min at the maximum speed. Carefully remove the supernatant solution and resuspend the precipitated RNA in 50 μL nuclease-free water. ~30 μg mRNA will be synthesized. Take an aliquot and adjust the concentration to 250 ng/μL for microinjection.

3.2.2. Microinjection

The *Tol2* transposon system is introduced into zebrafish cells by microinjection of a transposon donor plasmid and the transposase mRNA into fertilized eggs.

1. Prepare the transposon-donor plasmid DNA with the QIAfilter plasmid maxi kit. Then purify DNA by phenol–chloroform extraction, precipitate with 1/10 vol of 3 M sodium acetate

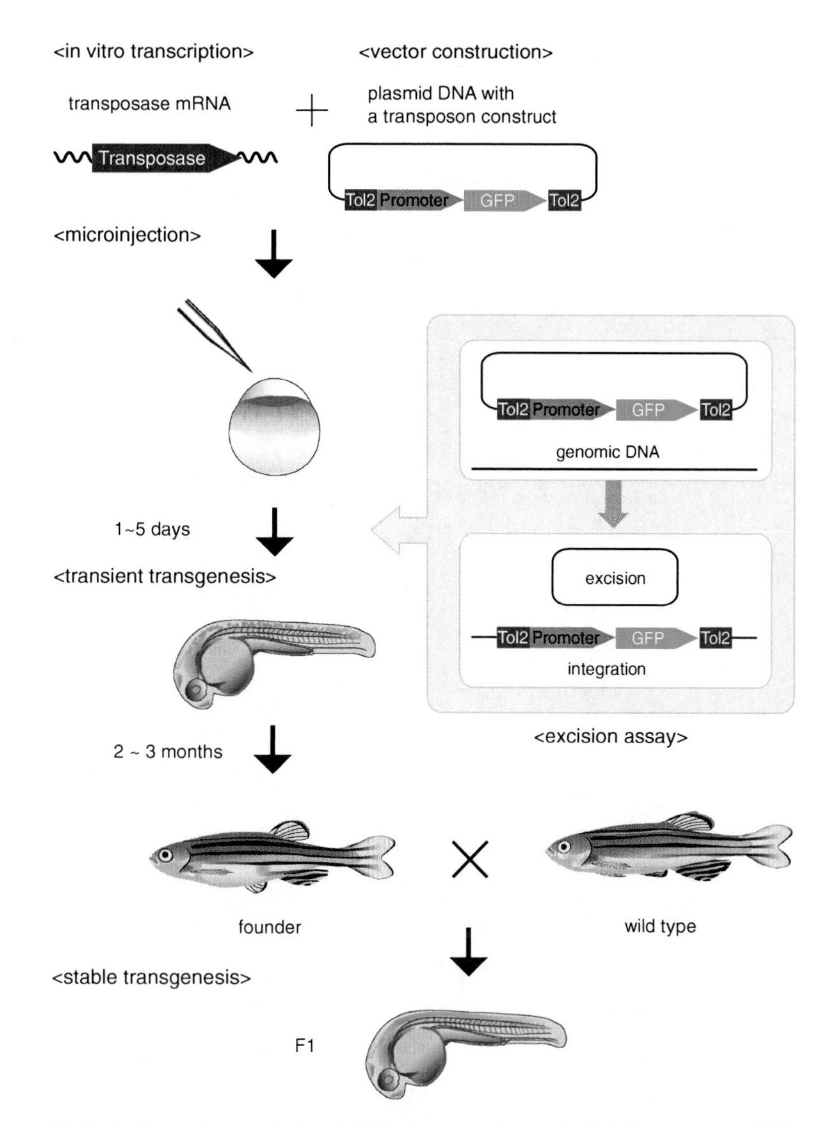

Fig. 3. A scheme for transient and stable transgenesis in zebrafish. Transposase mRNA synthesized by in vitro transcription and a plasmid DNA harboring the *Tol2* construct are coinjected into zebrafish fertilized eggs. The transposase protein synthesized from mRNA catalyzes excision of the *Tol2* construct from the plasmid and integration of the excised *Tol2* into the genome. When the GFP gene is connected to an appropriate promoter, mosaic GFP fluorescence (green spots) is observed in the injected embryo (transient transgenesis). The injected embryos are raised and crossed with wild-type fish. The integrated *Tol2* construct is transmitted to the F1 generation. F1 transgenic embryos show nonmosaic GFP expression (stable transgenesis).

of pH 5.2, and 3 vol of 100% ethanol, rinse once with 70% ethanol, and suspend in nuclease-free water at 250 ng/μL.

2. Place male and female zebrafish in mating boxes in the evening (**Fig. 4a**). Zebrafish mate and lay eggs in the next morning. Microinjection should be carried out at the one-cell stage, namely at 0–30 min post fertilization.

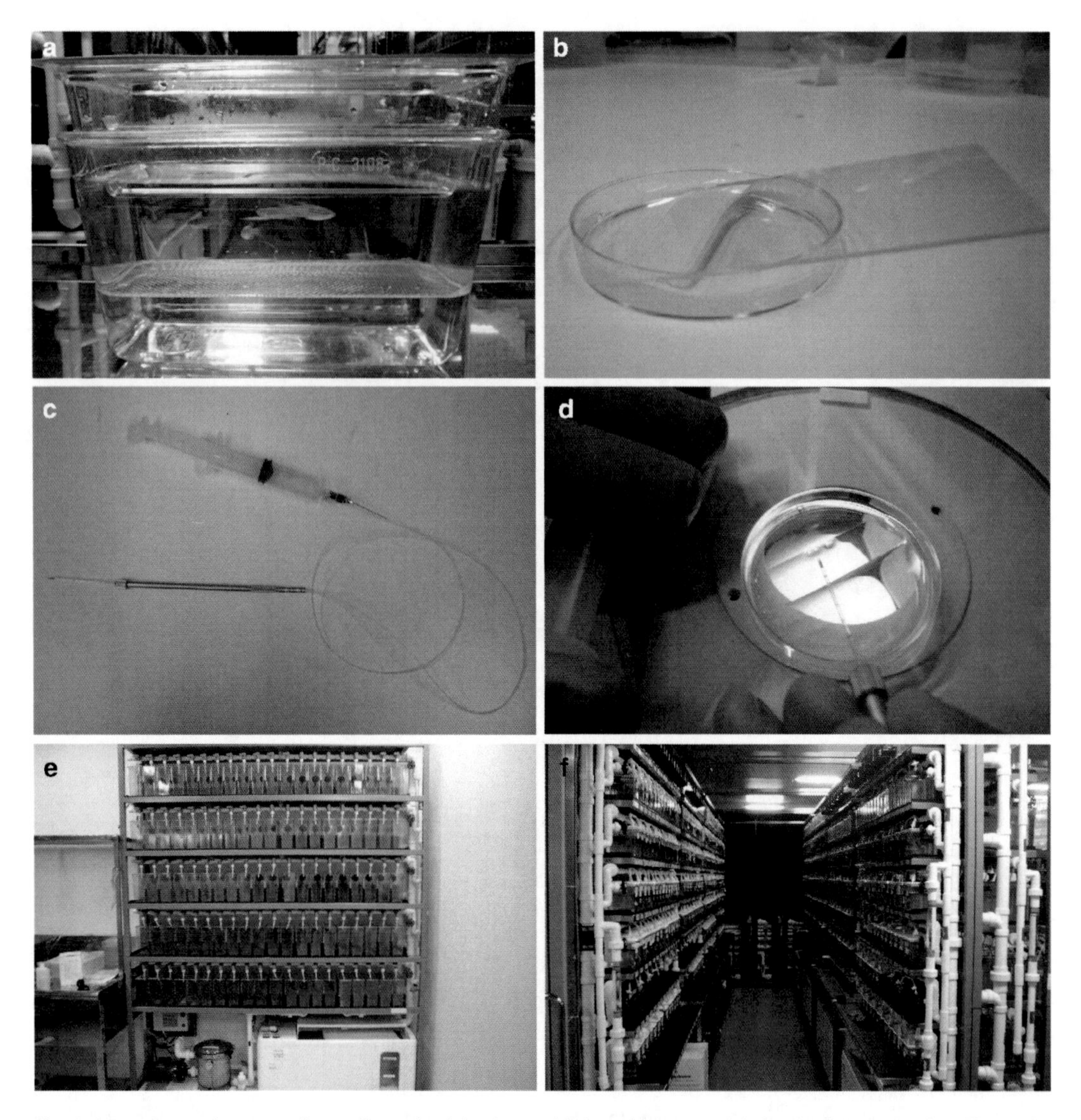

Fig. 4. Apparatus and systems for mating, microinjection and fish maintenance. (**a**). A zebrafish mating box. Two pairs of adult zebrafish are placed in a plastic tank with an insert with a sieve to separate eggs from their parents. (**b**). An agarose ramp for microinjection. Melted agarose is poured in a 60-mm Petri dish and then a glass plate is placed inside at an angle. (**c**). Microinjection apparatus. A glass capillary attached to a holder is connected to a syringe with a teflon tube. The DNA/RNA mixture is loaded into the capillary prior to attachment to the holder. (**d**) Microinjection under a dissecting microscope. The left hand keeps giving air pressure to the capillary. (**e**) A front view of a tank system used for maintenance of zebrafish. (**f**) A view of the fish room. The tank systems are aligned.

3. Make the injection ramp of 1% agarose with a glass plate (**Fig. 4b**). Prepare fine needles for microinjection by pulling a glass capillary, and cut the tip with a sharp blade.
4. Fill the DNA/RNA mixture into the glass capillary. Attach the capillary to a holder and connect to a 20-mL syringe with a Teflon tube (**Fig. 4c**).

5. Inject ~ 1 nL of the mixture (measure the approximate volume by observing the diameter of the injected blob by eye) into the cytoplasm **(Fig. 4d)**. Incubate the injected fish in a dish at 28°C.

3.2.3. Excision Assay

Whether transposition occurred properly can be tested easily and quickly by a single-embryo excision assay *(35)* **(Fig. 5)**.

1. Transfer injected embryos at 10 hpf to wells in a PCR 8-strip tube. Remove as much water as possible. Add 50 μL of embryo lysis buffer and incubate at 50°C for 3 h to overnight. Inactivate the enzyme at 95°C for 5 min.
2. Mix 1 μL of the sample with 2 μL 10× reaction buffer, 2 μL 25 mM $MgCl_2$, 1.5 μL dNTP mix (2.5 mM each), 1 μL of 20 μM exR and exL primers (*see* **Note 1**), 0.2 μL High-Fidelity Taq polymerase, and 11.3 μL water. Perform PCR; 94°C for 2 min and 35 cycles of 94°C for 30 s, 55°C for 30 s and 72°C for 30 s.

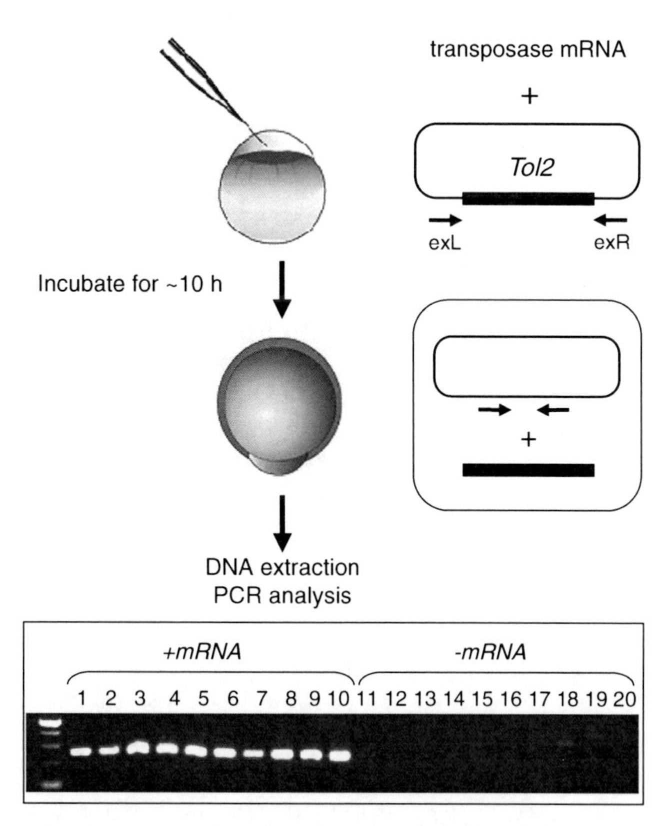

Fig. 5. A scheme for excision assay. In the injected embryos, the *Tol2* construct on a donor plasmid is excised and the excision site is repaired and religated. The exL and exR primers that are located at both sides of *Tol2* amplify short PCR products when excision and religation occurred properly. The gel shows an example of the excision assay. PCR products are detected in embryos injected with both transposase mRNA and a donor plasmid (*lanes 1–10*), but not in embryos injected only with a plasmid (*lanes 11–20*).

3. Check 10 μL of the PCR sample by agarose gel electrophoresis (1.5–2% agarose). The excision product is detected from embryos injected both with the transposon donor plasmid and the transposase mRNA, but not from embryos injected without the mRNA.

3.2.4. Transient Transgenesis: An Application to Analysis of a Genomic cis-Regulatory Element

A gene or a reporter gene injected into fertilized eggs is expressed during embryogenesis, allowing their activities to be analyzed by observing the injected embryos. The *Tol2* transposon system facilitates such transient transgenesis analyses since a *Tol2* construct injected with transposase mRNA is integrated in the genome of somatic cells with high efficiencies. Therefore reliable analyses of *cis*-regulatory elements as well as labeling and manipulation of a large number of cells during embryogenesis have been made possible. For example, when the activities of tissue-specific enhancer fragments were analyzed, ~20% of the injected embryos showed nonmosaic reporter expression *(26)*.

1. Prepare the pT2H-fgf8DCR-hspGFP plasmid DNA and perform microinjection.
2. Incubate the injected embryos on a dish at 28°C. Observe GFP expression under a fluorescent microscope at day 1 and time points of interest (**Fig. 2b**).

3.2.5. Stable Transgenesis

In comparison to transient transgenesis, stable transgenesis (**Fig. 3**) gives rise to fish with more reliable and reproducible expression of a transgene. Also, transgenes can be inherited by offspring in a Mendelian fashion. *Tol2*-mediated transgenesis is a highly efficient method to create stable transgenic fish since 50–70% of injected fish transmit genomic insertions of the injected *Tol2* construct to the next generation *(24, 29)*. Furthermore, expression of transgenes created by *Tol2*-mediated transgenesis persists from generation to generation.

1. Incubate the injected fish in a dish at 28°C for the first 5 days. Then transfer the fish to a fish room. Start feeding fish with paramecia for the next 10 days and then with brine shrimp and Tetramin flakes. Fish tanks and systems are commercially available for the maintenance of fish (**Fig. 4e, f**).
2. Raise the injected fish to sexual maturity. This usually takes 2.5–3 months. Cross the injected fish (founder fish) with wild-type fish in mating boxes.
3. Analyze the offspring. When the transgene contains a fluorescent marker such as GFP, observe its expression under a fluorescent stereoscope. Pick up GFP-positive embryos and raise them (**Fig. 2c**).
4. Alternatively, perform PCR on DNA extracted from pooled F1 embryos; transfer ~50 F1 embryos (24–36 hpf) into a microtube, lyse them with 250 μL of DNA extraction buffer

at 50°C overnight, purify DNA by phenol–chloroform extraction, precipitate with 1/10 vol of 3 M sodium acetate and 3 vol of 100% ethanol, rinse once with 70% ethanol, and suspend in 50 μL TE; 10 mM Tris-HCl of pH7.5, 1 mM EDTA. Use 1 μL of the DNA sample for PCR of 94°C for 2 min and 35 cycles of 94°C for 30 s, 56°C for 30 s, and 72°C for 30 s using transgene-specific primers. When a PCR-positive F1 pool is found, raise their siblings and analyze them individually at the adult stage for the presence of the transgene sequence.

5. We highly recommend analysis of the F1 fish by Southern blot hybridization. F1 fish often carry multiple transposon insertions, and this may complicate analysis of their phenotypes. If all of the F1 fish analyzed carry multiple insertions, cross the fish with the smallest number of transposon insertions with wild-type fish, raise F2 offspring, and analyze them by Southern blot hybridization to identify fish with single insertions (*see* **Subheading 3.4.1**).

3.3. The Tol2-Mediated Gal4 Gene Trap and Enhancer Trap Methods in Zebrafish

Tol2-mediated gene trap and enhancer trap methods have been developed *(29-31)*. The gene trap construct contains a splice acceptor and the GFP gene. When it is integrated in the genome and the splice acceptor traps an endogenous transcript, GFP is expressed in places where the endogenous gene is expressed *(29)*. The enhancer trap construct contains the zebrafish *hsp70* promoter as a minimal promoter and the GFP gene. When it is integrated in the genome and the *hsp70* promoter is activated by a chromosomal enhancer, GFP is expressed under the control of the enhancer *(31)*. These methods have further advanced by incorporating with the Gal4-UAS system *(32–34)*. Namely, transgenic fish expressing Gal4 in specific cells and tissues are created by gene trap and enhancer trap strategies and crossed with transgenic fish carrying a reporter or an effector gene downstream of the Gal4 recognition sequence UAS (**Fig. 6**). The T2KSAGFF gene trap construct contains a splice acceptor from the rabbit β-globin gene and the Gal4FF gene, which encodes a modified version of the Gal4 protein *(34)* (**Fig. 1c**). The T2KhspGFF enhancer trap construct contains the *hsp70* promoter and the Gal4FF gene (**Fig. 1c**). The Gal4 gene trap and enhancer trap methods are useful to visualize, modify, and ablate specific cell types during vertebrate development.

3.3.1. The Gal4 Gene Trap and Enhancer Trap Screens

1. Perform microinjection using a plasmid harboring T2KSAGFF or T2KhspGFF as described earlier. Raise the injected fish to sexual maturity (2.5–3 months).
2. Cross the injected fish with UAS:GFP reporter fish that contain the EGFP gene downstream of five tandem repeats of UAS (5X UAS) (**Fig. 6a**). GFP is expressed when and where

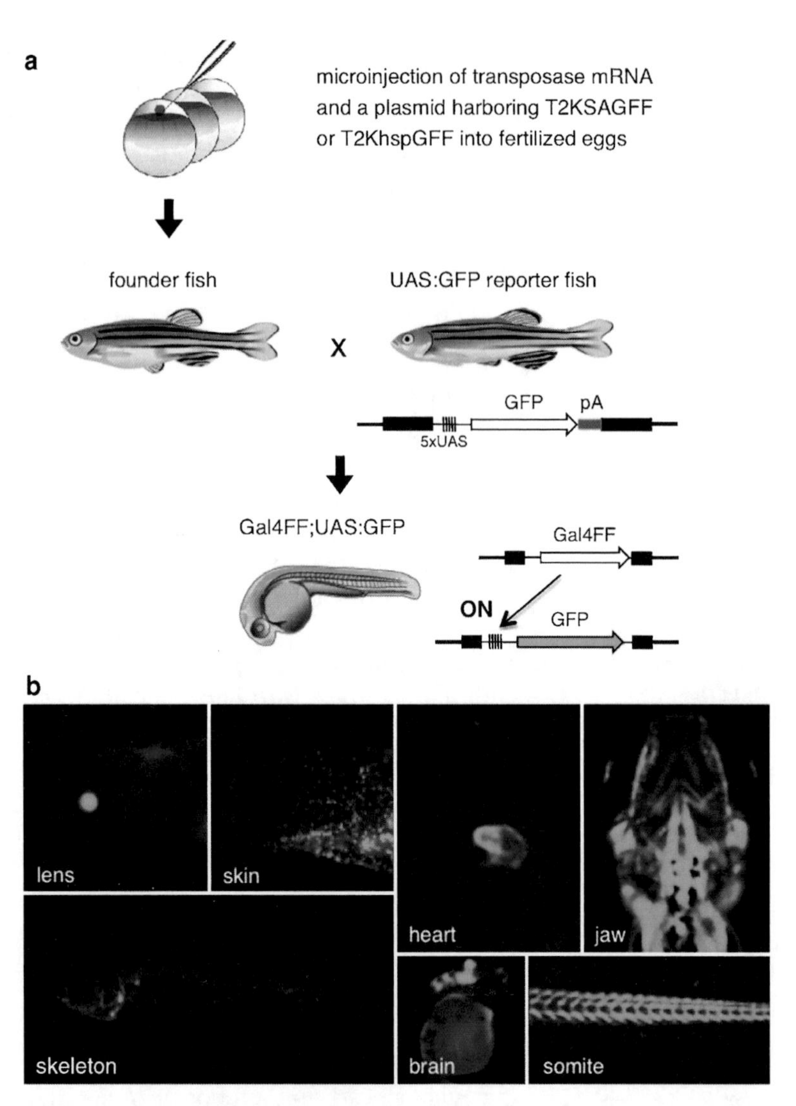

Fig. 6. *Tol2*-mediated Gal4 gene and enhancer trap screens. (**a**) A scheme for Gal4 gene and enhancer trapping. T2KSAGFF (gene trap) and T2KhspGFF (enhancer trap) are injected into fertilized eggs with transposase mRNA. Injected fish are raised and mated with homozygous UAS:GFP reporter fish. Double transgenic F1 embryos that express Gal4FF can express GFP (ON). (**b**) Various GFP expression patterns that represent Gal4FF expression are identified in T2KSAGFF;UAS:GFP double transgenic embryos.

Gal4FF is expressed. Place the F1 embryos on a 9-well glass depression plate and analyze the GFP expression patterns under a fluorescent stereoscope at different time points (e.g., 24 hpf, 48 hpf, 72 hpf, and 5 dpf).

3. Pick up double transgenic embryos that show GFP expression patterns of interest (**Fig. 6b**). GFP-positive F1 embryos may be obtained at the frequency of one per two injected fish for T2KSAGFF and more than one per one injected fish for T2KhspGFF. Raise them to adulthood.

4. In some cases, insertions of the gene trap or enhancer trap construct disrupt transcription, splicing, or translation of endogenous genes *(31, 36)*. To identify such insertional mutations, cross heterozygous male and female fish that carry the same insertion. Mutant phenotypes may be identified in a quarter of their offspring, namely homozygous embryos. To find a gene affected by the insertion, genomic DNA surrounding the integration site should be analyzed by Southern blot hybridization, adaptor-ligation PCR, or inverse PCR (*see* **Subheading 3.4**).

3.3.2. Construction of UAS Reporter and Effector Fish

The T2MUASMCS vector contains a multicloning site (MCS) between 5X UAS and the polyA signal **(Fig. 1d)** *(34)*. This vector is used to create UAS reporter and effector transgenic fish.

1. Insert any gene of interest at the MCS of pT2MUASMCS.
2. Inject the plasmid DNA with the transposase mRNA into fertilized eggs. Raise the injected fish, cross them with wild-type fish, and identify transgenic fish in the F1 progeny by PCR analysis (*see* **Subheading 3.2.5**).

3.3.3. Targeted Gene Expression with the Gal4-UAS System and Its Application to Genetic Studies of Vertebrate Neural Circuits

The Gal4-UAS system is applicable to targeted gene expression. The UAS:TeTxLC:CFP effector fish carry a fusion of the tetanus toxin light-chain gene (TeTxLC) and the CFP gene downstream of 5X UAS **(Fig. 1d)**. TeTxLC cleaves a vesicle membrane protein synaptobrevin and thereby blocks neurotransmitter release from synaptic vesicles *(37)*. When gene and enhancer trap fish that express Gal4FF in subpopulations of neurons are crossed with the UAS:TeTxLC:CFP effector fish, activities of Gal4FF-expressing neurons are reduced or abolished *(34)*. The expression of the TeTxLC:CFP fusion protein is detected by CFP fluorescence and immunohistochemistry. The activity of the toxin can be detected by a behavioral analysis, the touch response assay.

Targeted Expression of TeTxLC:CFP in Gal4FF Gene Trap and Enhancer Trap Fish

1. Cross enhancer trap (hspGGFF27A) and gene trap (SAGFF31B and SAGFF36B) fish lines that express Gal4FF in specific populations of neurons with the UAS:TeTxLC:CFP fish. These *et* and *gt* fish are kept as double transgenic for the Gal4FF and UAS:GFP transgenes. Use males for the Gal4FF fish and females for the UAS:TeTxLC:CFP effector fish to avoid maternal Gal4FF and GFP expression. Pick embryos that show CFP but not GFP fluorescence at 24 hpf. CFP fluorescence observed at 48 hpf is shown in **Fig. 7a**. These embryos are double transgenic for the Gal4FF transgene and UAS:TeTxLC:CFP.
2. Fix the embryos with 4% PFA/PBS for 2 h at room temperature. Then wash with PBS-TX for 5 min four times and once for 30 min.

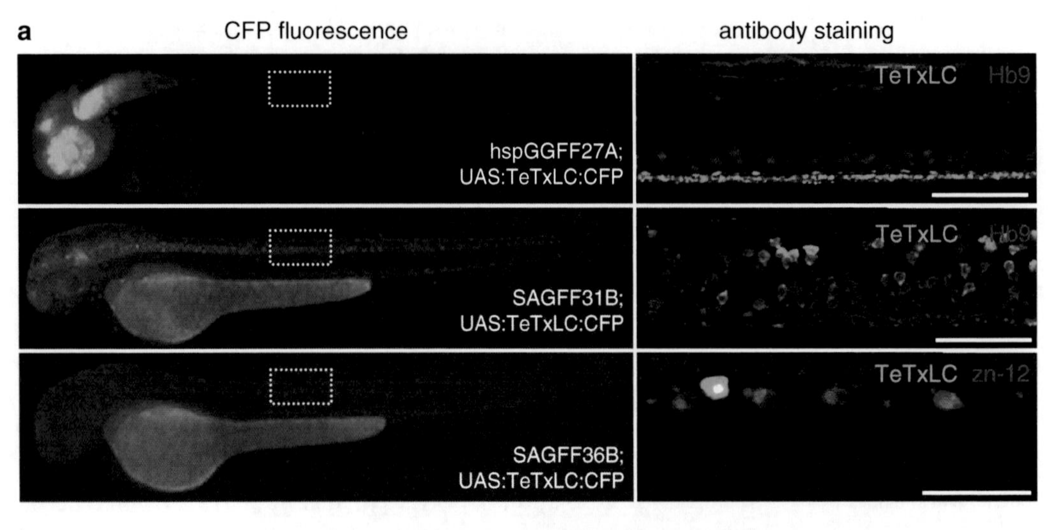

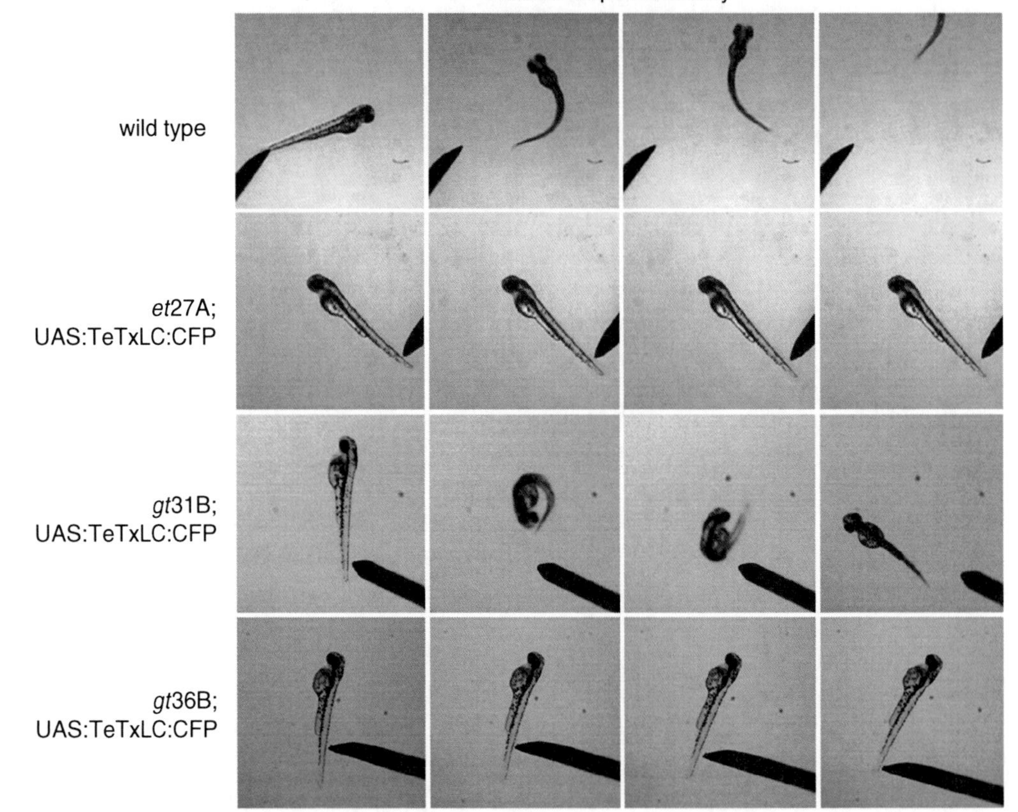

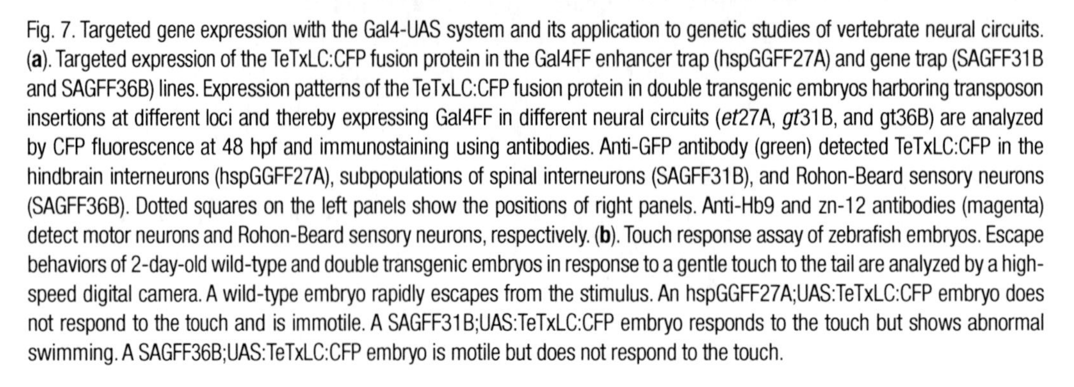

Fig. 7. Targeted gene expression with the Gal4-UAS system and its application to genetic studies of vertebrate neural circuits. (**a**). Targeted expression of the TeTxLC:CFP fusion protein in the Gal4FF enhancer trap (hspGGFF27A) and gene trap (SAGFF31B and SAGFF36B) lines. Expression patterns of the TeTxLC:CFP fusion protein in double transgenic embryos harboring transposon insertions at different loci and thereby expressing Gal4FF in different neural circuits (*et*27A, *gt*31B, and gt36B) are analyzed by CFP fluorescence at 48 hpf and immunostaining using antibodies. Anti-GFP antibody (green) detected TeTxLC:CFP in the hindbrain interneurons (hspGGFF27A), subpopulations of spinal interneurons (SAGFF31B), and Rohon-Beard sensory neurons (SAGFF36B). Dotted squares on the left panels show the positions of right panels. Anti-Hb9 and zn-12 antibodies (magenta) detect motor neurons and Rohon-Beard sensory neurons, respectively. (**b**). Touch response assay of zebrafish embryos. Escape behaviors of 2-day-old wild-type and double transgenic embryos in response to a gentle touch to the tail are analyzed by a high-speed digital camera. A wild-type embryo rapidly escapes from the stimulus. An hspGGFF27A;UAS:TeTxLC:CFP embryo does not respond to the touch and is immotile. A SAGFF31B;UAS:TeTxLC:CFP embryo responds to the touch but shows abnormal swimming. A SAGFF36B;UAS:TeTxLC:CFP embryo is motile but does not respond to the touch.

3. Incubate the embryos in 1% BSA/PBS-TX for 1 h. Add a diluted primary antibody (1/500 dilution for rabbit polyclonal anti-GFP antibody to detect TeTxLC:CFP, 1/25 dilution for mouse monoclonal anti-Hb9 antibody to detect motor neurons, and 1/250 dilution for mouse monoclonal zn-12 antibody to detect Rohon-Beard sensory neurons) to the sample and incubate at 4°C overnight.
4. Remove the primary antibody and wash the sample with PBS-TX for 5 min three times and then for 20 min three times.
5. Incubate the sample in 1% BSA/PBS for 30 min. Add a diluted secondary antibody (1/1,000 dilution for goat antirabbit Alexa Fluor 488 and goat antimouse Alexa Fluor 633, Molecular Probes) and incubate at room temperature for 2 h.
6. Remove the secondary antibody and wash the sample with PBS-TX for 5 min three times and then for 20 min four times. Soak the sample in 30% glycerol, mount on a slide glass, and analyze under a confocal microscope (**Fig. 7a**).

Inhibition of Neuronal Functions by Targeted Expression of TeTxLC:CFP in Gal4FF Gene Trap and Enhancer Trap Fish

1. Place the Gal4FF;UAS:TeTxLC:CFP double transgenic embryos at 48 dpf in a plastic dish.
2. Touch the tail gently with a needle three times. Take images by using a high-speed digital video camera. At this stage, wild-type embryos can respond to the gentle touch in the tail and escape rapidly from the stimulus. The double transgenic embryos in which activities of distinct spinal neural circuits are inhibited display distinct behavioral abnormalities (**Fig. 7b**).

3.4. Genomic Analysis of Tol2 Insertions

It is important to analyze integration sites of the *Tol2* construct in gene trap and enhancer trap fish lines. First, genes expressed in specific cells and tissues may be identified. These should include important developmental genes. Second, by performing comparative analyses of surrounding genomic sequences against other animals, enhancers that regulate such specific expression may be identified.

3.4.1. Analysis of Tol2 Insertions by Southern Blot Hybridization

Transgenic F1 fish created by the *Tol2* transposon system often carry multiple insertions. Therefore, we recommend analyzing the GFP-positive F1 fish by Southern blot hybridization to determine the number of transposon insertions (**Fig.** 8a). When all of the fish analyzed carry multiple insertions, fish with the smallest number of insertions should be outcrossed to wild-type fish to obtain fish with single insertions in the next generation.

1. Anesthetize adult fish with tricaine. Cut tail fins with a blade, submerge in 200 μL of DNA extraction buffer, incubate at 50°C for 3 h to overnight, and dissolve them completely.

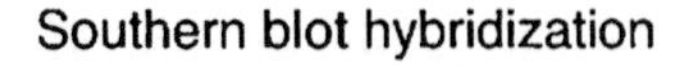

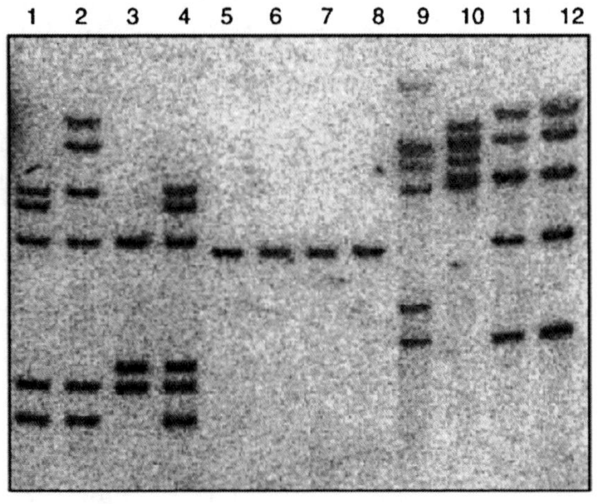

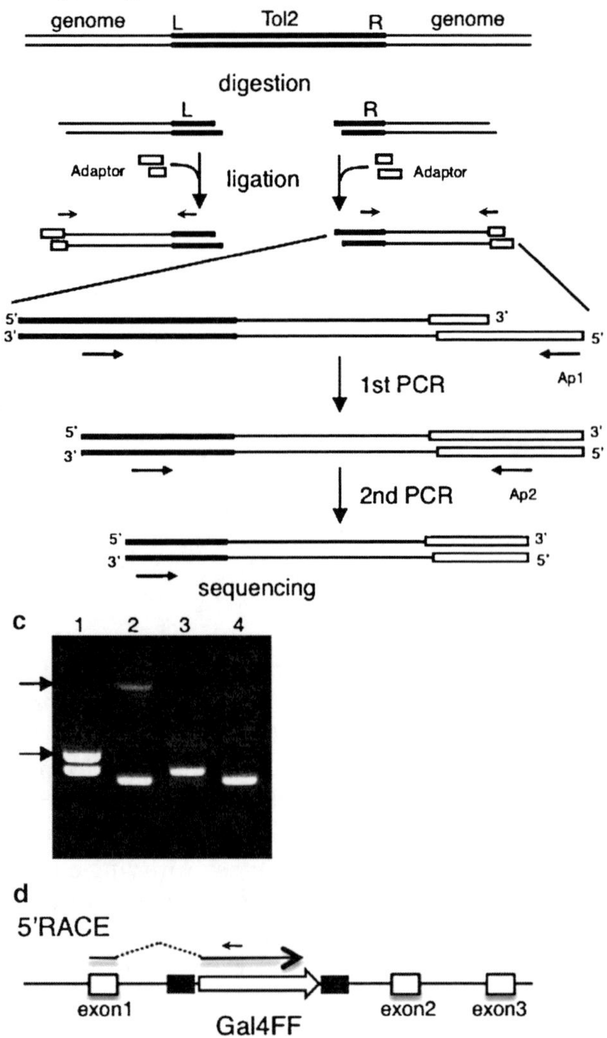

Fig. 8. Genomic analyses of *Tol2* insertions. (**a**). An example of Southern blot hybridization analysis of F1 fish using the ^{32}P-labeled Gal4FF probe. *Lanes 1–4*, *5–8*, and *9–12* are F1 fish from same injected fish. Many fish harbor multiple insertions and some fish harbor single insertions. Note that F1 fish from the same injected fish that show the same GFP expression patterns contain some insertions in common. (**b**). A scheme for adaptor-ligation PCR method. Genomic DNA (*thin line*) containing the *Tol2* sequence (*thick line*) is digested with an appropriate restriction enzyme, ligated to an adaptor (boxes), and amplified by two rounds of PCR using nested primers shown by *arrows*. (**c**). An example of the result obtained from adaptor-ligation PCR. *Lane 1*: PCR products of the left junction from a Gal4FF;UAS:GFP fish. *Lane 2*: PCR products of the right junction from a Gal4FF;UAS:GFP fish. *Lane 3*: A PCR product of the left junction from the UAS:GFP fish. *Lane 4*: A PCR product of the right junction from the UAS:GFP fish. *Arrows* indicate PCR products unique to the Gal4FF transgene. (**d**). 5′ RACE analysis of the trapped gene. A gene whose transcript is trapped by an insertion of the gene trap construct can be identified by performing 5′ RACE.

2. Purify DNA by phenol–chloroform extraction, then precipitate with 1 mL ethanol, rinse once with 70% ethanol, and suspend in 50 μL TE. Measure DNA concentrations by a spectrometer.
3. Digest 5 μg of the genomic DNA with *Bgl*II in 30 μL reaction solution at 37°C for 1 h. Perform electrophoresis by using 1% TAE-agarose gel and take a photo of the gel under UV illuminator.
4. Soak the gel in 0.1 N HCl for 15 min, rinse with deionized water, soak in 0.5 N NaOH for 30 min, rinse with deionized water, and soak in 10× SSC.
5. Place the gel in a vacuum transfer apparatus with Hybond-N+ presoaked with 10× SSC. Perform transfer according to the manufacturer's instructions. After transfer, rinse the membrane in 2× SSC and dry completely at 50°C for 1 h to overnight.
6. Amplify the Gal4FF DNA fragment from T2KSAGFF by PCR using Gal4FF-f2 and Gal4FF-r2. Purify the band from the gel with QIAquick gel extraction kit. Label DNA with ^{32}P-dCTP by using Prime It-II labeling kit. Purify the labeled DNA with a minispin column.
7. Soak the membrane in 25 mL preheated hybridization buffer on a tray and together with the buffer transfer into a hybridization tube. Incubate in a hybridization oven at 65°C for 1 h.
8. Remove hybridization buffer from the tube, and add 10 mL preheated hybridization buffer containing the labeled DNA probe. Perform hybridization at 65°C for overnight.
9. Wash the membrane with ~100 mL wash solution at 65°C for 30 min twice. Transfer the membrane on a tray and rinse with in 0.1× SSC. Wrap the membrane in a plastic wrap, and place on an imaging plate (or an X-ray film) in an X-ray cassette. After exposure for 3 h to overnight, the image can be analyzed by using BAS2500.

3.4.2. Identification of Tol2 Integration Sites by Adaptor-Ligation PCR

Once fish with a single insertion are isolated, it is relatively easy to clone the genomic DNA surrounding the *Tol2* insertion by adaptor–ligation PCR **(Fig. 8b)** (*see* **Note 3**). First, genomic DNA is digested with an appropriate restriction enzyme and ligated with an adaptor. Then junction fragments of *Tol2* and genomic DNA are amplified by using a primer in the *Tol2* sequence and a primer in the adaptor sequence (*see* **Note 4**). It should be noted that, since GFP-positive Gal4 fish are double transgenic for the Gal4FF insertion and the UAS:GFP transgene, which was also created by *Tol2*-mediated transgenesis **(Fig. 1d)**, primers located in the *Tol2* sequence can amplify junction fragments from both transgenes. The junction DNA may also be cloned also by inverse PCR *(38)*.

1. Digest 1 μg of genomic DNA with *Mbo*I in 20 μL reaction solution at 37°C for 1 h. Inactivate the enzyme at 70°C for 15 min. Ligate 2 μL of the DNA sample with the GATC adaptor in a ligation mixture at 16°C for 3 h or overnight.
2. Inactivate the ligase at 70°C for 15 min. Dilute the sample to 1/10 by adding 180 μL of H_2O. Use 2 μL of the sample for the first PCR (30 cycles of 94°C for 30 s, 57°C for 30 s, 72°C for 1 min) in 50 μL 1× PCR buffer containing 1 μM Ap1 and 175L-out or 150R-out.
3. Use 1 μL of 1/10 diluted first PCR product for the second PCR (30 cycles of 94°C for 30 s, 57°C for 30 s, 72°C for 1 min) in 50 μL 1× PCR buffer containing 1 μM Ap2 and 150L-out or 100R-out.
4. Analyze 10 μL of the sample using 1.5% agarose–TAE gel electrophoresis **(Fig. 8c)**. Purify a DNA band that is unique to the Gal4 transgenic fish from the gel with QIAquick gel extraction kit according to the manufacturer's instructions. The final volume will be ~40 μL.
5. Use 4 μL of the DNA sample for sequencing with 100L-out, 100R-out, and BigDye terminator cycle sequencing kit according to the manufacturer's instruction. Mix the sample with 1/10 vol of 3 M sodium acetate and 3 vol of ethanol. Incubate the sample at room temperature for 5 min, centrifuge at 20,400 g for 20 min at 4°C, rinse with 70% ethanol, dry and suspend in 20 μL of HiDi formamide. Analyze the sample using the ABI PRISM 3130xl DNA sequence analyzer.
6. Analyze the identified DNA sequences by using Ensembl and NCBI databases to find candidate genes that are expressed similarly to the GFP expression or are affected by the *Tol2* insertion.

3.4.3. Discovery of the Trapped Gene by 5′ RACE

An unknown endogenous transcript spliced with GFP or Gal4 in the gene trap lines can be discovered by performing 5′RACE *(29, 39)* **(Fig. 8d)**.

1. Transfer ~50 GFP-positive embryos to a teflon tissue grinder. Add 1 mL of Trizol reagent, homogenize, and incubate for 5 min at room temperature.
2. Add 0.2 mL chloroform and mix vigorously by hand for 15 s. Incubate the sample at room temperature for 2–3 min and centrifuge at 13,000 g for 15 min at 4°C.
3. Transfer these upper aqueous phase to a new tube. Add 0.5 mL isopropyl alcohol, mix, incubate at room temperature for 10 min, and centrifuge at 9,100 g for 10 min at 4°C. Rinse the precipitate with 1 mL of 75% ethanol. Dry the precipitated RNA briefly, dissolve in ~50 μL RNase-free water by pipetting, and incubate for 10 min at 55–60°C.

4. Synthesize cDNA using 5′RACE system for rapid amplification of cDNA ends according to the manufacturer's instructions. Use ~5 μg of total RNA as a template (*see* **Note 5**) and Gal4-r3 as a primer.
5. Use 5 μL of dC-tailed cDNA for the first PCR (30 cycles of 94°C for 30 s, 55°C for 30 s, 72°C for 2 min) in 50 μL 1× PCR buffer containing 0.4 μM Gal4-r2 primer and 0.4 μM AAP primer from the kit. Then, perform the second PCR using Gal4-r1 and AUAP primer from the kit.
6. Analyze the 5′RACE product using 1.5–2% agarose-TAE gel electrophoresis. Purify and clone DNA bands with TOPO TA cloning kit. Prepare plasmid DNA using the QIAprep spin miniprep kit and perform sequencing.

4. Notes

1. Forward and reverse primers for excision assay should be different when a plasmid containing a *Tol2* construct has a different backbone. In such a case, primers for excision assay have to be newly designed according to the backbone sequence.
2. Amplification of large genomic fragments (>5 kb) by PCR is often challenging when genomic DNA is used as a template, and in some cases may not work. In such a case, we recommend purchasing BAC or PAC clones (ImaGenes GmbH, Germany) that cover the genomic region of interest and using them as templates for PCR.
3. Adaptor-ligation PCR (and also inverse PCR) may fail when the *Tol2* insertion has occurred within genomic loci containing highly repetitive sequences.
4. It is necessary to design different adaptors when different restriction enzymes are used. Also, it is necessary to design different sets of primers when different types of constructs are used.
5. When 5′RACE does not work using total RNA, prepare polyA+ RNA and use it for 5′ RACE.

Acknowledgments

We thank postdoctoral fellowships from the Japan Society for the Promotion of Science (MLS and KA) and the Takeda Research Foundation (MLS) and from National Institute of Genetics (HK), and the National BioResource Project and grants from the Ministry of Education, Culture, Sports, Science and Technology of Japan.

References

1. Kimmel, C. B., Ballard, W. W., Kimmel, S. R., Ullmann, B., and Schilling, T. F. (1995) Stages of embryonic development of the zebrafish. *Dev Dyn* **203**, 253–310
2. Haffter, P., Granato, M., Brand, M., Mullins, M. C., Hammerschmidt, M., Kane, D. A., Odenthal, J., van Eeden, F. J., Jiang, Y. J., Heisenberg, C. P., Kelsh, R. N., Furutani-Seiki, M., Vogelsang, E., Beuchle, D., Schach, U., Fabian, C., and Nüsslein-Volhard, C. (1996) The identification of genes with unique and essential functions in the development of the zebrafish, *Danio rerio*. *Development* **123**, 1–36
3. Driever, W., Solnica-Krezel, L., Schier, A. F., Neuhauss, S. C., Malicki, J., Stemple, D. L., Stainier, D. Y., Zwartkruis, F., Abdelilah, S., Rangini, Z., Belak, J., and Boggs, C. (1996) A genetic screen for mutations affecting embryogenesis in zebrafish. *Development* **123**, 37–46
4. Amsterdam, A., Nissen, R. M., Sun, Z., Swindell, E. C., Farrington, S., and Hopkins, N. (2004) Identification of 315 genes essential for early zebrafish development. *Proc Natl Acad Sci U S A* **101**, 12792–12797
5. Nasevicius, A. and Ekker, S. C. (2000) Effective targeted gene 'knockdown' in zebrafish. *Nat Genet* **26**, 216–220
6. Xu, Q., Alldus, G., Holder, N., and Wilkinson, D. G. (1995) Expression of truncated Sek-1 receptor tyrosine kinase disrupts the segmental restriction of gene expression in the *Xenopus* and zebrafish hindbrain. *Development* **121**, 4005–4016
7. Hammerschmidt, M., Bitgood, M. J., and McMahon, A. P. (1996) Protein kinase A is a common negative regulator of Hedgehog signaling in the vertebrate embryo. *Genes Dev* **10**, 647–658
8. Xu, Q., Alldus, G., Macdonald, R., Wilkinson, D. G., and Holder, N. (1996) Function of the Eph-related kinase rtk1 in patterning of the zebrafish forebrain. *Nature* **381**, 319–322
9. Wada, H., Iwasaki, M., Sato, T., Masai, I., Nishiwaki, Y., Tanaka, H., Sato, A., Nojima, Y., and Okamoto, H. (2005) Dual roles of zygotic and maternal Scribble1 in neural migration and convergent extension movements in zebrafish embryos. *Development* **132**, 2273–2285
10. Xiao, T., Roeser, T., Staub, W., and Baier, H. (2005) A GFP-based genetic screen reveals mutations that disrupt the architecture of the zebrafish retinotectal projection. *Development* **132**, 2955–2967
11. Stuart, G. W., McMurray, J. V., and Westerfield, M. (1988) Replication, integration and stable germ-line transmission of foreign sequences injected into early zebrafish embryos. *Development* **103**, 403–412
12. Amsterdam, A., Lin, S., and Hopkins, N. (1995) The *Aequorea victoria* green fluorescent protein can be used as a reporter in live zebrafish embryos. *Dev Biol* **171**, 123–129
13. Long, Q., Meng, A., Wang, H., Jessen, J. R., Farrell, M. J., and Lin, S. (1997) GATA-1 expression pattern can be recapitulated in living transgenic zebrafish using GFP reporter gene. *Development* **124**, 4105–4111
14. Higashijima, S., Okamoto, H., Ueno, N., Hotta, Y., and Eguchi, G. (1997) High-frequency generation of transgenic zebrafish which reliably express GFP in whole muscles or the whole body by using promoters of zebrafish origin. *Dev Biol* **192**, 289–299
15. Lin, S., Gaiano, N., Culp, P., Burns, J. C., Friedmann, T., Yee, J. K., and Hopkins, N. (1994) Integration and germ-line transmission of a pseudotyped retroviral vector in zebrafish. *Science* **265**, 666–669
16. Gaiano, N., Allende, M., Amsterdam, A., Kawakami, K., and Hopkins, N. (1996) Highly efficient germ-line transmission of proviral insertions in zebrafish. *Proc Natl Acad Sci U S A* **93**, 7777–7782
17. Ellingsen, S., Laplante, M. A., Konig, M., Kikuta, H., Furmanek, T., Hoivik, E. A., and Becker, T. S. (2005) Large-scale enhancer detection in the zebrafish genome. *Development* **132**, 3799–3811
18. Raz, E., van Luenen, H. G., Schaerringer, B., Plasterk, R. H., and Driever, W. (1998) Transposition of the nematode *Caenorhabditis elegans* Tc3 element in the zebrafish *Danio rerio*. *Curr Biol* **8**, 82–88
19. Fadool, J. M., Hartl, D. L., and Dowling, J. E. (1998) Transposition of the mariner element from *Drosophila mauritiana* in zebrafish. *Proc Natl Acad Sci U S A* **95**, 5182–5186
20. Kawakami, K., Koga, A., Hori, H., and Shima, A. (1998) Excision of the Tol2 transposable element of the medaka fish, Oryzias latipes, in zebrafish, *Danio rerio*. *Gene* **225**, 17–22
21. Davidson, A. E., Balciunas, D., Mohn, D., Shaffer, J., Hermanson, S., Sivasubbu, S., Cliff, M. P., Hackett, P. B., and Ekker, S. C. (2003) Efficient gene delivery and gene expression in zebrafish using the *Sleeping Beauty* transposon. *Dev Biol* **263**, 191–202
22. Kawakami, K. (2005) Transposon tools and methods in zebrafish. *Dev Dyn* **234**, 244–254

23. Kawakami, K. (2007) Tol2: a versatile gene transfer vector in vertebrates. *Genome Biol* **8** Suppl 1, S7
24. Urasaki, A., Morvan, G., and Kawakami, K. (2006) Functional dissection of the Tol2 transposable element identified the minimal cis-sequence and a highly repetitive sequence in the subterminal region essential for transposition. *Genetics* **174**, 639–649
25. Balciunas, D., Wangensteen, K. J., Wilber, A., Bell, J., Geurts, A., Sivasubbu, S., Wang, X., Hackett, P. B., Largaespada, D. A., McIvor, R. S., and Ekker, S. C. (2006) Harnessing a high cargo-capacity transposon for genetic applications in vertebrates. *PLoS Genet* **2**, e169
26. Fisher, S., Grice, E. A., Vinton, R. M., Bessling, S. L., Urasaki, A., Kawakami, K., and McCallion, A. S. (2006) Evaluating the biological relevance of putative enhancers using Tol2 transposon-mediated transgenesis in zebrafish. *Nat Protoc* **1**, 1297–1305
27. Kwan, K. M., Fujimoto, E., Grabher, C., Mangum, B. D., Hardy, M. E., Campbell, D. S., Parant, J. M., Yost, H. J., Kanki, J. P., and Chien, C. B. (2007) The Tol2kit: a multisite gateway-based construction kit for Tol2 transposon transgenesis constructs. *Dev Dyn* **236**, 3088–3099
28. Villefranc, J. A., Amigo, J., and Lawson, N. D. (2007) Gateway compatible vectors for analysis of gene function in the zebrafish. *Dev Dyn* **236**, 3077–3087
29. Kawakami, K., Takeda, H., Kawakami, N., Kobayashi, M., Matsuda, N., and Mishina, M. (2004) A transposon-mediated gene trap approach identifies developmentally regulated genes in zebrafish. *Dev Cell* 7, 133–144
30. Parinov, S., Kondrichin, I., Korzh, V., and Emelyanov, A. (2004) Tol2 transposon-mediated enhancer trap to identify developmentally regulated zebrafish genes in vivo. *Dev Dyn* **231**, 449–459
31. Nagayoshi, S., Hayashi, E., Abe, G., Osato, N., Asakawa, K., Urasaki, A., Horikawa, K., Ikeo, K., Takeda, H., and Kawakami, K. (2008) Insertional mutagenesis by the Tol2 transposon-mediated enhancer trap approach generated mutations in two developmental genes: tcf7 and *synembryn-like*. *Development* **135**, 159–169
32. Davison, J. M., Akitake, C. M., Goll, M. G., Rhee, J. M., Gosse, N., Baier, H., Halpern, M. E., Leach, S. D., and Parsons, M. J. (2007) Transactivation from Gal4-VP16 transgenic insertions for tissue-specific cell labeling and ablation in zebrafish. *Dev Biol* **304**, 811–824
33. Scott, E. K., Mason, L., Arrenberg, A. B., Ziv, L., Gosse, N. J., Xiao, T., Chi, N. C., Asakawa, K., Kawakami, K., and Baier, H. (2007) Targeting neural circuitry in zebrafish using GAL4 enhancer trapping. *Nat Methods* **4**, 323–326
34. Asakawa, K., Suster, M. L., Mizusawa, K., Nagayoshi, S., Kotani, T., Urasaki, A., Kishimoto, Y., Hibi, M., and Kawakami, K. (2008) Genetic dissection of neural circuits by Tol2 transposon-mediated Gal4 gene and enhancer trapping in zebrafish. *Proc Natl Acad Sci U S A* **105**, 1255–1260
35. Kawakami, K. and Shima, A. (1999) Identification of the Tol2 transposase of the medaka fish *Oryzias latipes* that catalyzes excision of a nonautonomous Tol2 element in zebrafish Danio rerio. *Gene* **240**, 239–244
36. Kotani, T. and Kawakami, K. (2008) *Misty somites*, a maternal effect gene identified by transposon-mediated insertional mutagenesis in zebrafish that is essential for the somite boundary maintenance. *Dev Biol* **316**, 383–396
37. Schiavo, G., Benfenati, F., Poulain, B., Rossetto, O., Polverino de Laureto, P., DasGupta, B. R., and Montecucco, C. (1992) Tetanus and botulinum-B neurotoxins block neurotransmitter release by proteolytic cleavage of synaptobrevin. *Nature* **359**, 832–835
38. Kawakami, K. (2004) Transgenesis and gene trap methods in zebrafish by using the Tol2 transposable element. *Methods Cell Biol* 77, 201–222
39. Kotani, T., Nagayoshi, S., Urasaki, A., and Kawakami, K. (2006) Transposon-mediated gene trapping in zebrafish. *Methods* **39**, 199–206

Chapter 4

Generation of Transgenic Frogs

Jana Loeber, Fong Cheng Pan, and Tomas Pieler

Summary

The possibility of generating transgenic animals is of obvious advantage for the analysis of gene function in development and disease. One of the established vertebrate model systems in developmental biology is the amphibian *Xenopus laevis*. Different techniques have been successfully applied to create *Xenopus* transgenics; in this chapter, the so-called meganuclease method is described. This technique is not only technically simple, but also comparably efficient and applicable to both *Xenopus laevis* and *Xenopus tropicalis*. The commercially available endonuclease *I-SceI* (meganuclease) mediates the integration of foreign DNA into the frog genome after coinjection into fertilized eggs. Tissue-specific gene expression, as well as germline transmission, has been observed.

Key words: *Xenopus laevis*, Transgenesis, Meganuclease, *I-SceI*, Germline transmission, Tissue-specific promoter

1. Introduction

For half a decade, *Xenopus* has served as an attractive model system in developmental biology. It is easy to produce large numbers of embryos which can easily be manipulated by microinjection. However, the lack of efficient methods to create transgenic animals has for a long time been a serious drawback for the use of this system.

In 1996, Kroll and Amaya developed a protocol to produce nonmosaic transgenic *Xenopus laevis (1)*; decondensed sperm nuclei were incubated with linearized plasmid DNA and restriction enzyme, followed by coinjection into unfertilized eggs. Unfortunately, this so-called REMI method is technically quite

Elizabeth J. Cartwright (ed.), *Transgenesis Techniques*, Methods in Molecular Biology, vol. 561
DOI 10.1007/978-1-60327-019-9_4, © Humana Press, a part of Springer Science+Business Media, LLC 2009

demanding for the inexperienced user. It requires special equipment, which might not be available in every *Xenopus* lab, and the survival rates of sperm nuclei-injected embryos generated using this method are relatively low depending on the user's experience. Thus, a lot of eggs have to be injected to produce a sufficient number of transgenic embryos.

In 2005, two new transgenesis techniques for *Xenopus* were published. One makes use of a phage integrase-mediated integration of injected plasmid into genomic DNA *(2)*. So far, germline transmission of the resulting transgenes has not been demonstrated. Our lab has successfully modified a transgenesis protocol for *Xenopus laevis* which was originally established for Medaka fish *(3)*. In this protocol we make use of a special endonuclease, the meganuclease *I-SceI* from *Saccharomyces cerevisiae(4)*. Injection of *I-SceI* restriction digest reaction mix into one-cell embryos facilitates integration of the transgene DNA into the genome of the early embryo, and the transgene DNA can be efficiently transmitted to the F1 offspring. Importantly, the preparation of the embryos and the injection procedure itself do not principally differ from normal mRNA microinjection protocols.

2. Materials

2.1. Plasmid Preparation

1. Plasmid DNA (100 ng/μl in H_2O) (*see* **Note 1**). The expression cassette of choice has to be flanked by two I-SceI recognition sites. A recommended vector is the I-SceI-pBSII SK plasmid (GenBank accession number: DQ836146). The plasmid preparation should be as clean as possible. Avoid any contaminants. The best way to achieve this is using a commercial midi- or maxiplasmid isolation kit (e.g., from Qiagen).
2. I-SceI, 10 × I-SceI reaction buffer, 10 × BSA (New England Biolabs).
3. 37° C Incubator or heat block.

2.2. Preparation of X. laevis Embryos

1. *Xenopus laevis* males and females (Nasco, Fort Atkinson, WI).
2. Human chorionic gonadotropin (HCG) (Sigma). Prepare a 2,000 U/ml solution in water, aliquot, and store at –20° C.
3. Single-use syringe, 1 ml, for HCG injection.
4. Tricaine methanesulfonate (Sigma). Prepare 1 l of a 0.2% solution in water and store at 4° C.
5. Scissors, forceps, scalpel, Petri dishes (35, 60, 94 mm).

6. 1 × MBS: 88 mM NaCl, 2.4 mM $NaHCO_3$, 1 mM KCl, 10 mM Hepes of pH 7.6, 0.82 mM $MgSO_4$, 0.41 mM $CaCl_2$, 0.66 mM KNO_3. Prepare a 5 × stock and autoclave. Store at room temperature.
7. 0.1 × MBS. Store at room temperature.
8. 2% L-cysteine hydrochloride: Dissolve in H_2O and titrate to pH 8.0 with NaOH (Sigma). Prepare fresh on the day of injection.

2.3. Microinjection

1. Microinjection needles. Prepare them from borosilicate glass capillaries with 1 mm outside diameter and 0.58 mm inside diameter (Harvard Apparatus Ltd, Edenbridge, UK) using a microneedle puller (Narishige, Tokyo, Japan).
2. Microloader tips (Eppendorf).
3. Injection buffer (1 × MBS, 1% Ficoll). Prepare fresh from 5 × MBS and 10% Ficoll stocks. Precool to 12.5° C.
4. Microscope slide. On this slide the embryos will be arranged for injection. For better handling you can use tape to fix a pipette tip at one side of the slide. The tip will serve as a handle. Alternatively, embryos can be placed in a Petri dish with 800-μm nylon mesh (e.g., Neolab, Heidelberg, Germany) at the bottom. The mesh can be fixed with drops of chloroform. Let the chloroform completely evaporate before use.
5. Pressure-pulsed microinjector (World Precision Instruments, Sarasota, FL).
6. Cooling plate with cooling thermostat (Lauda, Lauda-Königshofen, Germany). Precool to 12.5° C.
7. Stereomicroscope.

2.4. Screening of Positive Embryos

1. 0.1 × MBS.
2. 0.01% Tricaine methanesulfonate (MS-222). Prepare a 20% solution in DMSO. This stock is stored at –20° C and diluted 1:2000 in 0.1 × MBS before use.
3. Plastic Pasteur pipette with cut tip.
4. Petri dishes.
5. Fluorescence microscope (e.g., SZX10, Olympus), with a mercury lamp device (U-RFL-T, Olympus).

2.5. Tadpole Husbandry

1. Plastic or glass tank, 2–30 l (e.g., IKEA).
2. Powdered growth food (e.g., Sera Micron, Sera, Heinsberg, Germany or leaf powder).
3. Sinking granular food (e.g., Novo Grano Mix, JBL, Neuhofen and Sterlet Sticks, Tetra Pond, Melle, Germany).
4. 0.1 × MBS.

5. Tap water, drinking quality. If the water is chlorinated, it is highly recommended to let it sit overnight before use.

3. Methods

The most critical parameter in this protocol is time. The earlier the DNA is delivered into the embryo, the higher the chance to achieve early and, as a consequence of this, nonmosaic integration. Thus, it is necessary to be well prepared and to work quickly. Injections should be done within the first 30–90 min after fertilization. We recommend injecting only one construct per batch of embryos.

3.1. Meganuclease Plasmid Preparation

The meganuclease digest reaction should be set up as soon as the frog starts to lay eggs.

Set up the I-SceI digest as follows:

0.5 μl I-SceI buffer (10 ×)

1.0 μl BSA (10 ×)

0.5–1.0 μl DNA (100 ng/μl)

1.0 μl I-SceI enzyme (20 U/μl)

6.5–7.0 μl H_2O

Incubate at 37° C for 40 min prior to injection.

3.2. Preparation of Xenopus laevis Embryos

3.2.1. Stimulation of Female Frogs

1. On the day before the injection the frogs have to be primed with human chorionic gonadotropin (HCG). Inject approximately 900 U subcutaneously into the dorsal lymph sac of the frog. Keep the frog at 16° C (*see* **Note 2**).

3.2.2. Testis Preparation

1. On the morning of the injection narcotize a male frog in 0.2% MS222 for 20–30 min until it does not show any responses.
2. Kill the frog by decapitation.
3. Dissect the testes with sharp scissors and forceps. With an incision into the abdominal wall the abdomen can be opened and the inner organs become visible. The testes are two small, white bean-like organs lying in the back of the abdomen. Cut them out carefully and immediately put them into ice-cold 1 × MBS in a Petri dish. Remove any remaining blood or connective tissue (*see* **Note 3**). Submerge one-third of a testis in 1 ml ice-cold 1 × MBS in a small Petri dish (35 mm) and mince with a scalpel. Alternatively mince in a microcentrifuge tube with a plastic pestle. To resuspend the testis thoroughly

pipette the mix through a yellow tip which has the end of the tip cut off. Keep this preparation on ice.

3.2.3. In Vitro Fertilization

1. Bring the female frogs to room temperature. Their cloacae should be red.
2. When they start laying, gently massage the eggs out of the animal. To do this remove the frog from the tank, carefully dry the cloaca, and hold the frog over a fresh Petri dish. Carefully massage the back and the belly of the frog and harvest the eggs into the dish.
3. Remove any water in the dish before fertilization.
4. Pipette 100 μl of testis suspension into the Petri dish and immediately dilute 1:10 with 900 μl water (*see* **Note 4**). Mix the sperm quickly and thoroughly with the eggs using a blue pipette tip.
5. After 10 min cover the batch with 0.1 × MBS.
6. 20 min after fertilization the embryos must be dejellied. Transfer them into a beaker and add 30 ml 2% L-cysteine. Carefully swirl the embryos for 3–5 min. The jelly coat is completely removed when the embryos have direct contact with each other. Monitor the status of the embryos frequently during the dejellying process and immediately remove the cysteine when the dejellying is completed (*see* **Note 5**).
7. Wash the embryos intensively with 200–400 ml precooled 0.1 × MBS.
8. Transfer the embryos into precooled injection buffer and keep them in an incubator at 12.5° C until injection.

3.3. Microinjection

1. When the meganuclease digest reaction is ready, load the whole reaction mix into a microinjection needle using a microloader tip.
2. Open the needle with fine forceps and adjust the injection volume to 4 nl.
3. Arrange approximately 50 embryos in a row on the microscope slide and remove the buffer with a Pasteur pipette.
4. Turn the animal poles upside with a shortened microloader tip. In this way it is easier to inject at the right site with appropriate speed.
5. Inject 4 nl of the meganuclease reaction mix between the sperm entry site and the animal pole (*see* **Note 6**). The embryos can be injected up to 90 min post fertilization at 12.5° C (*see* **Note 7**).
6. After injection incubate the embryos at 12.5° C in injection buffer until they have reached the four- to eight-cell stage. Then exchange the buffer for 0.1 × MBS and remove any dead embryos. Only a few embryos should share a dish (around 100 embryos in a 94-mm dish). They must not touch each other.

7. Keep the embryos at 12.5° C overnight, and then transfer them 18° C the next day.
8. During the first week of development wash the embryos twice a day with 0.1 × MBS (*see* **Note 8**).

3.4. Screening of Positive Embryos

The time point of screening depends on the onset of the promoter activity. If possible, the screening should be done later rather than too early in order to discriminate between mosaic and nonmosaic integration and gene expression from the nonintegrated plasmid (*see* **Note 1**). The efficiency of the method may vary greatly depending on the time point of injection and the quality of the embryos as well as the *I-SceI* meganuclease. Efficiencies from 3 to 20% of uniform, nonmosaic transgene expression have been reported *(4–6)*.

1. Carefully collect the embryos with a plastic Pasteur pipette with a cut tip, (depending on the size of the embryos) and transfer to a Petri dish with 0.02% MS222 in 0.1 × MBS.
2. Narcotize them for 3–4 min and screen for reporter gene expression under the fluorescence microscope. If the expression of the reporter is very strong, a UV filter can be used in order to avoid damage of the animals due to an overdose of UV exposure (*see* **Note 9**).
3. Sort the positive embryos into a new dish and raise them to the desired stage.

3.5. Tadpole Husbandry

1. Keep the tadpoles at room temperature (22° C) in a static container. They are very susceptible to any chemical contamination, especially chlorine. Hence it is recommended to keep them in 0.1 × MBS buffer as long as possible (up to stage 47–50). Tap water can be used instead, but the water quality should be carefully controlled. If chlorinated, the water should be stored in an open container over night before use.
2. Clean the container at least once per week. Exchange approximately one-third of the dirty water for clean chlorine-free water or 0.1 × MBS. Additionally, any excrements or food leftovers should be removed daily in order to keep the water clean.
3. Feed tadpoles daily. Resuspend a small amount of powdered food in 1–2 ml water in a small glass or plastic container. Add the suspension to the tadpole tank and mix carefully. Adjust the amount of food depending on the stage and number of tadpoles per tank. Tadpoles will stop feeding during metamorphosis, i.e., when the forelimbs appear. After completion of metamorphosis the froglets immediately start to eat small granular food (*see* **Note 10**).

4. Notes

1. As a reporter, the use of fluorescent proteins is recommended, since they allow for fast, easy, and noninvasive screening. The promoter should be expressed during embryogenesis to enable early screening. It is important to consider that nonintegrated promoters can also be active; hence, reporter activity might come from the plasmid. Rescreening at a later time point of development would therefore be necessary. Another problem to be considered is a mosaic expression pattern. If the DNA is not injected early enough the chances of late and nonuniform integration rise. The territory of the promoter activity should be large enough to visualize a homogeneous pattern of expression.We have obtained good results making use of the elastase promoter which is active in the exocrine pancreatic tissue *(7)*. The pancreas arises from vegetal cells. Since the DNA is injected into the animal pole region, nonintegrated plasmids should not give rise to reporter gene expression. Moreover, the pancreatic bud at late tadpole stages is large enough, so that uniform and mosaic expression can easily be discriminated.The γ-crystallin 1 promoter which was also described in several studies allows screening from stage 40 *(8)*. However, it is a very potent promoter; hence, transcription might also occur directly from the unintegrated plasmid. We have occasionally observed embryos which show a decreasing area of GFP expression in the lens due to plasmid dilution during development beyond stage 40.
2. Depending on the quality of the hormone, the frogs will start to lay eggs 12–15 h after priming. It is recommended to prime two frogs for one injection day.
3. In 1 × MBS the testes can be kept for up to 4 days at 4° C, but it has to be considered that the quality will drop with time. Therefore freshly isolated testis is always preferred.
4. In high salt the sperm is not motile and thus does not use its stored ATP. Dilution will lead to a decrease of the salt concentration and an activation of the sperm. Hence, it is important to mix the sperm with the eggs immediately after dilution.
5. If the embryos are exposed to the cysteine solution for too long, they will be severely damaged and eventually die.
6. The preferred injection site is between the animal pole and the sperm entry site. In this way the DNA comes as close as possible to the pronuclei. The sperm entry site can be observed by pigment accumulation at one side of the animal pole. If it is not visible, inject into the center of the animal pole.

7. The earlier the DNA is injected, the higher is the chance of integration before the first cleavage. It is advised to separate the embryos injected early from the ones injected at a later point in time. The early embryos should have a higher transgenesis rate than the late-injected ones.
8. Embryos tolerate DNA injection to a much lesser extent than injection of RNA. Instant removal of dead embryos and cell debris is necessary for the rest to develop normally.
9. Take only 10–20 embryos at one time. The embryos will suffer from extended exposure to MS222 and UV light. The number of embryos taken should also depend on the visibility of the reporter gene expression site. If the reporter gene is driven by the CMV or actin promoter, ubiquitous expression is quickly visible, whereas embryos with tissue-specific promoter (e.g., γ-crystallin or elastase promoter) might have to be turned to a certain side in order to see expression. The embryos should be transferred to fresh 0.1 × MBS as soon as the examination is done.
10. It is normal that some animals grow faster than others. When the frogs are still small, they can share the tank with big tadpoles. However, the froglets may feed on tadpoles.

Acknowledgments

This work was supported by funds from the Deutsche Forschungsgemeinschaft to T.P.

References

1. Kroll, K.L. and Amaya, E. (1996) Transgenic Xenopus embryos from sperm nuclear transplantations reveal FGF signaling requirements during gastrulation. *Development*, **122**, 3173–3183.
2. Allen, B.G. and Weeks, D.L. (2005) Transgenic *Xenopus laevis* embryos can be generated using phiC31 integrase. *Nat Methods*, **2**, 975–979.
3. Thermes, V., Grabher, C., Ristoratore, F., Bourrat, F., Choulika, A., Wittbrodt, J. and Joly, J.S. (2002) I-SceI meganuclease mediates highly efficient transgenesis in fish. *Mech Dev*, **118**, 91–98.
4. Pan, F.C., Chen, Y., Loeber, J., Henningfeld, K. and Pieler, T. (2006) I-SceI meganuclease-mediated transgenesis in Xenopus. *Dev Dyn*, **235**, 247–252.
5. Ogino, H., McConnell, W.B. and Grainger, R.M. (2006) Highly efficient transgenesis in Xenopus tropicalis using I-SceI meganuclease. *Mech Dev*, **123**, 103–113.
6. Ogino, H., McConnell, W.B. and Grainger, R.M. (2006) High-throughput transgenesis in Xenopus using I-SceI meganuclease. *Nat Protoc*, **1**, 1703–1710.
7. Beck, C.W. and Slack, J.M. (1999) Gut specific expression using mammalian promoters in transgenic *Xenopus laevis*. *Mech Dev*, **88**, 221–227.
8. Offield, M.F., Hirsch, N. and Grainger, R.M. (2000) The development of *Xenopus tropicalis* transgenic lines and their use in studying lens developmental timing in living embryos. *Development*, **127**, 1789–1797.

Chapter 5

Pronuclear DNA Injection for the Production of Transgenic Rats

Jean Cozzi, Ignacio Anegon, Valérie Braun, Anne-Catherine Gross, Christel Merrouche, and Yacine Cherifi

Summary

In the absence of germ-line-competent ES cells in the rat, pronuclear microinjection remains an essential tool to generate transgenic rat models. However, DNA microinjection procedures first developed for mouse do not provide scientists with satisfying results when applied to rat. Here we describe optimized procedures for rat with a special focus on rat embryo production, in vitro incubation and culture, DNA pronuclear injection, and the transfer of embryos into foster females.

Key words: Microinjection, Transgenic rat, Ovarian cycle synchronization, Isoflurane, Sprague–Dawley, Embryo transfer

1. Introduction

DNA microinjection is a straightforward method to obtain transgenic animals. The procedure consists of the injection of a transgene into the male pronucleus of a one-cell stage fertilized embryo. Injected embryos are reimplanted into foster mothers to obtain offspring harboring one or several copies of the transgene in their germ cells. Breeding of the transgenic founder animals enables stable transmission of the transgene to the progeny. In the absence of germ-line-competent ES cells in the rat, DNA pronuclear injection remains one of the most powerful technologies to generate transgenic rat models of high relevancy *(1)*. Pronuclear injection was developed in the mouse in the early 1980s and successfully translated to the rat in the 1990s *(2, 3)*. Successful translation

Elizabeth J. Cartwright (ed.), *Transgenesis Techniques,* Methods in Molecular Biology, vol. 561
DOI 10.1007/978-1-60327-019-9_5, © Humana Press, a part of Springer Science+Business Media, LLC 2009

of this gene transfer technology to the rat has been delayed because several features of the rat physiology and embryology differ from the mouse. As an example, the conventional superovulation regimen that has proved to be successful in the mouse is often ineffective in generating high-quality embryos in the rat *(4)*. In addition, a sequential culture system is required to achieve embryonic preimplantation in vitro development beyond the two-cell stage *(5)*. Finally, rat single-cell embryos exhibit highly flexible pronuclear and plasma membranes that render transgene injection more difficult and increase the rate of cell lysis. Robust and reliable production of transgenic rats thus requires the use of dedicated procedures. The procedures described here have been mostly developed and validated in the Sprague–Dawley (SD) strain (outbred). The percentage of injected eggs that develop into transgenic founders ranges from 1 to 2%. Refinement of these protocols may be required if transgenic models are to be developed in other strains.

2. Materials

2.1. Zygote Production

2.1.1. Estrus Synchronization

Rat estrus can be synchronized using luteinizing hormone releasing hormone agonist (LHRHa) *(6, 7)* (*see* **Note 1**).

1. Luteinizing hormone releasing hormone agonist (LHRHa) (Sigma): Dissolve 1 mg in 5 ml PBS, 0.1% BSA. Store aliquots at –20°C for up to 2 months.
2. Sprague–Dawley (SD) females, 8–12 weeks of age.
3. Adult males of proven fertility.
4. 1-ml syringe with hypodermic needle.

2.1.2. Zygote Isolation and Preparation for Microinjection

1. Understage illumination stereomicroscope equipped with heated stage.
2. Surgical tools: surgical scissors, fine forceps, iris forceps, and iris scissors.
3. 37°C, 5% CO_2 incubator.
4. 35-mm (35 × 10 mm) culture dishes.
5. Mouthpiece, tubing, and microcapillary holder.
6. Transfer pipettes: borosilicate glass capillaries pulled by hand over a Bunsen burner and broken at an adequate internal diameter using a diamond cutter (*see* **Table 1**).
7. M2 (Sigma) for handling zygotes outside of the incubator.

Table 1
Characteristics of glass pipettes

Pipette type	Use	Length of pulled tip (mm)	Outer diameter (μm)	Inner diameter (μm)	Tip angle	Capillary required
Transfer	Collecting and moving zygotes	400	400–500	300–400	-	Without filament, length = 100 mm, OD = 1.0 mm, ID = 0.78 mm
Holding	Maintaining zygotes during pronuclear injection	200	80–100	15–20	30–40°	
Reimplanta-tion	Transferring injected zygotes into oviduct	300–400	140–200	120–180	-	
Injection	DNA injection	200	1	<1	-	with filament, length = 100 mm, OD = 1.0 mm, ID = 0.78 mm

Table 2
Formulation of modified rat 1-cell embryo culture medium (mR1ECM)

Ingredient	mR1ECM
NaCl	110 mM
KCl	3.2 mM
$CaCl_2$	2.0 mM
$MgCl_2$	0.5 mM
$NaHCO_3$	25.0 mM
Sodium lactate	10.0 mM
Sodium pyruvate	0.5 mM
Glucose	7.5 mM
BSA[a]	4.0 mg
Glutamine	0.1 mM
EAA[b]	2% (v/v)
NEAA[c]	1% (v/v)
Osmolarity	290 mOsM

[a] Embryo-tested fraction V powder (Sigma)
[b] Minimal essential medium (MEM) amino acid solution
[c] MEM nonessential amino acid solution

8. mR1ECM media (modified rat embryo culture medium): for incubation and in vitro culture of rat one-cell preimplantation embryos at 37°C in an atmosphere of 5% CO_2 (*see* **Note 2**). (*See* **Table 2** for formulation.) Reconstitute the medium in double-distilled or higher-purity water and sterilize using a 0.22-μm sterile-filter. Store in 5-ml polypropylene bottles at 4°C for up to 1 month. Use a fresh bottle of medium on each experimental day.
9. Mineral oil, embryo tested (*see* **Note 3**).
10. Embryo-tested bovine testis hyaluronidase (Sigma) stock. Dissolve 30 mg in 3 ml M2 to give 10 mg/ml. Store in 50-μl aliquots at –20°C for several months. Use at a final concentration of 300 μg/ml.
11. Mated SD rat females, 8–12 weeks of age.

2.2. Preparation of DNA

1. DNA preparation kit from Qiagen (Qiaquick® Gel Extraction Kit).
2. Ultrafree-MC spin column (Merck).
3. Slide-A-Lyzer Dialysis kit (Pierce).
4. DNA electrophoresis setup.
5. UV transillumination table.
6. Photometer to measure DNA concentration at 260/280 nm.
7. 1× TAE buffer.
8. Microinjection buffer: 8 mM Tris–HCl (pH 7.4), 0.15 mM EDTA (*see* **Note 4**).
9. Agarose, e.g., UltraPure™-agarose (Invitrogen).
10. Ethidium bromide solution.

2.3. DNA Microinjection

1. Vibration-damped table.
2. Work station including:
   - Inverted microinjection microscope with DIC optics (Leica DM IRB or equivalent) equipped with 10× eyepieces, 5× and 40× objectives.
   - Heated stage.
   - Manual microinjector for holding oocytes (Cell Tram oil, Eppendorf).
   - 2× micromanipulator (TransferMan NK2, Eppendorf).
   - Microinjector (Femtojet, Eppendorf).
3. Holding and microinjection capillaries: Borosilicate glass capillaries pulled using a pipette puller (e.g., Flaming/brown P97/IVF, Sutter instrument Co.). Holding capillaries are modified using a microforge (manufacturers include Narishige and De Fonbrune) to heat polish the tip and adjust the internal diameter.
4. Microloader (Eppendorf).
5. Microscope slides.
6. Purified DNA at 3–5 ng/µl.

2.4. Reimplantation of Injected Zygotes

2.4.1. Preparation of Foster Recipients

1. 8–12-week old females.
2. Sprague–Dawley vasectomized rats (*see* **Note 5**).

2.4.2. Anesthesia

1. Isoflurane anesthesia setup (*see* **Fig. 1** and **Note 6**).
2. Isoflurane.
3. Buprenorphine.
4. Warming pad.

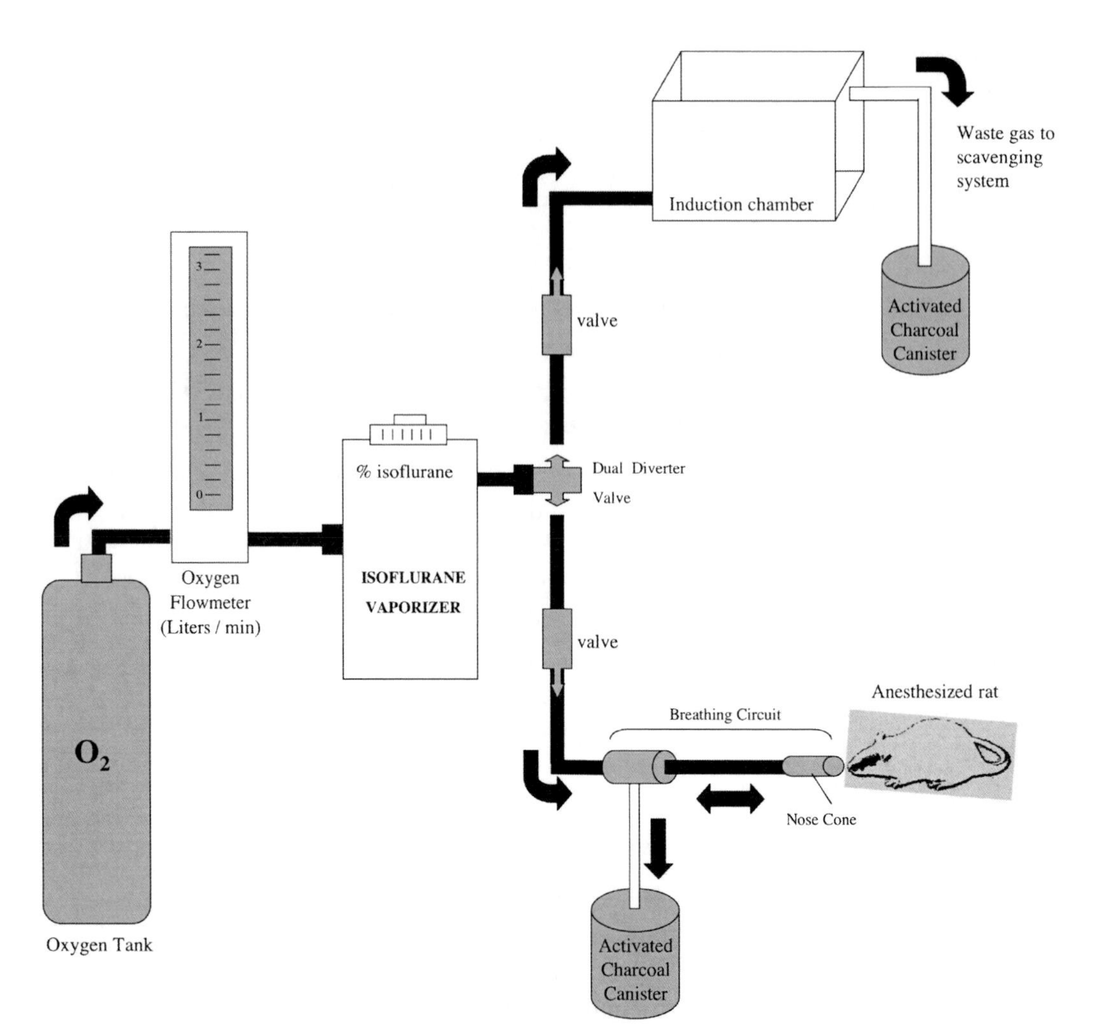

Fig. 1. Scheme of basic anesthetic delivery system using isoflurane. O_2 flows through the vaporizer and enables the adjustment of the percentage of isoflurane and picks up isoflurane vapors. O_2-isoflurane mix flows through the breathing circuit and into the rat lungs, or alternatively through the induction chamber for preanesthesia of the animal. Scavenging devices (activated charcoal canisters) remove any excess anesthetic gases in order to prevent room pollution. The scheme is not depicted to scale.

2.4.3. Embryo Transfer

1. Surgical tools: 2× Dumont #5 forceps, surgical forceps, iris forceps, iris scissors, serafine clamp, vessel clamp, wound clips, and applier. All instruments should be clean and sterilized with 70% ethanol or heat sterilized prior to use.
2. Surgical suture.
3. Surgical gauze or Kimwipe tissues.
4. Embryo transfer capillary and mouth pipetor.
5. Fiber optic light source.
6. Warming pad or red lamp.

2.5. Genotyping

1. Wizard Genomic Purification kit (Promega).
2. Incubator at 55°C.

3. Methods

Timelines of the estrus synchronization, females mating, zygote collection, DNA injection, and transfer of embryos to recipient females at one-cell and two-cell stages are presented in **Fig. 2**.

3.1. Zygote Production

3.1.1. Estrus Synchronization

Zygote production is a critical step in the overall rat gene transfer procedure. Zygote production efficiency is dependent on various factors such as the rat strain, the age and weight of the animals, the timing of hormone delivery, embryo collection, the quality of hormonal preparation, the doses of hormones, the housing condition of the animals (light/dark cycle duration), and the seasonal effect. Most laboratories have used superovulation regimens that allow the production of a high number of embryos from a minimum number of animals. However, reliable methods that yield large numbers of viable oocytes with normal developmental potential are not easy to set up in the rat. The major disadvantages of hormonal stimulation include great variability in ovarian response between females, poor quality of embryos, and low rate of fertilization. In addition, superovulation protocols are most efficient when applied to immature females. However, low mating frequency is not uncommon when mating immature females with adult males *(8)*. All these issues can be avoided by

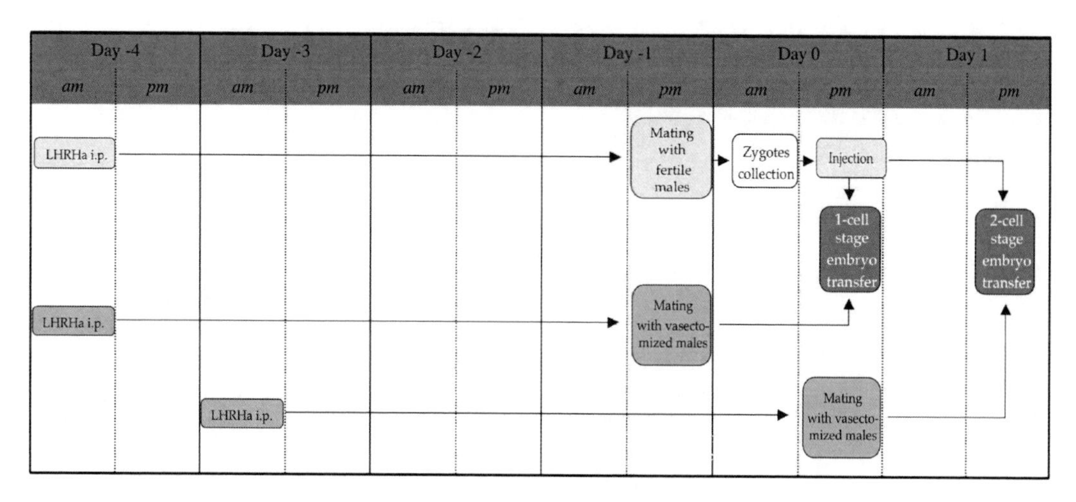

Fig. 2. Timelines of pronuclear injection procedure from estrus synchronization to embryo reimplantation into foster females.

using adult females as zygote donors. We describe here a simple protocol using adult rat females for the production of fertilized oocytes that have been synchronized but have a nonstimulated ovarian cycle (*see* **Note 7**). The timing of experiments indicated is valid for animals kept under a 12:12-h light-dark cycle.

1. Inject 0.2 ml of LHRHa preparation (i.p. 0.04 mg/female) on the morning of day –4.
2. Mate females on the afternoon of day –1 with fertile males in a 1:1 ratio in single cages overnight.
3. Kill females with a copulation plug (*see* **Note 8**) between 12 and 2 pm on day 0 in order to collect zygotes.

3.1.2. Zygote Isolation and Preparation for Microinjection

The procedure for zygote collection and preparation for pronuclear injection is similar to that used in the mouse.

1. Prepare the dishes required for zygote collection, preparation, and incubation before pronuclear injection (*see* **Fig. 3**).
2. Kill donor females using CO_2.

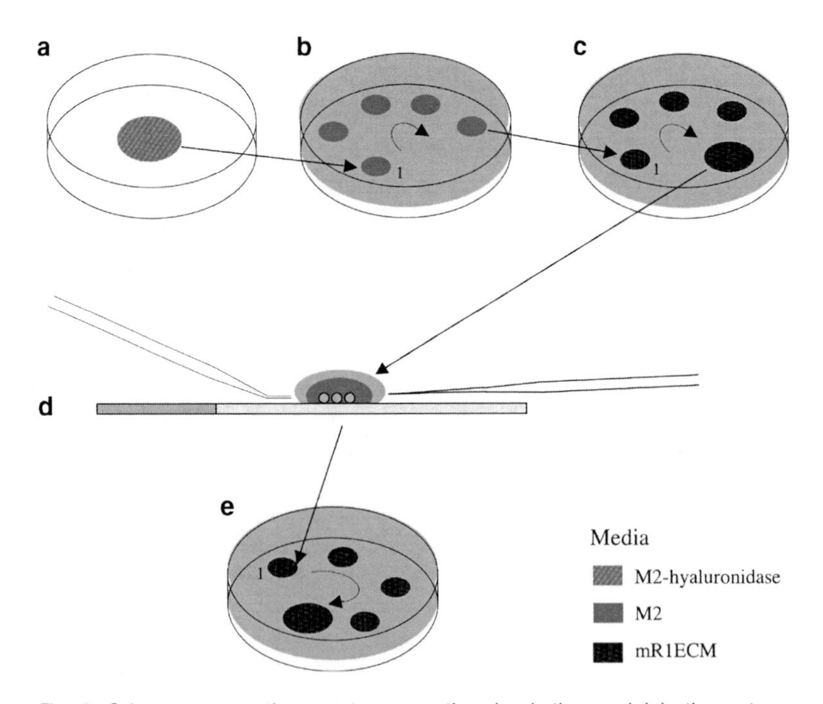

Fig. 3. Scheme representing zygote preparation, incubation, and injection setups. (**a**) Zygotes are freed from cumulus cells in a 200-μl drop of M2 containing hyaluronidase. (**b**) Free cumulus zygotes are successively rinsed in five 10-μl M2 drops under oil. (**c**) Zygotes are successively rinsed in four 10-μl mR1ECM drops and incubated in 50 μl of the same medium under oil before pronuclear injection. (**d**) Zygotes are microinjected with DNA in a 10-μl M2 drop overlaid by mineral oil on a glass slide. (**e**) Injected zygotes are rinsed in 10-μl mR1ECM drops and incubated in 50 μl of the same medium prior to reimplantation into the oviduct of a recipient pseudopregnant female.

3. Place the rats on their back on absorbent paper and rinse thoroughly with 70% ethanol. Incise the peritoneum to expose the inner organs.
4. Collect the oviducts on both sides and place them in M2 medium. The rest of the procedure is performed on a warming plate using prewarmed media (37°C) to avoid cold shock. A transfer pipette is used to move zygotes from one drop to another.
5. Under a stereoscopic microscope (15–20 × magnification) transfer one oviduct at a time in a 200-μl drop of M2 containing hyaluronidase at 300 mg/ml and dilacerate the ampulla (swollen part of the oviduct) using a pair of fine forceps. Gently squeeze the oviduct to help the oocyte–cumulus masses to flow out.
6. Leave the oocyte–cumulus masses in the same drop until complete dispersion of the cumulus (5–10 min).
7. Rinse zygotes by serial transfer in five 10-μl drops of fresh M2.
8. Repeat the procedure for the remaining oviducts.
9. Cumulus granulosa cells that remain after hyaluronidase treatment can be stripped by pipetting eggs through transfer pipettes 90–100 μm in diameter (*see* **Note 9**).
10. Observe zygotes at 300–400× magnification to assess pronuclei size. The time of the day to perform microinjection should be empirically determined according to the pronuclei size in maturing zygotes. The pronuclei should be large enough and clearly visible (*see* **Fig. 4a**). If the pronuclei are too small then incubate zygotes for an extra period of time.
11. Transfer all collected zygotes to a 50-μl drop of equilibrated culture medium mR1ECM under mineral oil in a 35-mm dish, washing once to remove M2 medium.

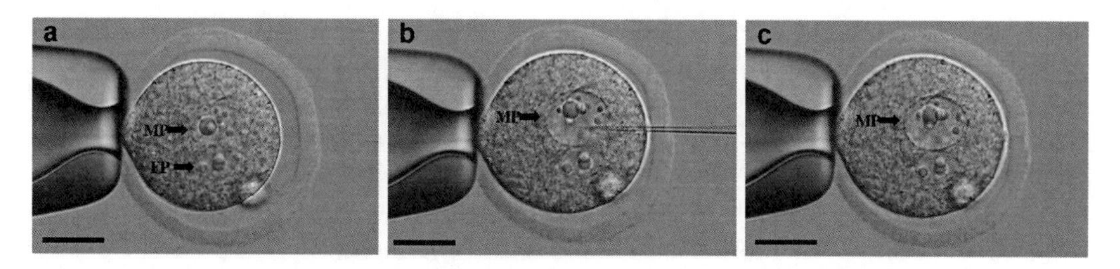

Fig. 4. Rat DNA pronuclear microinjection Pronuclear injection is performed under 400× magnification (40× objective). (**a**) Male pronucleus (MP) is positioned such as it is in the equatorial plane of the oocyte and relatively close to the injection needle. (**b**) DNA injection induces male pronucleus swelling. The pronucleus size should increase by 1.5–2-fold. (**c**) Rat zygote after retrieval of the injection pipette. Scale bars = 20 μm.

12. Place the dish containing the zygotes in a 37°C, 5% CO_2, humidified incubator.
13. Incubate zygotes for at least 30 min before pronuclear injection.

3.2. Preparation of DNA

The purity of the transgene DNA used for microinjection is critical for the successful production of transgenic founder rats. DNA impurities in the form of bacterial endotoxins or organic contaminants can severely decrease the integration efficiency of transgene DNA or the viability of microinjected embryos. The procedure described here applies to conventional plasmid vectors. The expression vector should be isolated from the vector backbone prior to microinjection since the presence of plasmid sequences may interfere with transgene expression.

1. Digest 40 μg of DNA with appropriate restriction enzymes to release the insert.
2. Run 200 ng of the DNA digest on an agarose gel to test for complete digestion. Increase digestion time if needed.
3. Load the digested sample, a control DNA, and size markers onto a standard preparative agarose gel in 1× TAE buffer to separate the transgene from the backbone DNA. The DNA sample and the control DNA should be loaded sufficiently apart for ease of **step 4**.
4. After electrophoresis cut the gel vertically to separate markers and the control DNA from the sample.
5. Stain the gel fragment containing the marker and control DNA in ethidium bromide.
6. Appose the two gel fragments. Using the stained markers as reference, locate the region of the nonstained gel containing the transgene and cut it out.
7. Purify the DNA from the gel using QIAquick columns and quantify the recovered DNA.
8. On an agarose gel, verify the purity and determine the concentration of the DNA.
9. Undiluted DNA can be stored at –80°C.
10. On the day before injection (D-1) microdialyze the filtrate using floating microdialysis disk. Dialysis is performed against 2 l of microinjection buffer for 2 h at room temperature. The buffer is then renewed and dialysis is conducted for an additional 2 h.
11. Store the sample at 4°C overnight.
12. On the day of injection (D0) dilute DNA in microinjection buffer to 3–5 ng/μl and ultrafiltrate the DNA using a 0.22-μm Ultrafree-MC spin column (*see* **Note 10**).

13. On an agarose gel, verify the purity and determine the concentration of the DNA (*see* **Note 11**).
14. Dilute DNA to 1–5 μg/ml for microinjection using injection buffer.
15. Centrifuge the diluted DNA at 13,000 × *g* for 5 min at 4°C and keep on ice until use. When loading the capillary, take the DNA from the upper phase.

3.3. Pronuclear Injection of DNA

To preserve the developmental ability of the rat embryos the microinjection should be performed in prewarmed M2 medium (to avoid cold shock) on a heating microscope stage.

1. Prepare the injection slide as described in **Fig. 3d**.
2. Place the microcapillary for holding zygotes into the connector piece.
3. Fill the injection capillary using a microloader with 2 μl of DNA suspension and insert into the connector piece.
4. Place the injection slide under a stereoscopic microscope.
5. Transfer about 20 zygotes into the upper part of the drop.
6. Position the holding and injection microcapillaries on the stage according to the arrangement described in **Fig. 3d**. Set a low point to prevent breakage of capillary.
7. Clean the injection capillary following recommendations of the manufacturer of the microinjection device.
8. Using the holder pipette move one noninjected zygote to the center of the drop. Rotate the zygote so that the male pronucleus (largest pronucleus) is in alignment with the opening of the holder pipette (same focus plane). Take care that the female nucleus is located below or above **(Fig. 4a)**.
9. For pronuclear injection, bring the capillary tip in focus with the nuclear membrane of the male pronucleus **(Fig. 4a)**. Penetrate the cytoplasm and the male pronucleus taking care not to touch the nucleoli. Inject DNA until the volume of the male nucleus has nearly doubled **(Fig. 4b, c)**. Rapid swelling of the male pronucleus, with the nuclear membrane becoming clearly visible, is evidence of successful injection (*see* **Note 12**).
10. Withdraw the needle smoothly from the oocyte (*see* **Note 13**). It should take about 5 s from zona penetration to injection needle withdrawal.
11. Move the injected zygotes to the bottom of the drop in order to avoid a mix up of injected and uninjected embryos (*see* **Note 14**).
12. Repeat the procedure until all zygotes are injected.
13. Rinse the injected zygotes four times in equilibrated culture medium to remove HEPES before incubation.

3.4. Reimplantation of Injected Zygotes

Outbred strain females which show very good reproductive performance should be preferentially used as surrogate mothers for the microinjected eggs (e.g., Sprague–Dawley). Reimplantation of 30 injected zygotes per recipient is commonly achieved on the same day of injection (D1), at the one-cell stage (*see* **Note 15**), or on the following day (D2), at the two-cell stage. The latter timing will allow observation of cleavage of injected zygotes and better selection of viable embryos for transfer. Bilateral embryo transfer is performed.

3.4.1. Preparation of Foster Recipients

Pseudopregnant females for reimplantation of microinjected eggs can be prepared using the same procedure described for zygote production; however, vasectomized rats must be used (*see* **Subheading 3.1.1**).

3.5. Anesthesia

The most reproducible, efficient, and safe procedure for embryo transfer combines isoflurane for anesthesia and buprenorphine for analgesia. Isoflurane provides better standardization of the depth and duration of anesthesia and more rapid postoperative recovery than other methods relying on injection of anesthetic compounds such as ketamine *(9)*. Buprenorphine ensures analgesia during the peri- and early postoperative periods. Successful surgical anesthesia depth is achieved in 100% of rats given isoflurane. In contrast, about 25% of rats given ketamine/xylazine do not achieve a surgical anesthesia depth within 10 min *(9)*.

1. Place the animal in the induction chamber.
2. For induction anesthesia: adjust the flowmeter to 1–1.5 l/m (oxygen as carrier gas) and the isoflurane vaporizer to 5%.
3. Inject buprenorphine subcutaneously about 2 min before surgery
4. Maintain anesthesia with 3% isoflurane by placing the nose of the preanesthetized rat in the nose cone connected to the breathing circuit.
5. To verify surgical depth of anesthesia before surgery, pinch the foot of the narcotized rat; absence of retraction reflex should be observed.
6. Isoflurane will be administered for approximately 12–15 min; 2 min prior to and until the end of the surgical procedure (*see* **Note 16**).

3.5.1. Embryo Transfer

1. Anesthetize a plugged recipient female as described earlier.
2. When the animal is properly anesthetized (lack of spontaneous movement to toe pinch), shave both sides of the flank. Rinse the shaved area with 70% alcohol.

3. Make a transverse skin incision of approximately 1 cm, about 1 cm down from the spinal cord just posterior to the last rib. Make a small incision in the body wall above the ovary. With blunt forceps, pick up the fat pad. Clip the serafine clamp to the ovarian fat pad and use to exteriorize the ovary, the oviduct, and the tip of the uterus. Use the clamp to position the uterus, ovary, and oviduct. The coiled oviduct within the transparent bursa should be clearly visible and should be orientated horizontally with the ampulla on the left side such that the reimplantation capillary can be easily inserted into the infundibulum.
4. Transfer the embryos to a drop of M2 medium and load a pipette for reimplantation with embryos as shown in **Fig. 5** (*see* **Note 17**). The pipette is first filled with light paraffin oil to just past the shoulder (*see* **Note 18**). The viscosity of the oil allows one to pick up or blow out the embryos with greater control.
5. Using Dumont #5 forceps gently tear a hole in the bursa membrane at the point between the ovary and the oviduct where the infundibulum is located. Take great care to avoid rupturing the capillaries of the bursa membrane (*see* **Note 19**). Grasp the edge of the infundibulum with one pair of forceps while inserting the reimplantation capillary into the aperture and slowly blow the zygotes into the oviduct until the first air bubble behind the embryos has entered the ampulla.
6. Gently nurse the uterus, oviduct, and ovary back inside the body wall. Close the body wall with three to four sutures and the apposed skin edges with three to four sterile wound clips.
7. Keep the rat warm using a red lamp or warming pad until narcosis ends and return the rat to the animal facilities.

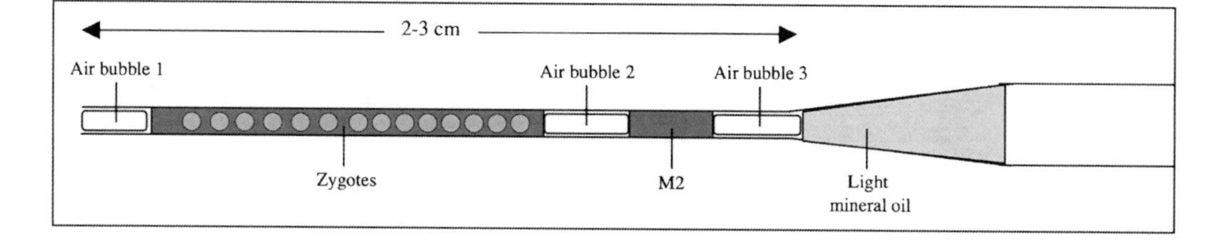

Fig. 5. Embryo transfer pipette organization. The internal diameter of the capillary is slightly larger than a single embryo thus allowing the embryos to be packed within a minimal volume of medium. Mineral oil ensures better control over embryo transfer. Air bubbles, which are easily seen during reimplantation, enable the operator to ensure that all embryos have been injected into the oviduct.

3.6. Genotyping

1. Clip tail biopsy, 5–10-mm fragment, from 10-day-old pups.
2. Extract genomic DNA from tail biopsies using the Wizard Genomic DNA purification kit according to the manufacturer's instructions.
3. Estimate the DNA concentration.
4. Submit extracted DNA to PCR and/or Southern blot analysis for transgene detection according to standard procedures.

4. Notes

1. This protocol is very efficient with up to 95% of rats being in estrus on the scheduled day. It is thus unnecessary to stage LHRHa treated rats for estrus.
2. mR1ECM is specifically for the culture of preimplantation rat embryos within a 5% CO_2 atmosphere. Outside the incubator its pH changes quickly. Manipulation of oocytes (washing steps and injection) should therefore always be performed in M2 medium.
3. Mineral oil may contain contaminants deleterious to embryonic development. Use embryo-tested mineral oil preferentially from Sigma.
4. When preparing the injection buffer, use disposable plastic free of traces of detergents. Use reagents designated solely for tissue culture work to avoid any trace of contaminants. Solutions should be filtered with a 0.2-μm filter to remove particles which may clog the injection needle.
5. Vasectomized rats can be bought from animal suppliers or prepared in house using 2–3 month old males. Keep track of plugging performance for best results. Vasectomized males should be replaced approximately every 10 months.
6. It is assumed that prolonged and repetitive exposure to anesthetic gases can be toxic to procedure area personnel. Ensure that all individuals responsible for anesthesia are properly trained. Maintain equipment in good working order and have it certified yearly to ensure optimal performance. Work in a well-ventilated area, ideally under a fume hood or hard-ducted biosafety cabinet.
7. On average for the SD strain 90% of females successfully mate allowing collection of 12 fertilized oocytes per female.
8. Females tend to lose the vaginal plug a few hours after copulation. It is thus advisable to check for plugs early in the morning (7–8 am) or use dedicated mating cages that enable plug recovery thereby greatly facilitating identification of mated females (e.g., "E" type cut-out base section complete cage for rat, Charles River Lab.).

9. The pipette should be slightly larger than a zygote. A pipette that is too narrow may cause zygote lysis.
10. A DNA concentration that is too high will cause developmental arrest of embryos, but a concentration that is too low decreases the number of transgenic founders.
11. DNA concentration is best determined by preparing serial dilutions of lambda DNA of known concentration (25, 50, 100, 200, 400 ng of lambda DNA) and using them as standards. Run the lambda DNA and the DNA to be microinjected on a gel to determine DNA concentration of the diluted DNA for microinjection. DNA concentration may be assessed in parallel by fluorimetry.
12. The pronucleus membrane is extremely flexible. In case of resistance insert the pipette further to the opposite side of the pronucleus to facilitate penetration.
13. If contact between the nucleolus and the injection pipette has occurred a string of DNA may formed while retrieving the pipette from the oocyte. The pipette may then become sticky. Do not hesitate to change the pipette in that case.
14. Oocyte trauma during injection usually results in cell lysis within the next 15 min. At least 80% of injected zygotes should survive the injection procedure. Low survival rate is frequently caused by injection pipettes that are too large or by nucleolus damage.
15. One-cell embryo reimplantation should be performed after allowing injected embryos to recover for 1–3 h in mR1ECM culture medium.
16. Anesthesia through inhalation of halogenated gases causes significant body heat loss. Prevent heat loss until the animal recovers using red lamp or heating-pad devices. The animal should recover within 5 min postoperatively.
17. It is crucial to avoid cold shock, so the time between loading of zygotes into the reimplantation pipette and embryo transfer into the oviduct should be as short as possible.
18. Care must be taken to avoid injecting oil into the oviduct because it may dramatically reduce litter size by interfering with the active ciliary-driven egg transport along the oviduct.
19. The rat ovarian bursa is much tougher than in the mouse and is highly vascular. Bleeding is thus virtually unavoidable. To perform optimal embryo transfer a few drops (approximately 20–30 μl each) of epinephrine are sprinkled over the bursa to reduce bleeding. Alternatively, it is possible to transfer embryos through a hole created with a 30-G needle in the oviduct segment upstream of the ampullae. When following this latter procedure it is important to ensure that the embryos are blown toward the ampulla and not the infundibulum.

Acknowledgments

The author would like to thank staff from the embryology and molecular biology department for their technical help in writing the document.

References

1. Cozzi, J., Fraichard, A. and Thiam, K. (2008) Use of genetically modified rats for translational medicine. *Drug Discov Today*, **13**(11–12), 488–494.
2. Mullins, J.J., Peters, J. and Ganten, D. (1990) Fulminant hypertension in transgenic rats harbouring the mouse Ren-2 gene. *Nature*, **344**, 541–544.
3. Hammer, R.E., Maika, S.D., Richardson, J.A., Tang, J.P. and Taurog, J.D. (1990) Spontaneous inflammatory disease in transgenic rats expressing HLA-B27 and human beta 2m: an animal model of HLA-B27-associated human disorders. *Cell*, **63**, 1099–1112.
4. Armstrong, D.T. and Opavsky, M.A. (1988) Superovulation of immature rats by continuous infusion of follicle-stimulating hormone. *Biol Reprod*, **39**, 511–518.
5. Miyoshi, K., Funahashi, H., Okuda, K. and Niwa, K. (1994) Development of rat one-cell embryos in a chemically defined medium: effects of glucose, phosphate and osmolarity. *J Reprod Fertil*, **100**, 21–26.
6. Vickery, B.H. and McRae, G.I. (1980) Synchronization of oestrus in adult female rats by utilizing the paradoxical effects of an LH-RH agonist. *J Reprod Fertil*, **60**, 399–402.
7. Filipiak, W.E. and Saunders, T.L. (2006) Advances in transgenic rat production. *Transgenic Res*, **15**, 673–686.
8. Hamilton, G.S. and Armstrong, D.T. (1991) The superovulation of synchronous adult rats using follicle-stimulating hormone delivered by continuous infusion. *Biol Reprod*, **44**, 851–856.
9. Smith, J.C., Corbin, T.J., McCabe, J.G. and Bolon, B. (2004) Isoflurane with morphine is a suitable anaesthetic regimen for embryo transfer in the production of transgenic rats. *Lab Anim*, **38**, 38–43.

Part II

Transgenesis in the Mouse

Chapter 6

Cell-Type-Specific Transgenesis in the Mouse

James Gulick and Jeffrey Robbins

Summary

Since the early 1980s, when the first transgenic mice were generated, thousands of genetically modified mouse lines have been created. Early on, Jaenisch established proof of principle, showing that viral integration into the mouse genome and germline transmission of those exogenous sequences were possible (Proc Natl Acad Sci USA 71:1250–1254, 1974). Gordon et al. (Proc Natl Acad Sci USA 77:7380–7384, 1980) and Brinster et al. (Cell 27:223–231, 1981) subsequently used cloned genes to create "transgenic constructs" in which the exogenous DNA was randomly inserted into different sites in the mouse genome, stably maintained, and transmitted through the germline to the progeny. The utility of the process quickly became apparent when a transgene carrying the metallothionein-1 (Mt-1) promoter linked to thymidine kinase was able to drive expression in the mouse liver when promoter activity was induced by administration of metals. In an attempt to find stronger and more reliable promoters, viral promoter elements from SV40 or cytomegalovirus were incorporated. However, while these promoters were able to drive high levels of expression, for many applications they proved to be too blunt an instrument as they drove ubiquitous expression in many, if not all cell types, making it very hard to discern organ-specific or cell-type-specific effects due to transgene expression. Thus the need to find cell-type-specific promoters that could reproducibly drive high levels of transgene expression in a particular cell type, e.g., cardiomyocyte, became apparent. One such example is the α myosin heavy-chain (MHC) promoter, which has been used extensively to drive transgene expression in a cardiomyocyte-specific manner in the mouse. This chapter, while not written as a typical methods section, will describe the necessary components of the α myosin promoter. In addition, common problems associated with transgenic mouse lines will be addressed.

Key words: Mouse, Cardiomyocyte, Transgenesis, Heart, Muscle

1. Introduction

Traditionally, classical genetics has involved the process of randomly mutating an organism's genetic complement (DNA or RNA) either by radiation or mutagenic substances, followed by

Elizabeth J. Cartwright (ed.), *Transgenesis Techniques*, Methods in Molecular Biology, vol. 561
DOI 10.1007/978-1-60327-019-9_6, © Humana Press, a part of Springer Science+Business Media, LLC 2009

a systematic search for a phenotype and the associated mutation. With the development of transgenesis the process was reversed, with the genetic sequence of interest known and the transgene incorporating either the normal sequence or one in which a known mutation was made *(1–3)*. Through micromanipulation, the modified DNA is injected into a fertilized mouse egg, where the modified DNA is inserted into the mouse genome by nonhomologous recombination. The number of transgene sequences (copy number) and the site of insertion into the host genome are random, although using specially engineered constructs *(4, 5)* this potential limitation can be overcome. Transgenic mice have allowed researchers to study inherited diseases, define metabolic and signaling pathways, delineate structure–function relationships in critical proteins, and determine the function(s) of newly discovered sequences. For transgenesis to be successful several technical factors need to be considered, such as the purity and concentration of the DNA to be injected, the skill of the injector, and the rigor with which animal husbandry is practiced. However, of paramount importance is the transgenic construct itself, which consists of four components: (1) The promoter, which drives transcription of the transgene to be expressed. (2) The cDNA insert, which contains the sequence encoding the peptide to be expressed. (3) The untranslated region and polyadenylation (polyA) signal lying 3′ to the cDNA and (4) The plasmid vector. Each of these components will be considered in detail in **Subheading 3**.

2. Materials

1. Mouse genomic DNA or (purchased) Bacterial Artificial Chromosomes (BACs) containing the gene of interest.
2. Restriction enzymes.
3. Klenow enzyme.
4. Taq polymerase with proofreading ability designed for long templates or GC-rich regions.
5. Plasmid vectors.
6. DNA ligase.
7. Gene-specific primers with cloning sites incorporated into the 5′ end.
8. Reverse transcriptase or cDNA clone.
9. Competent bacterial cells.
10. Qiaex gel extraction kit.

3. Methods

3.1. Choosing a Promoter

The choice of the promoter that will be used to drive expression of the construct is a critical one. Ideally, a suitable promoter has already been defined and one can "clone by phone." Promoters of widely varying specificity and strength have been used ranging from promoters that can drive expression in all tissues and at all developmental stages, to those that are restricted in their patterns of expression to a single cell type at a particular developmental stage. The purpose of the particular experiment will dictate which promoter is used but, in general, the more precisely expression is controlled, the more easily the investigator can interpret the resulting phenotype. This chapter will focus on precisely manipulating the output of the cardiomyocyte and thus, the murine α myosin heavy-chain (MHC) promoter will be used as an example to help illustrate the process of building a transgene *(6)*.

Each promoter contains a unique combination of DNA sequences (*cis* regulatory elements) which, taken together, are responsible for recruiting specific protein complexes that activate or silence the expression of the downstream gene. The availability of these complexes is responsible for the developmental stage and tissue or cell-type-specific expression of each gene. While there is no set length of DNA that defines a promoter, certain sequences are required to establish a minimal eukaryotic functional promoter *(7, 8)*. At about 25–35 base pairs (bp) upstream of the first exon is a *cis* element defined by the sequence <TATAA>, known as a TATA or Hogness box. About 50-bp upstream of this element is the CAAT box with a consensus sequence of <CCAAT>. A schematic diagram of the murine α MHC promoter, showing the location of the CAAT and TATAA elements with respect to the start site of transcription, illustrates these points **(Fig. 1a)**. Additional transcriptional control elements are also shown: these consist of small sequence cassettes that confer tissue-specific, developmental-stage-specific, and hormone responsive (thyroid responsive) transcriptional modulation. Other elements that may be found in many promoters include enhancers (which confer high levels of transcriptional activity), insulators (which confer immunity from either enhancing or silencing effects of the chromosomal context in which the gene or transgene is inserted), and repressors, although it is rare that all these elements have been defined completely for a single promoter.

If a promoter has not been previously characterized, as a first step a detailed search of Pubmed (http://www.ncbi.nlm.nih.gov/pubmed/) should be performed for the gene of interest. Often, a bioinformatics-based approach is useful and an analysis of potential binding sites may already exist. In addition, *Bioinformatic Harvester* (http://harvester.fzk.de/harvester/) is an

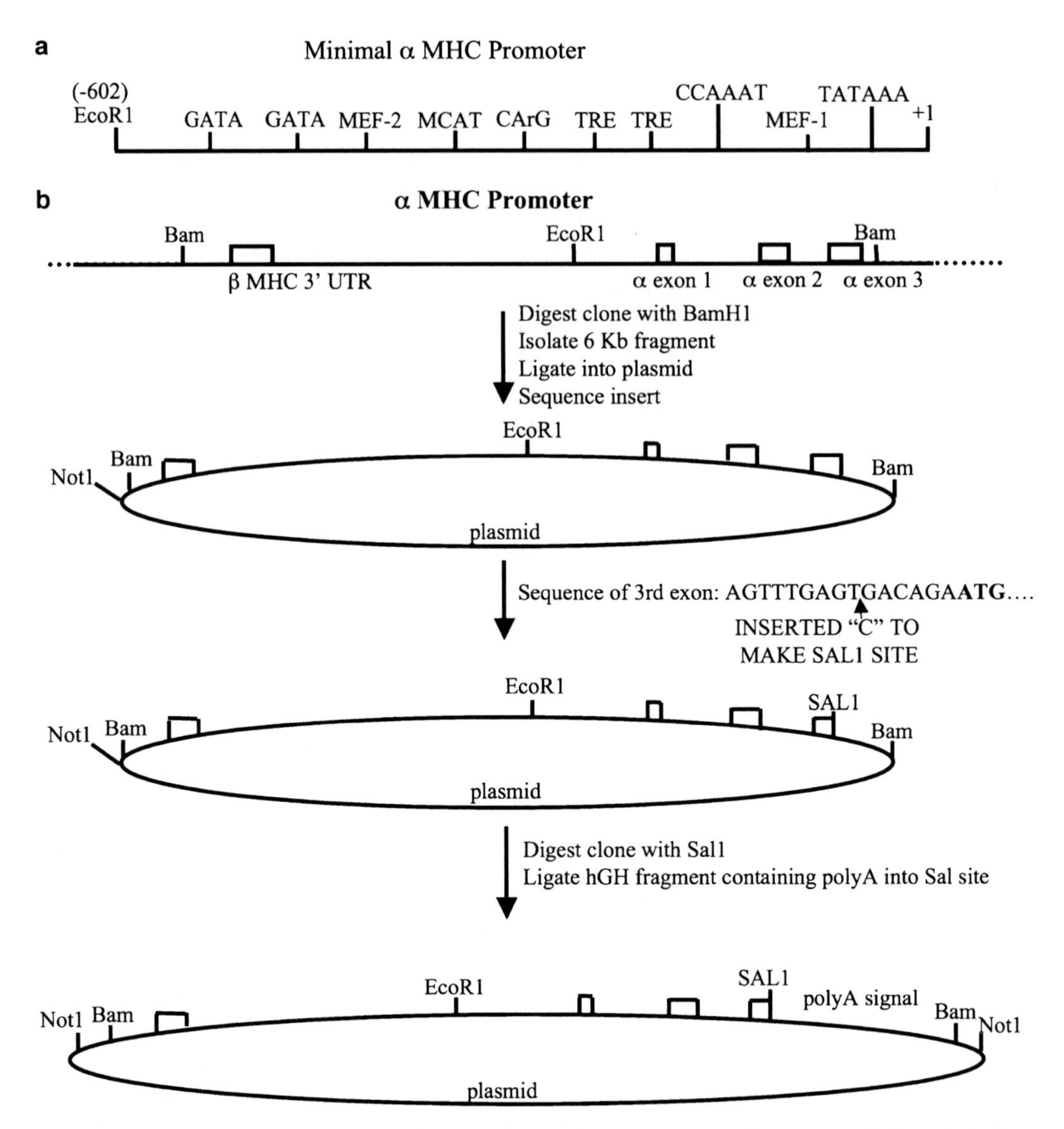

Fig. 1. The α myosin heavy-chain (MHC) promoter. (**a**) MHC minimal promoter, which is contained within a 600-bp fragment upstream of the start site of transcription. The TATAA and the CAAT *boxes* are shown as well as several muscle-specific regulatory elements needed for high levels of cell-type-specific expression. (**b**) The cloning strategy used to isolate the promoter is shown. The final construct contains a unique restriction site (Sal1) allowing for easy insertion of the cDNA. Polyadenylation is provided by the growth hormone (hGH) fragment, which contains ~100 bp of 3′ UTR between the Sal1 cloning site and the polyA signal, AATAAA. The construct also contains unique restriction sites on each end of the transgene, allowing for easy removal of the plasmid vector before injection.

excellent compiler of gene structure and function. With the complete sequence and extensive annotation of the mouse genome available, the DNA sequence surrounding the transcriptional start site can be (relatively) easily analyzed. By using the available sequence, a homology search, using *Blast* (http://www.ncbi.nlm.nih.gov/blast/Blast.cgi), against other species will reveal conserved regions that may define the most important potential regulatory elements that are necessary for cell-type controllable expression. Other bioinformatic-based approaches may also be useful in designing a new promoter. For example, the program *Trafac* (http://www.cincinnatichildrens.org/research/cores/informatics/labs/aronow/tools/trafac.htm) screens for potential protein-binding motifs and compares potential elements between different promoters and between species, emphasizing the conserved nature of the *cis* transcriptional elements.

Once the sequence of the putative promoter is selected, it can now be isolated from the mouse genome. Previously, promoter regions were isolated by screening and purifying bacteriophage or cosmid clones that contained mouse genomic DNA. As shown in **Fig. 1b**, the mouse α MHC promoter was contained within a single restriction fragment that spanned the intergenic region between the α and the β myosin genes. In addition to the extreme 3′ terminus of β MHC, the fragment encompassed the α MHC promoter and the first three noncoding exons. The noncoding exons were left in the promoter fragment to provide a substrate for splicing, which is important for RNA processing, transport, and stability: all transgenic constructs should include a splicing site of this nature, either at the 5′ or 3′ end, or both *(9, 10)*. Using PCR mutagenesis, a "C" was inserted into the third exon 5′ of the beginning of translation (ATG) to create a unique Sal1 site, which provides a convenient cloning site for inserting any cDNA to be expressed.

With current available technology, screening genomic libraries is rarely necessary and the sequence of interest is produced directly using PCR. As shown in **Fig. 2**, primers are made to the ends of the promoter region and the DNA amplified using a high-fidelity polymerase. For difficult-to-amplify DNA sequences, polymerase kits containing additives that "open" G/C-rich regions should be used. If the region is still difficult to PCR then a BAC (http://www.tigr.org/tdb/bac_ends/mouse/bac_end_intro.html) can

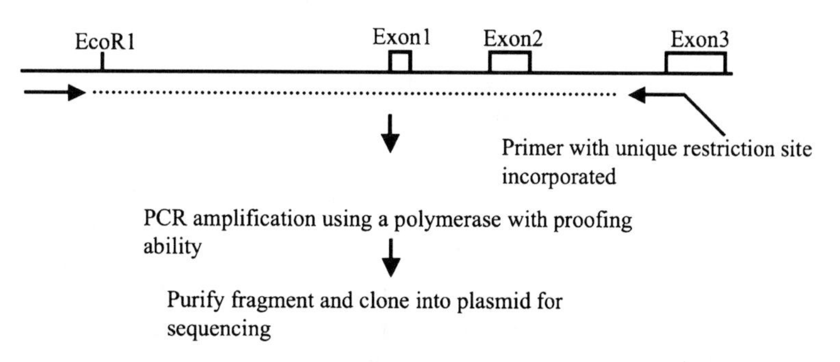

Fig. 2. PCR amplification of the MHC promoter. The unique Sal1 site was incorporated into the reverse primer. After amplification with a proofing reading polymerase, the fragment was cloned into a blunt-ended vector and the sequence verified before further use.

be ordered (http://clones.invitrogen.com/bacpacsearch.php), which contains a relatively small DNA with the genomic DNA of interest. By reducing the size of the DNA template used in the reaction, a cleaner PCR product can be obtained. Once the promoter fragment has been isolated it should be cloned into a plasmid vector. Most vectors used are high copy number plasmids, which is necessary since a large quantity of DNA is required by the Transgenic Facility for purification.

3.2. The cDNA Insert

The DNA to be expressed is subsequently inserted 3′ of the isolated promoter. The DNA usually consists of a double-stranded cDNA (a DNA sequence that contains only the exonic sequences) fragment cloned into a unique site behind the desired promoter, although some labs use genomic fragments that contain the coding region of the gene to be expressed, creating a minigene. In the case of a cDNA insert, PCR can be used to amplify the coding region from a cDNA template, removing as much of the 5′ and 3′ UTRs as possible. At the same time, unique restriction sites should be added onto the ends of the cDNA, allowing the fragment to be easily cloned downstream of the promoter. As an example, using the α MHC promoter, Sal1 restriction sites were added to the ends of the cDNA by PCR to aid in cloning. In those cases where the cDNA contains internal Sal1 sites, Xho1 restriction sites can be added since Sal 1and Xho 1 contain cohesive ends that can be ligated together. In those cases where the cDNA contains both internal Sal 1 and Xho 1 sites, the ends of the cDNA fragment, as well as the promoter, must be blunt ended with the use of Klenow before they can be ligated together.

In addition to incorporating restriction endonuclease sites onto the ends, the cDNA should contain sequences around the translational start site (AUG) that enhance translation. Kozak *(11)* performed a survey of several hundred translational start sites and discovered a consensus sequence of GCCA/GCC**AUG**G (the initiating methionine is in bold and underlined). To optimize translation of the transcribed cDNA, this consensus sequence should be incorporated into the construct whenever possible. Another necessary component of the cDNA is the inclusion of a termination signal at the 3′ end of the open reading frame. Any of the three stop codons can be used.

In our experience, the cDNA should contain as little of the gene's 5′ and 3′ UTRs as possible, although minimal UTRs should be present (**Subheading 3.3**). UTRs are known to contain regulatory sequences that may affect transcription and translation *(12)*. miRNAs (microRNAs) are also known to interact with sequences in the 3′ UTRs and can influence both mRNA stability and translational efficiency *(13)*. In one example of the importance of the flanking sequences, an entire cDNA with complete 5′ and 3′ UTRs was cloned into the α promoter, with no detectable protein being produced. Once the UTRs were removed from the transgene, the modified sequence resulted in detectable levels of protein.

In most cases, the cDNA PCR template can be obtained commercially or "cloned by phone." Instead of isolating RNA, buying a kit to reverse transcribe the RNA, cloning and sequencing the potential cDNA clone, and then sorting through all of the clones that represent splicing variants, a cDNA clone can be in your lab in a few days by simply placing an order, either through commercial sources or by contacting an academic laboratory.

3.3. The UTR and PolyA Signal

Several studies have shown that message stability is influenced by the 3′ UTR attached to the construct *(14)*. Transgene efficiency can be increased dramatically by using a 3′ UTR that prolongs mRNA half-life, which can increase the amount of protein produced. In the case of the α MHC promoter, the human growth hormone 3′ UTR was used. A 600-bp fragment that contains about 100 bp of 3′ UTR as well as a polyA signal was placed 3′ of the cDNA cloning site. Another useful 3′ UTR/polyA fragment is derived from the SV40 virus. The SV40 polyA fragment also contains an intron, which may be needed to confer a splicing event during processing of the primary transcript (**Subheading 3.1**), if your promoter lacks an intron. Most commercially available plasmids use the SV40 polyA fragment, providing an easy source for ligating the appropriate signal 3′ to your cDNA as the transgenic construct is built.

3.4. The Plasmid Vector

All the aforementioned components are then inserted into a plasmid backbone. The vector should be a high copy number plasmid and contain an antibiotic resistance marker for easy selection. In our hands, the kanamycin resistance gene is preferable, as ampicillin can be unstable during bacterial growth. A high copy number is preferable as this lowers the total culture volume needed to produce the approximately 100 μg of construct usually requested by the Transgenic Injection Facility. However, if the construct contains repetitive sequences that have the potential to recombine during growth, a high copy number plasmid should be avoided and a low copy number plasmid backbone used instead, such as pBR322 and its derivatives.

3.5. Purification of the Transgene for Injection

The transgene, consisting of the promoter, cDNA, and polyA, needs to be digested away from the plasmid backbone. After digestion, the transgene needs to be purified. We normally digest 100 μg of DNA in a volume of 300 μl. The DNA fragments are separated on an agarose gel and the band containing the transgene isolated. The DNA is purified using standard column or extraction procedures: we have found that the Qiaex gel extraction method (Qiagen) provides DNA suitable for pronuclear injections. After extraction, the DNA should be resuspended in

Tris/EDTA (the exact concentration and pH requested by your transgenic injection facility may vary). The DNA must be filtered or passed through a spin column. This procedure will remove any contaminating matrix or beads, which might clog the microinjection pipette. In all cases, it is a good idea to check with your transgenic facility as they may have a preferred method of transgene purification (*see* **Note 1**). Many facilities offer a DNA purification service for an additional fee and we strongly recommend that the investigator use their services as it can avoid considerable delays or finger pointing later on if founders are not forthcoming. Some facilities may add an additional step for quality control and recommend a test injection with the DNA to see if the postinjected oocyte can undergo a few rounds of division, indicating that the DNA itself is not lethal to the cell.

If there is no backlog of requests at the Transgenic Facility, then tissue samples from potential founders (transgene-positive mice) can be expected in approximately 4–6 weeks. The facility will start DNA purification within days, followed by quantification and dilution of the DNA. Within 1–2 weeks, the DNA should be injected into the fertilized mouse eggs. In about 19 days the pups will be born and after 10 days they will be large enough for sample collection (usually an earclip). The sample is digested and PCR analysis used to check for the presence of the transgene.

Although these techniques and procedures are now well established, numerous issues can arise that will challenge successful execution of the experiments. These are detailed below (*see* **Note 2–15**).

4. Notes

1. *Use a successful Transgenic Facility.* Not all facilities have a successful track record or a high success for making transgenic mice. Even the best facilities have their bad days (or weeks) when success is limited. Some though may have more bad days than good and it is important that the investigator have a clear sense of the overall competence of the facility. Many academic institutions with very competent personnel offer offsite pricing and even though it may appear to be more expensive, the amount of time saved more than makes up for the price differential that may be incurred.
2. *No pups.* If no transgenic pups can be detected from approximately 200 oocyte injections, this may indicate that the transgene encodes a protein that is lethal during embryonic development. If this is the case then more eggs need to be

injected and the embryos analyzed during their development to determine the nature of the lethal event. To avoid lethality, either a different promoter that is tightly controlled such that expression occurs only after development, or an inducible system must be considered.

3. *Mosaic founders.* The transgenic facility will inject the DNA into the pronuclei of single-cell fertilized eggs. The DNA normally integrates at this stage and the transgene will be present in all cells of the mouse, including the germline. However, if integration occurs at a later stage when there are two or more cells, the resultant mouse may be mosaic for the transgenic DNA, with the transgene present (and expressed) in some of the cells, but not in others. It is possible that no, or very few, F1-generation transgenic mice will be born from the mosaic founder if those progenitor cells that form the germline did not acquire the DNA. If no transgenic offspring from a positive founder are obtained in the initial litter, we routinely screen at least three additional litters from the potential founders for positive offspring before terminating the line.
4. *Transgene expression levels and molecular torture.* Most transgenes have some degree of sensitivity to the surrounding chromosomal context and/or some degree of copy number-dependent expression. As such, not all lines containing the identical transgenic construct will express the transgene at the same level, as each line may contain varying (1→100) copies of the transgene inserted at different sites in the mouse genome. This can be useful as a range of protein expressions can illuminate the physiology of the resultant phenotype. However, nonphysiologically high levels of expression can be misleading, resulting in cell responses that are related simply to the abnormally high protein level rather than being directly related to the protein's function. If a mutated protein is being expressed this can be controlled by making transgenic lines in which the transgene consists of the normal protein being expressed at levels at or above the mutant polypeptide. In cases where this control cannot be done, the data from the transgenic line must be cautiously interpreted.
5. *Animal husbandry.* Most if not all injection facilities will use mice that have been raised in barrier rooms and housed in microisolator cages: these mice are referred to as being "clean." That is, they have been purchased from a reputable provider, or bred in house and have been tested serologically for viruses and pathogens. NIH mandates that transgenic mice made under the auspices of government support be freely shared. By keeping some of the resultant transgenic mice in a barrier, or clean room, mice can be easily exchanged between facilities without spreading disease. Additionally,

some animal models will require that the mice remain clean, since viral or bacterial contamination will complicate the phenotype. As examples, results will be compromised when studying liver function if the mice have hepatitis, or when studying digestive diseases and the mice have *helicobacter*.

6. *Sex-linked transmission*. The genomic integration of the transgene is a random event and occasionally the DNA inserts into the X or Y chromosomes. An insert in the Y chromosome will be obvious when no transgene-positive females are detected in the first- or second-generation offspring. An insertion into the X chromosome may be difficult to interpret if the transgene is silenced by x-linked inactivation in which some cells are expressing an active transgene, while other cells are not. These lines should be discarded.
7. *Double insertions*. Chromosomal integration most frequently occurs at a single site. In a few instances the transgene will insert into two distinct sites, possibly on different chromosomes. Sometimes the offspring will contain both inserts; at other times the pups may contain one or the other insert. This will become apparent when a single litter gives a wide range of expressions, or if too many positive pups are obtained (nonmendelian inheritance). Since maintaining the double insertion on one transgenic line is difficult, the two insertions should be bred out into two separate sublines.
8. *Insertional mutagenesis*. During the recombination events that underlie transgene insertion, genomic deletions may occur at the integration site. The deleted DNA may range from a few bases to a kilobase in size. Normally, as long as the transgenic mice are maintained as heterozygotes, the deletion will not be noticed as the mutation is only rarely dominant and one intact allele is still present. However, if the transgene is bred to homozygosity then the (potential) effects of the deletion may become apparent *(15)*, with a resultant, confounding phenotype. For this reason it is always a good idea to compare data using at least two distinct lines (in case the mutation is, in fact, dominant) and only use outbred transgenic offspring to ensure that the mouse is heterozygous for the transgene.
9. *Illicit splicing events*. In a few instances one is able to detect mRNA from the transgene, but no transgenic protein is produced. Using the α MHC promoter, after analysis by RT-PCR we determined that the RNA had been improperly spliced. Instead of the expected splicing between the second and third noncoding exons of the promoter (**Fig. 3**), the second exon was spliced into the cDNA, removing a third of the coding region. The third exon was later removed and the cDNA ligated directly to the second exon in an attempt

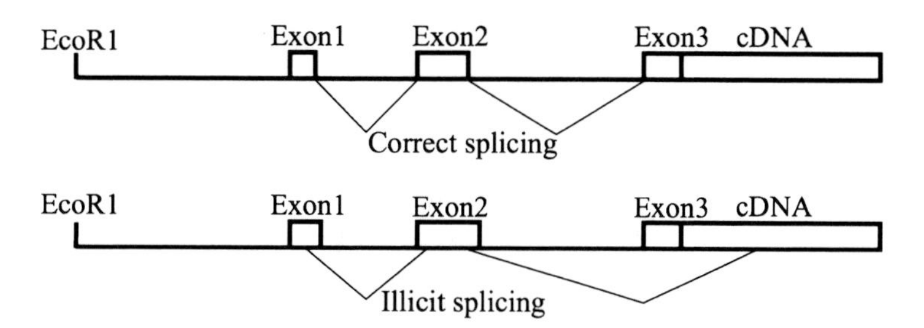

Fig. 3. Errant splicing. Most of the time the promoter splices correctly from exon 1 to exon 2, then to exon 3 with the entire cDNA sequence being incorporated into the mRNA. Occasionally, the splicing event occurs into the cloned cDNA, bypassing exon 3. In these cases a full-length protein product may not be produced or detected.

to eliminate the errant splicing. The new transgene spliced from the first exon into the cDNA, again removing most of the coding sequence. In another attempt we changed the sequence of the splice site in the cDNA. The construct still spliced incorrectly. Analysis of the cDNA sequence revealed several potential splice sites, which were eliminated where possible by changing the cDNA sequence. Most amino acids are encoded by more than one DNA triplet, allowing the sequence to be modified while conserving the proteins' amino acid sequence.

10. *Lack of a phenotype.* In some transgenics no phenotype is observed. Lack of a phenotype most often means that the investigator has not devised a suitable screening procedure or has a predetermined notion of what the phenotype should be. Occasionally, however, this can be due to a complete lack of protein function. This can arise for a number of reasons. The protein may not be trafficked correctly within the cell. If a "sorting" signal is left off of the transgene, such as a nuclear localization signal, then the protein may not reach the correct cellular subcompartment. Another possibility is that the transgene is post-transcriptionally controlled or the protein is post-translationally modified and subsequently degraded. The most frequent cause is due to a mutation occurring in the transgene, giving a truncated protein product or no protein at all. Your construct should be sequenced at all steps of development and care should be taken to derive the transgene from the mouse strain that will be the source of the oocytes used for injection, to minimize strain differences in the DNA that might result in subtle changes in protein sequence (*see* **Note 13**).
11. *Strain differences.* The Mouse Sequencing Consortium recently finished sequencing the genome of the C57BL/6J strain of mouse (http://www.genome.gov/10001859) with genetic analysis on fifteen of the most commonly used strains

currently underway. Strain variations are significant and must be considered as they can have a major impact on phenotype presentation. As an example, several strains of mice were characterized as to their cardiac function, with statistically significant differences *(16)*. In a transgenic model, strain differences were observed between C57 and FVB mice when a GATA4 mutation was introduced *(17)*, leading to variations in valve formation. Most journals will ask you to list which strains were used and the percentage of each if the strain is not pure. Some journals actually go so far as to demand that experiments be conducted in a pure, defined strain.

12. *Phenotypic consequences of reporter gene expression or epitope tagging.* Reporter genes are sometimes used in transgenesis as a way of easily assessing the cellular/tissue location of where a promoter is active or if a certain event occurs, such as genetic recombination. One common marker is GFP (Green Fluorescent Protein), which is often linked to a transgene, creating a fusion product. The GFP serves as a useful marker, allowing your protein to be tracked as it moves about the cell, or to quantify the levels of expression. An area of concern is that GFP can affect biological function, e.g., possibly altering the electrophysiology of cardiac cells *(18)*. Another protein that may cause a phenotype is a component of the commonly used binary system, the Cre protein in the *Cre-lox* system. Cre expression can, for example, cause dilated cardiomyopathy when expressed at high levels *(19)*. Another binary system, the tetracycline inducible promoter, uses a transactivator protein, which when expressed at high levels causes a cardiomyopathy *(20, 21)*.Frequently, epitope tags are added to the coding portion of the transgene, allowing the protein of interest to be tracked or identified within the cell by immunohistochemical means. These extra bases are usually incorporated onto either the N- or C-terminus of the protein. In some instances the tags can interfere with protein function by blocking protein-protein interactions or affecting intramolecular folding. Most tags contain charged amino acids or are highly structured, making them great epitopes but (potentially) terrible additions to your protein in terms of its normal function. A review by Maue *(22)* lists the most commonly used tags and their amino sequences, and describes the successes and failures of adding tags to various constructs.
13. *Phenotype drift.* Over time, an observed phenotype may change or drift within a colony of transgenic animals. For example, an investigator might report that all transgenic mice died at 6 months of age but, 2–4 years later, note that all are now living consistently for 9–12 months. One explanation

is that over time those mice that are the healthiest have a tendency to breed more frequently and for a longer period of time, skewing the colony's phenotype, while the sicker mice become progressively under-represented. There is no easy way to avoid this since the breeding ability of the mice is not known in advance, but the cautious investigator will only use breeding pairs that are less than 6–7 months and continuously monitor the phenotype of the colony.

14. *Transgene silencing*. Genes can be silenced by epigenetic changes, such as DNA methylation and histone acetylation. DNA becomes methylated when enzymes add methyl groups to the cytosine residue of CpG sequences *(23–25)*. This pattern of methylation is then inherited by the next generation of mice and can lead to transgene inactivation. One can minimize the problem by using endogenous sequences for all elements of transgene construction, but it remains a potential issue despite such precautions.

15. *Transgene instability*. Some genomic sequences are known to be unstable, especially those that contain repetitive elements. If repetitive elements are included in your construct or generated during the insertional event, rearrangements may occur, in some instances deleting the transgene and some of the surrounding chromosome.

References

1. Jaenisch, R. and Mintz, B. (1974) Simian virus 40 DNA sequences in DNA of healthy adult mice derived from preimplantation blastocysts injected with viral DNA. *Proc Natl Acad Sci USA*, **71**, 1250–1254.
2. Gordon, J.W., Scangos, G.A., Plotkin, D.J., Barbosa, J.A. and Ruddle, F.H. (1980) Genetic transformation of mouse embryos by microinjection of purified DNA. *Proc Natl Acad Sci USA*, **77**, 7380–7384.
3. Brinster, R.L., Chen, H.Y., Trumbauer, M., Senear, A.W., Warren, R. and Palmiter, R.D. (1981) Somatic expression of herpes thymidine kinase in mice following injection of a fusion gene into eggs. *Cell*, **27**, 223–231.
4. Bronson, S.K., Plaehn, E.G., Kluckman, K.D., Hagaman, J.R., Maeda, N. and Smithies, O. (1996) Single-copy transgenic mice with chosen-site integration. *Proc Natl Acad Sci USA*, **93**, 9067–9072.
5. Dupuy, A.J., Clark, K., Carlson, C.M., Fritz, S., Davidson, A.E., Markley, K.M., Finley, K., Fletcher, C.F., Ekker, S.C., Hackett, P.B. et al. (2002) Mammalian germ-line transgenesis by transposition. *Proc Natl Acad Sci USA*, **99**, 4495–4499.
6. Gulick, J., Subramaniam, A., Neumann, J. and Robbins, J. (1991) Isolation and characterization of the mouse cardiac myosin heavy chain genes. *J Biol Chem*, **266**, 9180–9185.
7. Pedersen, A.G., Baldi, P., Chauvin, Y. and Brunak, S. (1999) The biology of eukaryotic promoter prediction – a review. *Comput Chem*, **23**, 191–207.
8. Suzuki, Y., Tsunoda, T., Sese, J., Taira, H., Mizushima-Sugano, J., Hata, H., Ota, T., Isogai, T., Tanaka, T., Nakamura, Y. et al. (2001) Identification and characterization of the potential promoter regions of 1031 kinds of human genes. *Genome Res*, **11**, 677–684.
9. Proudfoot, N.J., Furger, A. and Dye, M.J. (2002) Integrating mRNA processing with transcription. *Cell*, **108**, 501–512.
10. Reed, R. (2003) Coupling transcription, splicing and mRNA export. *Curr Opin Cell Biol*, **15**, 326–331.
11. Kozak, M. (1987) An analysis of 5′-noncoding sequences from 699 vertebrate messenger RNAs. *Nucleic Acids Res*, **15**, 8125–8148.
12. Marano, R.J., Brankov, M. and Rakoczy, P.E. (2004) Discovery of a novel control element within the 5′-untranslated region of the vascular

endothelial growth factor. Regulation of expression using sense oligonucleotides. *J Biol Chem*, **279**, 37808–37814.

13. Lai, E.C. (2002) Micro RNAs are complementary to 3′ UTR sequence motifs that mediate negative post-transcriptional regulation. *Nat Genet*, **30**, 363–364.
14. Knouff, C., Malloy, S., Wilder, J., Altenburg, M.K. and Maeda, N. (2001) Doubling expression of the low density lipoprotein receptor by truncation of the 3′-untranslated region sequence ameliorates type iii hyperlipoproteinemia in mice expressing the human apoe2 isoform. *J Biol Chem*, **276**, 3856–3862.
15. McNeish, J.D., Thayer, J., Walling, K., Sulik, K.K., Potter, S.S. and Scott, W.J. (1990) Phenotypic characterization of the transgenic mouse insertional mutation, legless. *J Exp Zool*, **253**, 151–162.
16. Lerman, I., Harrison, B.C., Freeman, K., Hewett, T.E., Allen, D.L., Robbins, J. and Leinwand, L.A. (2002) Genetic variability in forced and voluntary endurance exercise performance in seven inbred mouse strains. *J Appl Physiol*, **92**, 2245–2255.
17. Rajagopal, S.K., Ma, Q., Obler, D., Shen, J., Manichaikul, A., Tomita-Mitchell, A., Boardman, K., Briggs, C., Garg, V., Srivastava, D. et al. (2007) Spectrum of heart disease associated with murine and human GATA4 mutation. *J Mol Cell Cardiol*, **43**, 677–685.
18. Sekar, R.B., Kizana, E., Smith, R.R., Barth, A.S., Zhang, Y., Marban, E. and Tung, L. (2007) Lentiviral vector-mediated expression of GFP or Kir2.1 alters the electrophysiology of neonatal rat ventricular myocytes without inducing cytotoxicity. *Am J Physiol Heart Circ Physiol*, **293**, H2757–H2770.
19. Buerger, A., Rozhitskaya, O., Sherwood, M.C., Dorfman, A.L., Bisping, E., Abel, E.D., Pu, W.T., Izumo, S. and Jay, P.Y. (2006) Dilated cardiomyopathy resulting from high-level myocardial expression of Cre-recombinase. *J Card Fail*, **12**, 392–398.
20. McCloskey, D.T., Turnbull, L., Swigart, P.M., Zambon, A.C., Turcato, S., Joho, S., Grossman, W., Conklin, B.R., Simpson, P.C. and Baker, A.J. (2005) Cardiac transgenesis with the tetracycline transactivator changes myocardial function and gene expression. *Physiol Genomics*, **22**, 118–126.
21. Sanbe, A., Gulick, J., Hanks, M.C., Liang, Q., Osinska, H. and Robbins, J. (2003) Reengineering inducible cardiac-specific transgenesis with an attenuated myosin heavy chain promoter. *Circ Res*, **92**, 609–616.
22. Maue, R.A. (2007) Understanding ion channel biology using epitope tags: progress, pitfalls, and promise. *J Cell Physiol*, **213**, 618–625.
23. Weng, A., Magnuson, T. and Storb, U. (1995) Strain-specific transgene methylation occurs early in mouse development and can be recapitulated in embryonic stem cells. *Development*, **121**, 2853–2859.
24. Boyes, J. and Bird, A. (1991) DNA methylation inhibits transcription indirectly via a methyl-CpG binding protein. *Cell*, **64**, 1123–1134.
25. Gundersen, G., Kolsto, A.B., Larsen, F. and Prydz, H. (1992) Tissue-specific methylation of a CpG island in transgenic mice. *Gene*, **113**, 207–214.

Chapter 7

Transgene Design and Delivery into the Mouse Genome: Keys to Success

Lydia Teboul

Summary

Mouse transgenesis by pronuclear injection generates random integration events resulting in variable overexpression efficiency. This chapter will outline current strategies for improved transgene design (using plasmids or bacterial artificial chromosomes) and delivery to the mouse genome. The choice of genetic background of embryos for microinjection will also be discussed.

Key words: Transgene, Mouse, Design, Delivery, Overexpression, Plasmid, BAC

1. Introduction

In the mouse overexpression models are used for numerous purposes including complementation studies, promoter-bashing experiments, time or tissue-specific recombination through Cre or related enzymes activity, and cell lineage tracing or deletion. Recent years have also seen the use of an array of reporter genes within transgene design: CAT, *luc*, various fluorescent proteins, *lacZ*, alkaline phosphatase, and modifications thereof for specific subcellular localisation, biosensors (such as calcium or cAMP sensors), and genes for lineage ablation (Diphtheria toxin, mutants of M2 protein of influenza A, etc.).

In mice, these overexpression models are generally obtained by microinjection of DNA constructs into the pronuclei of one-cell embryos *(1)*. This is a process which usually results in the random integration of several copies of the transgene.

Elizabeth J. Cartwright (ed.), *Transgenesis Techniques*, Methods in Molecular Biology, vol. 561
DOI 10.1007/978-1-60327-019-9_7, © Humana Press, a part of Springer Science+Business Media, LLC 2009

Such a strategy is often satisfactory for achieving the appropriate expression level and pattern for the experiment, provided that several transgenic founders are screened for transgene integrity and activity. There are several major advantages of this technology: it does not involve the building of large and complicated constructs, the production of transgenic founders is relatively fast, and it only occasionally results in chimeric animals.

However, the site of integration of the transgene in the genome (which is random) may contribute to transgene silencing or ectopic expression and, as with any genome modification, transgenesis can have unexpected consequences *(2)*. In addition, transgene rearrangement may result in the overexpression of a truncated or mutant protein. This is particularly important when introducing large constructs where coding sequences are spread over long distances as the construct is randomly broken upon integration.

Although there is no way to ensure intact integration of transgenes through pronuclear injection, it is possible to add features that will reinforce efficient expression (*see* Chapter 6 and **Subheading 2** of the present chapter for details).

Finally, it is essential that strategies for transgene detection and characterisation are worked out prior to microinjection of the construct. Screening of potential transgenic founders comprises checking transgene presence, integrity, and activity. It is furthermore recommended that transgenic colony maintenance periodically includes monitoring transgene activity as rearrangement and/or gene silencing may occur in time.

2. Transgene Design

Chapter 6 describes in detail the design of a plasmid-based construct, including the identification and isolation of regulatory elements and promoters from a Bacterial Artificial Chromosome (BAC). There are several refinements that one can consider when designing a plasmid-based transgene, as described later. The use of BACs, P1-derived Artificial Chromosomes (PAC), and Yeast Artificial Chromosomes (YAC) as vectors for transgenes is also described.

2.1. Refining Plasmid-Based Transgenes

Further improvements that can be included in plasmid-based strategies are presented in this section, and it highlights types of DNA sequences that have a negative effect on overexpression efficiency.

2.1.1. Additional Features

Improvements in the shape of intronic and Kozak sequences, MARs, insulators, IRES, or E2A cassettes can be introduced in the transgene design to try and ensure precise and robust expression of the coding sequence(s) of interest.

- *Intronic sequences* can positively or negatively affect gene/transgene expression as they may contain UTR or transcription regulatory elements as well as splicing sites (*see* Chapter 6).
- Many genes contain a *Kozak* translation initiation sequence upstream of ATG *(3)*. Although such a sequence is not absolutely required for initiation of translation to occur, its addition to a transgene design is a precaution that is well worth the effort.
- MARs (Matrix Attachment Regions) can shield a transgene from positive or negative effects of the site of integration on the transgene's expression *(4)*.
- *Insulators* are sequences that prevent enhancers activating transcription from a promoter, when situated between these two elements *(5)*. Because transgenes generally integrate in several copies, forming a direct repeat, one insulator placed at one extremity of a construct will result in most transgene copies being flanked on either site by an insulator.
- IRES (Internal Ribosome Entry Sites) are nucleotide sequences that allow translation initiation to take place inside the sequence of a messenger RNA (mRNA). Therefore such a sequence can be used to produce polycistronic transcripts. When using such a cassette, one should keep in mind that the 3′ coding sequences in the transcript are generally translated at a lower rate than the 5′ sequences (**ref.** *6* shows one example of IRES that was used to produce transgenes).
- *E2A* is a RNA sequence that induces ribosome skipping during the process of translation *(7)*. This event results in the production of two consecutive polypeptides from one transcript. The polypeptide translated from the sequence upstream of the E2A box retains the E2A sequence as a C terminal tag.

2.1.2. Unwanted Sequences

DNA sequences of non-eucaryotic origin such as plasmid backbone attract chromatin modification machineries, so gene expression is shut down in the vicinity of such sequences. It is therefore preferable, prior to microinjection, to cut out the transgene from the vector backbone, and isolate insert by separation on an agarose gel and subsequent gel purification. Restriction sites suitable for this should be planned at the stage of transgene design.

Similarly, one should consider codon optimised sequences for heterologous expression.

The functionality of a new construct should be ascertained in a cell culture system before microinjection, whenever possible.

2.2. Use of Larger Constructs in Transgenesis

Producing transgenic mice with large constructs such as BACs, PACs, or YACs is more challenging than it is with smaller-sized constructs. However, it is the strategy of choice for obtaining a tightly controlled pattern of expression. One should, check that the chosen vectors do not contain additional transcriptional units to those required for overexpression.

BACs covering the genomic region of choice can be identified on the Ensembl Genome Browser (http://www.ensembl.org/Mus_musculus/index.html) and are obtained from a variety of distributors. Libraries prepared from an array of genetic backgrounds are available. Therefore, one can take advantage of DNA sequence polymorphisms to design genotyping assays that will discriminate transgenes from cognate endogenous genes.

Furthermore, BACs can be modified to introduce mutations or include an additional coding sequence by homologous recombination in bacteria (also known as 'recombineering'). A versatile strategy for BAC recombineering involves in vitro homologous recombination between linear DNA (that can be produced by PCR) and circular constructs *(8)*.

Random breaks occur upon transgene integration, so insertions should be carefully checked for rearrangements if construct integrity is essential to the experiment.

3. Methods for Transgene Delivery into the Mouse Genome

There are several possible strategies for transgene delivery to the mouse genome:

- DNA microinjection into the pronucleus of one-cell embryos (described in further detail in Chapter 8).
- Co-injection of more than one construct: One can take advantage of the fact that co-injected transgenes tend to co-integrate at the same site in the genome *(9)*. Co-injecting two or more constructs will save potentially difficult cloning steps that would be required to physically link several inserts into the same vector. However, this method offers no control over the relative order and copy number of integration of the different co-integrants.
- Intracytoplasmic sperm injection has been used to facilitate transgenesis with 'genomic' constructs such as YACs *(10)*. This mode of injection is in itself a challenge compared to pronuclear injection and requires specific equipment and training.
- Infection is an alternative mean of transgene delivery. Doing away with the microinjection allows higher throughput capacity. This is to be balanced against the cargo size limitations

of viral particles and potential bio-safety issues with handling viruses *(11)*.

- ES cell transfection (as described in subsequent chapters of this book) is also used to create mouse overexpression models. This technique allows the production of large numbers of clones containing transgenes randomly integrated. Transgene overexpression can be characterised in vitro (both at the cellular and at the molecular levels) before the ES cells are introduced into embryos. In addition, this solution circumvents the difficulties associated with deploying the pronuclear microinjection technique. It does however require the specialist skills associated with ES cell manipulation and blastocyst injection.

There are situations where randomness of integration is a sub-optimal approach. In particular, this is the case where a very specific pattern of expression has to be achieved, and regulatory elements able to drive such expression have not been isolated. One may then consider alternative overexpression strategies such as knock-in gene targeting if an integration site that allows the desired pattern and level of expression is known. This applies also to situations where it is desirable to have a single copy of integration of the transgene. An example of this is the building of models of inducible overexpression that will involve in vivo deletion of a STOP or a reporter cassette by Cre driven recombination, although random integration by pronuclear injection has also been successfully applied in this case *(12)*.

4. Genetic Background of Embryos for Microinjection

It is crucial to consider the genetic background of the embryos to be used for DNA pronuclear microinjection as this can greatly influence the rate at which successful transgenic mice are produced. Morphology and mechanical properties of one-cell embryos vary depending on their genetic background. FVB/N embryos are the easiest to employ for transgenesis by pronuclear injection as (1) superovulation and mating of FVB/N mice reliably produces good yields and (2) the embryos are quite translucent and present large pronuclei *(13)*. However, this background can be problematic as it can display repetitive hyperactive behaviour.

Hybrids such as C57BL/6xCBA/Ca or C57BL/10xCBA/Ca are a convenient alternative for embryo production. An added benefit of using these is that females from the same breeding colony can be employed as oviduct transfer hosts.

Finally, some projects require a specific genetic background *(14)*. Performing pronuclear injection with inbred genetic backgrounds such as C57BL/6J is technically much more challenging.

There are two main difficulties in using such inbred genetic backgrounds for transgenesis: yields of embryos after superovulation and mating are generally poor and the visual and the mechanical properties of these embryos make them difficult to microinject. However, when carried out by experts it is feasible and will save lengthy and costly backcrossing.

References

1. Gordon, J. W., Scangos, G. A, Plotkin, D. J., Barbosa, J. A., and Ruddle, F. H. (1980) Genetic transformation of mouse embryos by microinjection of purified DNA. *Proc Natl Acad Sci USA*. 77, 7380–7384.
2. Matthaei, K. I. (2007) Genetically manipulated mice: a powerful tool with unsuspected caveats. *J Physiol*. **582**, 481–488.
3. Kozak, M. (1978) How do eucaryotic ribosomes select initiation regions in messenger RNA? *Cell*. **15**, 1109–1123.
4. McKnight, R. A., Shamay, A., Sankaran, L., Wall, R. J., and Hennighausen, L. (1992) Matrix-attachment regions can impart position-independent regulation of a tissue-specific gene in transgenic mice. *Proc Natl Acad Sci USA*. **89**, 6943–6947.
5. Chung, J. H., Bell, A. C., and Felsenfeld, G. (1997) Characterization of the chicken beta-globin insulator. *Proc Natl Acad Sci USA*. **94**, 575–580.
6. Jang, S. K., Kräusslich, H. G., Nicklin, M. J., Duke,G.M.,Palmenberg,A.C.,andWimmer,E. (1988) A segment of the 5′ nontranslated region of encephalomyocarditis virus RNA directs internal entry of ribosomes during in vitro translation. *J Virol*. **62**, 2636–2643.
7. de Felipe, P., Luke, G. A., Hughes, L. E., Gani, D., Halpin, C., and Ryan, M. D. (2006) E unum pluribus: multiple proteins from a self-processing polyprotein. *Trends Biotechnol*. **24**, 68–75.
8. Yu, D., Ellis, H. M., Lee, E. C., Jenkins, N. A., Copeland, N. G., and Court, D. L. (2000) An efficient recombination system for chromosome engineering in *Escherichia coli*. *Proc Natl Acad Sci USA*. **97**, 5978–5983.
9. Methot, D., Reudelhuber, T. L., and Silversides, D. W. (1995) Evaluation of tyrosinase minigene co-injection as a marker for genetic manipulations in transgenic mice. *Nucleic Acids Res*. **23**, 4551–4556.
10. Moreira, P. N., Pozueta, J., Pérez-Crespo, M., Valdivieso, F., Gutiérrez-Adán, A., and Montoliu, L. (2007) Improving the generation of genomic-type transgenic mice by ICSI. *Transgenic Res*. **16**, 163–168.
11. Hermens, W. T., and Verhaagen, J. (1998) Viral vectors, tools for gene transfer in the nervous system. *Prog Neurobiol*. **55**, 399–432.
12. Depaepe, V., Suarez-Gonzalez, N., Dufour, A., Passante, L., Gorski, J. A., Jones, K. R., Ledent, C., and Vanderhaeghen, P. (2005) Ephrin signalling controls brain size by regulating apoptosis of neural progenitors. *Nature*. **435**, 1244–1250.
13. Taketo,M.,Schroeder,A.C.,Mobraaten,L.E., Gunning, K. B., Hanten, G., Fox, R. R., Roderick, T. H., Stewart, C. L., Lilly, F., Hansen, C. T., et al. (1991) FVB/N: an inbred mouse strain preferable for transgenic analyses. *Proc Natl Acad Sci USA*. **88**, 2065–2069.
14. Lusis, A. J., Yu, J., and Wang, S. S. (2007) The problem of passenger genes in transgenic mice. *Arterioscler Thromb Vasc Biol*. **27**, 2100–2103.

Chapter 8

Overexpression Transgenesis in Mouse: Pronuclear Injection

Wendy J.K. Gardiner and Lydia Teboul

Summary

Pronuclear injection is a mode of DNA delivery for the production of transgenic animals. We present protocols for the preparation of DNA for microinjection from small or large plasmids and from BACs. We describe the production of one-cell mouse embryos, unfertile males, and pseudo-pregnant females. Finally, we detail the set up and procedures for the microinjection of one-cell embryos and their transfer into foster females.

Key words: Transgenic, Mouse, Embryo, Pronuclear, Injection, One cell, Oviduct, Transfer, Transgene

1. Introduction

Pronuclear injection (injection of a solution into one of the pronuclei of one-cell embryos) is a common mode of DNA delivery for the production of transgenic mice. Upon microinjection, the DNA integrates into the embryo's genome, giving rise to a transgenic founder. This is a relatively inefficient process, the success of which is highly dependent on the skills of the microinjectionist and on the quality of the DNA to be injected.

The acquisition of the pronuclear injection technique involves significant investments in both time and equipment. One should therefore consider the option of using the service of a facility specialising in transgenic production. When doing so, it is important to consider Service Level Agreements, as well as prices. Contracts

Elizabeth J. Cartwright (ed.), *Transgenesis Techniques,* Methods in Molecular Biology, vol. 561
DOI 10.1007/978-1-60327-019-9_8, © Humana Press, a part of Springer Science+Business Media, LLC 2009

can offer very different levels of guarantee, such as the injection and transfer of a number of embryos, or, the production of a number of transgenic founders. If, however, a large number of transgenic models are needed and control over the timeline for microinjection is crucial, one should consider setting up for transgenic production. This chapter, together with other important references *(1)*, offers some guidelines.

Typically, trainees require ten microinjection sessions before obtaining their first positive founders. In expert hands, and depending on the type of construct injected, one should expect from 2 (large constructs) to 30% of positive founders among the animals born from a microinjection experiment.

2. Materials

When setting up the laboratory ensure that it is dust free, the level of building vibrations is low (which is easier achieved at lower floor levels and away from traffic), and there is positive air pressure. Also take care that the injection set up is not placed under any venting outlet, as this will cause evaporation and disturbances to the injection chamber.

2.1. DNA Preparation

1. Maxiprep kit (Qiagen) (*see* **Note 1**).
2. Qiaquick DNA extraction kit (Qiagen).
3. Medical gauze.
4. BAC MIB (microinjection buffer):10 mM Tris–HCl of pH 7.5, 0.1 mM EDTA of pH 8.0, 100 mM NaCl; filter through 0.2-μ m filter.
5. β-agarase (New England Biolabs).
6. Plasmid MIB: 10 mM Tris–HCl of pH7.5; 0.15 mM EDTA, filter through 0.2-μ m filter.
7. Phenol of pH 8.0 and $CHCl_3$/isoamylic alcohol (24/1).
8. SpinX (Costar).
9. Spectrophotometer or Nanodrop (Thermo Fisher Scientific).
10. SlideALyser dialysis column (Pierce).

2.2. Embryo and Host Production, Oviduct Transfer, Scrotal Sac Vasectomy

1. CO_2 incubator (Set at 37°C and 5% CO_2).
2. Stereomicroscope with stage light.
3. Stereomicroscope with fibre optic (swan neck) light and optional motorised *Z*-axis.
4. Gas anaesthesia (Isoflurane) and Vetergesic (buprenorphine–HCl in 5% dextrose, *see* **Note 2**) for analgaesia (obtained from a veterinary practice).

5. Gaseous anaesthetic equipment, including vaporizer, scavenging unit with aldasorber O_2 supply, mouse chamber, and nosepiece attachments.
6. Surgery kit including fine and blunted forceps, fine scissors, micro serrefine bulldog clip, suture and needle holder, 9-mm wound auto clips and autoclip wound applier, 70% ethanol, Kettenbach Sugi sterile mounted swabs for absorption of blood.
7. PMS/Folligon (NHPP, USA) and HCG/Chorulon (W&J Dunlop, Dumfries, Scotland).
8. M2 medium, and PVP (MW 40,000) (Sigma-Aldrich).
9. Hyaluronidase (30 mg, Sigma-Aldrich). Dilute with M2 to 3 mg/ml, aliquot and freeze, defrost just before use.
10. KSOM medium (Metachem Diagnostics, Northampton, UK).
11. Embryo tested mineral oil (Sigma-Aldrich).
12. 35-mm cell culture dishes and four-well multiwells.
13. Graticule.
14. Borosillicate glass capillaries (*see* **Note 3**).
15. Bunsen burner.
16. Mouth aspirator (Sigma-Aldrich).

An example of embryo surgery area is shown in **Fig. 1**.

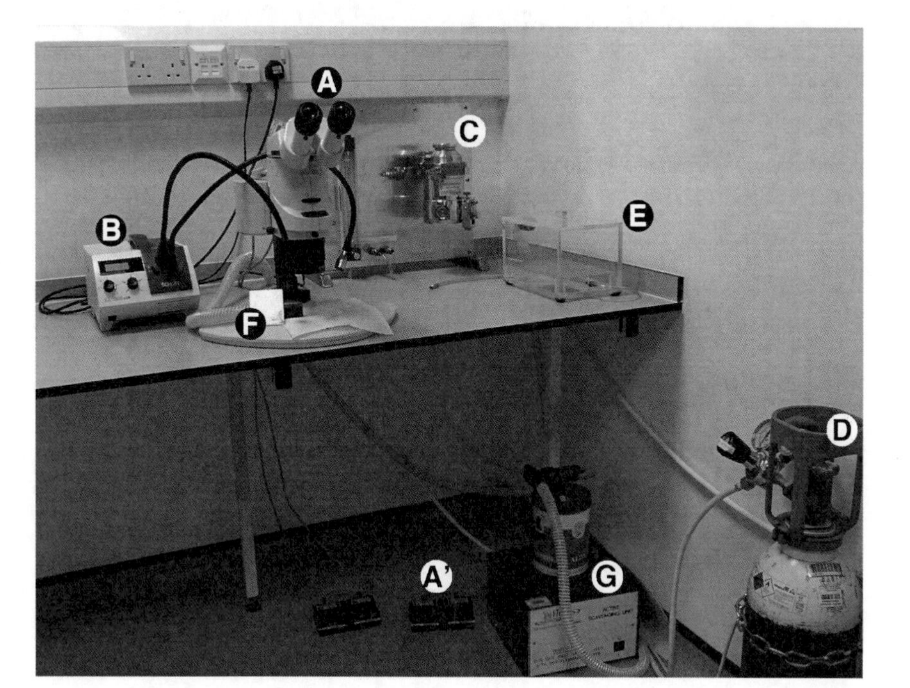

Fig. 1. Surgery area. (**A**) Motorised stereomicroscope with (**A′**) pedal switch, (**B**) fibre optic (swan neck) light, (**C**) vaporizer, (**D**) O_2 supply, (**E**) mouse anaesthesia chamber, (**F**) mouse nosepiece, (**G**) aldasorber (gas scavenging system). Picture by Kevin Glover.

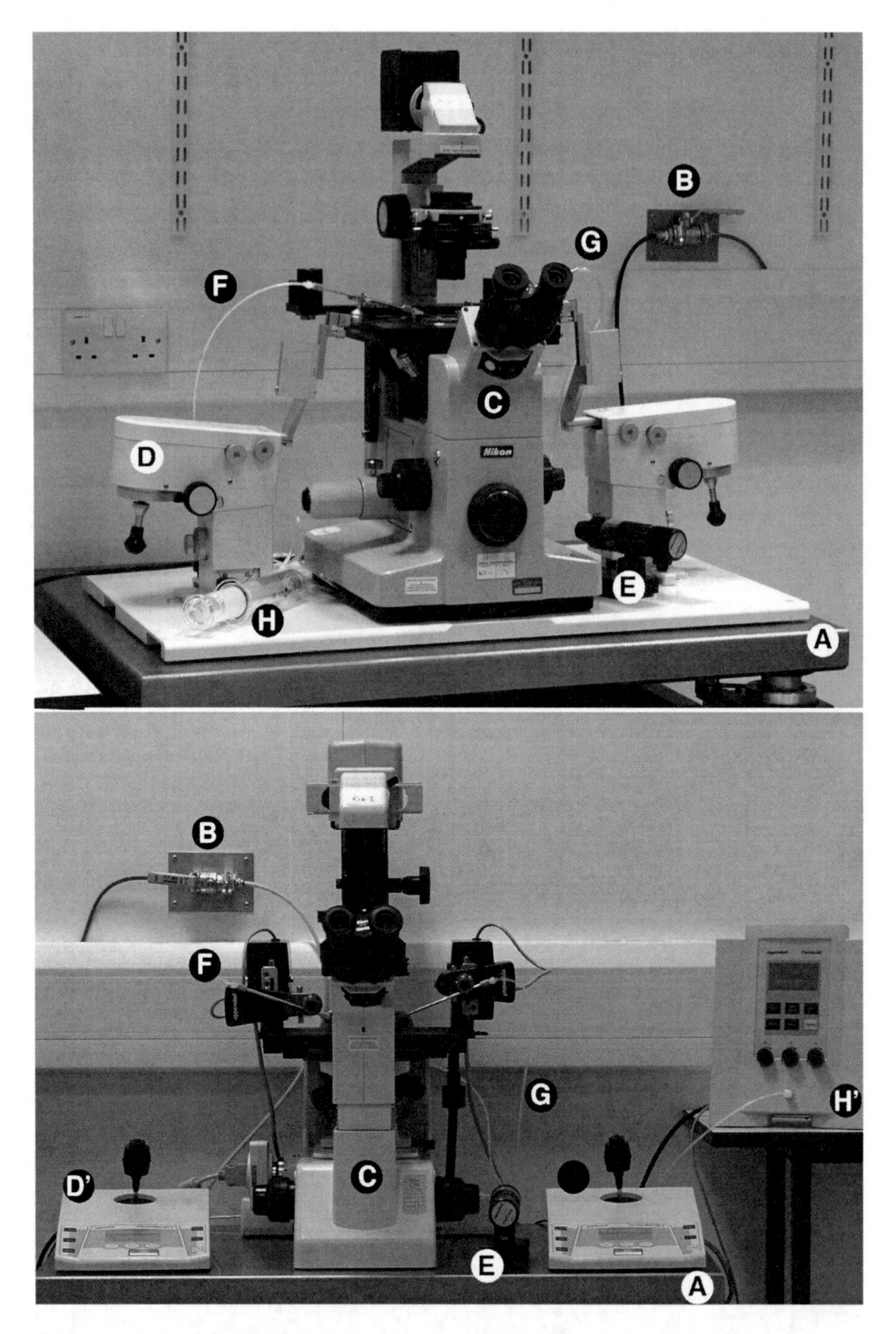

Fig. 2. Microinjection set up: *Top panel:* mechanical system. *Bottom panel:* electronic system. (**A**) Vibration isolation table, (**B**) compressed air supply for vibration isolation table, (**C**) inverted microscope, (**D**) mechanical, and (**D′**) electronic micromanipulator. Oil line for holding pipette: (**E**) cell tram oil, (**F**) tubing and holding pipette holder attached to micromanipulator. DNA injection device: (**G**) tubing and injection needle holder attached to micromanipulator, (**H**) syringe and tubing or (**H′**) Femtojet and tubing. Pictures by Kevin Glover.

2.3. Microinjection of One-Cell Embryos

1. Inverted microscope for microinjection with magnification X5 and differential interference contrast (DIC) for magnifications X10, X20, and X40. A long working distance between condenser and sample is useful.
2. Micromanipulators: mechanical (Micromanipulator R and L, Leica) or electronic (Transferman NK2, Eppendorf) (*see* **Note 4**).
3. Cell tram (Eppendorf) air or oil pipette holders (*see* **Note 5**).
4. DNA microinjection device: Eppendorf Femtojet Pump or a glass syringe adapted to a capillary holder (*see* **Note 6**).
5. Injection chamber: glass depression slide or a small Petri dish.
6. Vibration isolation table.
7. A good chair.
8. Electrode puller (Sutter Instrument).
9. Microforge (Narashige).

Examples of microinjection set up are shown in **Fig. 2**.

3. Methods

3.1. DNA preparation

3.1.1. Maxiprep

We prepare plasmids employing the Qiagen Maxiprep kit according to the manufacturer's protocol.

We prepare large constructs for microinjection according to Carvajal et al. *(2)*.

1. Prepare a maxiprep from 200-ml bacterial cultures according to the manufacturer's instructions, with the following modifications in the case of large constructs.
2. Resuspend the bacterial pellet in 50 ml of ice-cold P1 buffer. Incubate for 10 min on ice.
3. Gently add 50 ml of freshly prepared P2 buffer. Mix by gentle inversion. Incubate for 5–10 min at room temperature.
4. Add 50 ml of ice-cold N3 buffer. Mix by gentle inversion. Lay in a bed of ice with very gentle tilting for 20 min.
5. After centrifugation, filter the supernatant through several layers of gauze.
6. Load the supernatant on a column and wash following the manufacturer's instructions.
7. Elute DNA from the column using elution buffer pre-warmed to 55°C.

8. Precipitate the DNA according to the manufacturer's instructions.
9. Resuspend BAC or PAC DNA in BAC MIB overnight with very gentle agitation.

3.1.2. Preparation of Plasmid Insert for Microinjection

1. Digest 50 μg of plasmid with restriction enzymes that cut the insert away from the backbone and separate on an agarose gel (use low melting agarose for inserts larger than 10 kb).
2. *For inserts up to 10 kb:* Cut out the window of agarose that contains the insert and purify the DNA with a Qiaquick Gel Extraction kit (Qiagen) according to the manufacturer's instructions with the following modifications. Elute DNA from the column with plasmid microinjection buffer (plasmid MIB). Filter the DNA solution on SpinX column according to the manufacturer's instructions. Store at –20°C.
3. *For inserts larger than 10 kb:* Cut out the window of agarose that contains the insert and digest agarose with β-agarase according to the manufacturer's instruction. To eliminate residual agarose, add NaCl to obtain a final concentration of 0.5 M. Incubate on ice 15 min. Centrifuge at high speed for at least 15 min. Extract the supernatant with phenol pH 8.0 and with CHCl3/isoamylic alcohol (24/1). Precipitate the DNA by adding 2 volumes of isopropanol and incubating at –20°C for at least 1 h. Centrifuge at high speed for at least 30 min. Wash the pellet with 80% ethanol. Dissolve DNA in 20 μl BAC MIB.
4. Carefully evaluate the DNA concentration by running an aliquot on an agarose gel with several dilutions of standard DNA of relevant size and compare the signal intensity. Alternatively evaluate the DNA concentration using the Nanodrop system.
5. Dilute the DNA with plasmid MIB to an appropriate concentration (*see* **step 1** of **Subheading 3.4**) before microinjection.

3.1.3. Preparation of Large Constructs (such as BAC or PAC) for Microinjection

We inject BAC and PAC constructs in their circular form to minimise needle blockage during the microinjection process.

1. Optional: Further purify the DNA by $CsCl_2$ banding (*see* **Note 7**).
2. Dialyse overnight at 4°C against BAC MIB using a SlideA-Lyser dialysis column (Pierce) according to the manufacturer's instructions.
3. Carefully evaluate the DNA concentration by spectrophotometry or by Nanodrop. Store at 4°C.
4. Dilute the DNA with BAC MIB to an appropriate concentration (*see* paragraph 3.3.4.1) before microinjection (*see* **Note 8**).

3.2. Embryo and Host Production

3.2.1. Superovulation and Matings

Optimal superovulation conditions depend on mouse strain and environment (light cycle, type of housing). They will need to be worked out by trial and error at each animal house.

The following is a typical protocol for mice subject to a 12-h dark/light cycle regime with lights on at 7 am, and lights off at 7 pm.

1. Inject pregnant mare serum (PMS, Folligon) at 10 am, intraperitoneally. This acts as a follicle-stimulating hormone inducing oocyte maturation to occur.
2. 42–48 h later inject human chorionic gonadotropin (HCG, Chorulon), intraperitoneally. This acts as a lutenising hormone leading to rupture of mature follicles. HCG injection has to occur before the endogenous lutenising hormone is released; this is usually 15–20 h after the middle of the second dark period post-PMS injection (*see* **Note 9**).
3. Singly mate the females to the appropriate males after the HCG injection, or later in the same day. It is essential to use non-virgin young males (no older than 6–7 months for most strains) and to monitor the stud male fertility rates.
4. Harvest the embryos the next morning (0.5 day post coitum (dpc)), when a copulation plug has been observed as proof of mating.

3.2.2. Embryo Harvest

1. Kill the female by cervical dislocation.
2. Lay the female on its back with the legs splayed, and spray 70% ethanol on to the abdomen.
3. Hold the skin up at the lower part of the abdomen with forceps and make a cut.
4. Holding the skin between index finger and thumb, tear back the skin in both directions, towards and away from the head.
5. Cut through the peritoneum with scissors and move the viscera up towards the head so that the reproductive tract can be seen.
6. Locate the uterine horns, oviduct, and the ovary; the oviduct is coiled and is in between the uterine horn and the ovary.
7. Grasp the uterine horn just before the oviduct with forceps and pull up away from the abdomen, trim away the membranes and any fat attached, then cut the oviduct away from the ovary and then the uterine horn.
8. Place the oviduct in a 35-mm Petri dish containing approximately 3 ml of M2 media at room temperature.
9. Repeat removal of the oviduct for the other side of the female.

3.2.3. Embryo Preparation for Injection

1. Equilibrate the KSOM by placing into the CO_2 incubator for approximately 30 min.
2. Take M2 out of fridge to reach room temperature and thaw the hyaluronidase.
3. Prepare media dishes: 1 × 35 mm small Petri dish containing four small drops of M2, 2 × 35 mm small Petri dishes filled with approximately 3 ml of M2, 1 × 35 mm small Petri dish containing four small drops of KSOM, 1 × 35 mm dish with two small KSOM drops covered with mineral oil and placed into the incubator.
4. Place each oviduct into a clean dish containing 3 ml of M2 and tear out the cumulus mass (embryos surrounded by cumulus cells) from the swollen ampulla of each oviduct using fine forceps.
5. Drop the hyaluronidase solution (3 mg/ml) using a handling pipette onto the cumulus masses (this will separate the embryos from the cumulus cells) and wait a few minutes. The hyaluronidase is harmful to the embryos if left in the solution too long; therefore, incubate just long enough for the cumulus cells to detach from the embryos.
6. Using a mouth aspirator and a handling pipette wash the embryos through at least four drops/washes of M2 medium; pick up the embryos with a mouth aspirator and handling pipette and leave debris and abnormal embryos behind in each wash. **Figure 3** shows normal (a), lysed (b), fragmented (c), and unfertilised (d) embryos.
7. Wash the embryos through at least another four drops/washes of KSOM medium; pick up the embryos with a mouth aspirator and handling pipette and leave debris behind in each wash.
8. When the embryos are clean, transfer them to a dish containing KSOM medium covered with mineral oil and place into the incubator (the oil prevents evaporation and contamination).

3.2.4. Host Production

1. Females for oviduct transfer: Females with a good breeding capacity should be selected as hosts for microinjected embryos. For successful oviduct transfer the oviduct morphology of the strain should also be taken in consideration, as it will influence the level of difficulty of the surgery. CD-1 is a widely used strain.
2. Infertile males: T(7;19)145H (often called T145H) is a spontaneous mouse mutant. Most homozygous males carrying the translocation T(7;19)145H are recognised by brown eyes opposed to pink; these T(7;19)145H males have normal mating behaviour but are infertile and, therefore, they can be conveniently used to generate pseudo-pregnant females. These mutants can be used with at least comparable efficiency as

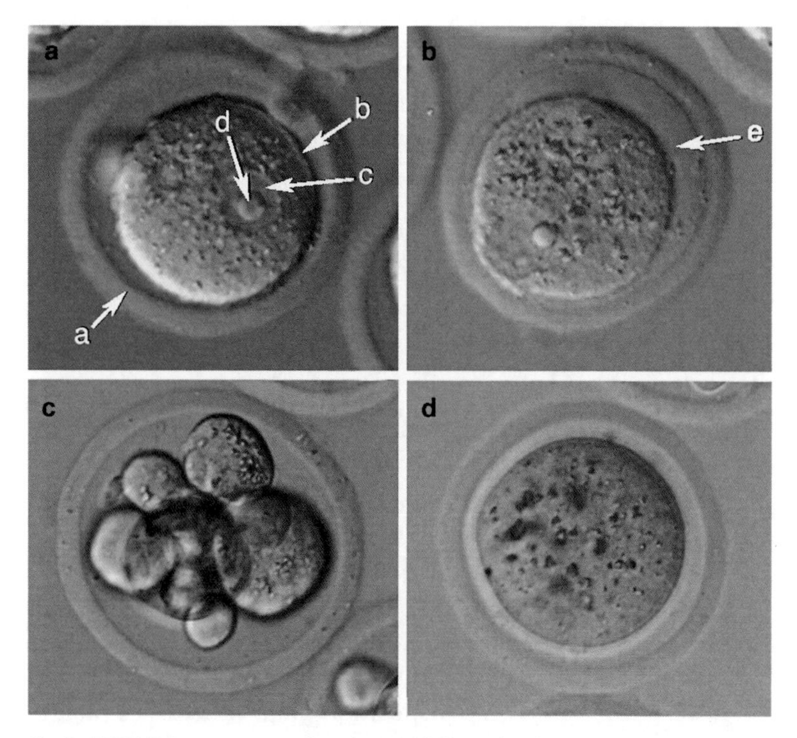

Fig. 3. C57BL/6J mouse one-cell embryos. (**a**) Normal embryo: (a) zona pellucida, (b) plasma membrane, (c) pronucleus (one of two), (d) nucleoli. (**b**) Lysed embryo: cytoplasm content has leaked into the perivitelline space (e). (**c**) Fragmented embryo. (**d**) Unfertilised embryo. Pictures by Tony Bell and Jane Vowles.

vasectomised males as they show good longevity and performance *(3)*. The strain can be obtained from the MRC Harwell, UK (http://www.har.mrc.ac.uk/services/fesa/).

If no genetically unfertile males are available scrotal vasectomy is the preferred alternative to produce infertile males.

3.2.5. Scrotal Sac Vasectomy and Pseudo-pregnancy

The following method is for using gaseous rather than injectable anaesthetic.

1. Put male into an anaesthetic chamber and anaesthetise the mouse (e.g. 2 l/min of O_2 and 5% of isoflurane; this is size and strain dependent), until the mouse has lost consciousness and its breathing is shallow. Check for lack of pedal reflex.
2. Remove mouse and administer analgaesic (0.1 ml Vetergesic).
3. Place mouse onto its back with its head to the left and into the anaesthetic nose piece ensuring the mouse's face is fully in the nose piece.
4. Turn down the anaesthetic dose (e.g. 1 l/min O_2 and 2% Isoflurane).

5. If breathing becomes to shallow turn the isoflurane down and if the breathing becomes fast then turn isoflurane up. Check for lack of pedal reflex before starting and check the breathing pace regularly.
6. Flush the scrotal area with 70% ethanol.
7. Make a small incision in the scrotal sac skin.
8. Make a small incision in the peritoneum.
9. Locate the vas deferens (with blood vessel running alongside, *see* **Fig. 4**) and cauterise a small section out with flamed forceps.
10. Sew the peritoneum using dissolvable suture.
11. Sew the skin using suture.
12. Remove mouse from the nose piece and place it into a warmed box and monitor. The male should be up and about within 5 min.
13. After 10 days, test mate by mating and then checking the female for a copulation plug the next morning. Leave the male with the female for 5–10 days, checking each day for a copulation plug. If there is a plug then record and separate.
14. Kill the female 10 days after the copulation plug was detected and check for the presence of embryos within the uterus. If there are no embryos present, the vasectomy was successful.
15. If the mating test is passed then the males can be used for producing pseudo-pregnant recipients.
16. Mate the verified unfertile males with females of suitable breeding age and check for copulation plugs the next morning.
17. If the female has a plug she is regarded as 0.5 dpc and can be used as a recipient for transferred embryos.

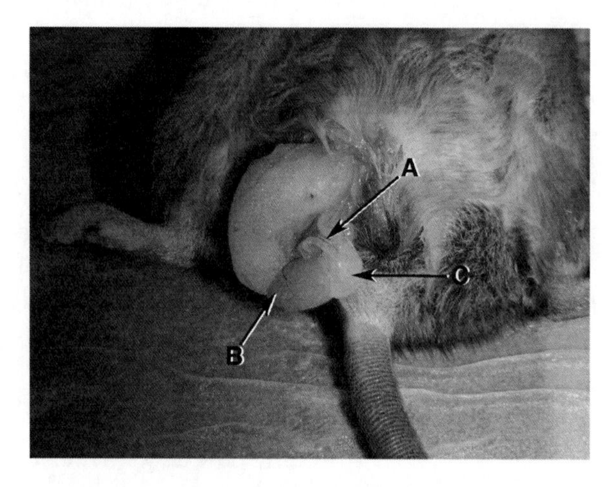

Fig. 4. Scrotal vasectomy. (**a**) Vas deferens, (**b**) testis, (**c**) epididymis.

3.3. Microinjection of One-Cell Embryos

3.3.1. DNA Concentration

Optimal concentrations for DNA microinjection depend on the microinjectionist's style. Typically, small inserts (up to 12–14 kb) are injected at a concentration ranging from 2 to 8 ng/μl, whereas larger constructs are injected at a concentration ranging from 0.5 to 2 ng/μl. The higher the concentration, the higher the post-injection embryonic lethality will be. One should choose a concentration that does not kill more than 80% of embryos after microinjection and recovery.

3.3.2. Injection

1. Prepare the injection chamber (slide or Petri dish) with a small drop of M2 media covered with mineral oil to stop evaporation.
2. Place the slide/dish onto the microscope and focus on the edge of the media using X4 objective.
3. Load the embryo-holding pipette into the holder; embryo-holding pipettes should have an inner diameter of approximately 100 μm and an opening of 10 μm.
4. Wash embryos in a drop of M2 and place the embryos in the chamber using a handling pipette and let them sink to the bottom. Only transfer the number of eggs that can be injected within 45 min to the microinjection set up. Only use the embryos where two pronuclei can be seen.
5. Fill the injection needle by dipping its back in DNA solution and load the needle into the holder. For slow-loading pipettes, it is acceptable to leave them standing in a microfuge tube containing the DNA solution, until a meniscus can be observed at the tip of the pipette.
6. Lower the holding pipette and injection needle until they are in the same plane of focus as the embryos.
7. Move to X20 magnification and break off the tip of the injection needle by brushing it against the holding pipette.
8. Hold onto one embryo with the holding pipette so one pronucleus is towards the injection needle side and therefore easily accessible for injection. Check embryo morphology, including the presence of two pronuclei. Line up the focus on one pronucleus with the tip of the injection needle. Focus on the pronucleus using the microscope dial and focus the needle using the manipulator dial.
9. Inject the pronucleus closest to the needle for less damage to the embryo.
10. Insert the needle through the four walls of the zona pellucida, cell membrane, cytoplasm, and nuclear membrane into either pronucleus (do not touch the nucleoli) and inject with DNA, this should result in a visible swelling; when this occurs remove the needle quickly. On piercing the embryo the membranes should relax rapidly allowing injection. If

the needle is too blunt penetration is harder and the membrane will not pierce and relax back into place. If you can see a bubble appear rather than swelling of the pronucleus on injecting DNA, the injection needle has not penetrated through all walls. Drive the needle further into the embryo and fractionally backwards and attempt to re-inject. Change the injection needle if it becomes blocked or blunted in the injection process.

11. Move injected embryos to a separate area of the chamber. If the embryo has been damaged during injection it will start to lyse and it will die.
12. When all embryos on the chamber are injected, load them in a handling pipette attached to a mouth aspirator and wash them in a drop of KSOM and place into the incubator dish and obtain more embryos for injection. Repeat until all embryos are injected. The amount of embryos that are put in the chamber will depend on the speed of the microinjectionist. It must be noted that it is harmful to the embryos to be left in the chamber too long; start with ten embryos, then increase this number as experience grows.
13. Let embryos recover in a CO_2 incubator for at least 30 min before transfer into recipient females. Check for lysis; the rate of lysis will depend on a number of factors including injection technique, quality of injection needle and of the DNA injected.

3.4. Oviduct Transfer

The embryos can be transferred into the oviduct of a 0.5-dpc host after 30 min of recovery in a CO_2 incubator. Alternatively, they can be cultured overnight and transferred the next morning into a 0.5-dpc female host.

Pseudo-pregnant females should be prepared by mating a female to a sterile male the previous afternoon and verifying the presence of a copulation plug the next morning, if there is a plug the female is 0.5 dpc. The embryos will fare better in vivo, so same-day transfer is logical. However, next-day transfer prevents transfer of embryos that may be dying and in a working day gives more time for microinjection.

Up to 15 embryos can be transferred into each oviduct of a CD-1 recipient female. CD-1 females are widely used as hosts as they are good mothers, large numbers of pups are born, and they also have a large infundibulum (entry to the oviduct), which makes surgery easier. If an inbred or hybrid strain is used as the host then reduce the number of embryos transferred to 8–10 per side. Check the natural number of pups usually born from the host strain you are using, but remember that all the embryos transferred will not implant.

When first learning to carry out embryo transfer it is best to start with transferring into one oviduct instead of both, but as experience grows embryos can be transferred to both sides, increasing the chance of success. Using both sides prevents using more females but one sided is easier, quicker, and the female is under anaesthetic for less time, enhancing the chances of a successful pregnancy.

Depending on how the female is positioned during surgery right-handed people may find the right-hand side of the mouse easier, and conversely left-handed people will find the left side of the female easier to transfer into.

Records should be kept of the numbers of embryos harvested, injected, transferred, pups born, and transgenic mice produced in order to troubleshoot any potential problems.

This method is for using gaseous rather than injectable anaesthetic.

1. Wash embryos in one drop of M2, load embryos into a pipette thus: aspirate a small amount of M2, then two small air bubbles, then embryos (close together), and then another small air bubble. The pipette should only be slightly larger than the embryos, e.g. 120–150 μm.
2. Put the recipient female into the anaesthetic chamber and anaesthetise the mouse (e.g. 2 l/min of O_2 and 5% of isoflurane; this is size and strain dependent), until the mouse has lost consciousness and its breathing is shallow.
3. Remove the mouse and administer analgaesic (0.1 ml of Vetergesic).
4. Place the mouse onto its front with the head to the left and into the anaesthetic nose piece ensuring the mouse's face is fully in the nose piece.
5. Turn down anaesthetic dose (e.g. 1 l/min of O_2 and 2% of isoflurane).
6. If breathing becomes to shallow turn the isoflurane down and if the breathing becomes fast turn isoflurane up. Check for no pedal reflex and check breathing rate regularly.
7. Flush the back area with 70% ethanol.
8. With blunt-ended forceps and fine dissection scissors make a small incision (no bigger than 1 cm) in the skin running along the spine, in line with the top of the hind legs.
9. Tease the epidermis away from the underlying peritoneum on one side of the body.
10. Move the incision sideways down and locate the white fat pad of the reproductive tract, which is visible through the peritoneum. Grasp the peritoneum with the fine forceps and make an incision in the peritoneum above the fat pad; avoid

cutting any large blood vessels with the tip of the fine scissors. Widen the opening by introducing closed scissors in the incision and opening them in the cavity. Take care not to nick any internal organs.

11. Locate and grasp the fat pad and pull out the ovarian fat pad with attached reproductive tract; attach a micro serrefine clip to the fat pad to stop it from retreating back into the body and drape the clip and reproductive tract over the back of the mouse.
12. Locate the ovary, oviduct, and uterus; then using fine forceps tear open the bursa membrane that is lying over the ovary and oviduct. Take care not to tear near any vessels; if a vessel is torn mop up any excess blood with a swab.
13. Place the female under a stereomicroscope and locate the infundibulum (opening of the oviduct) which lies in between the oviduct coils and the ovary. Insert one prong of the fine forceps into the infundibulum to ensure access for the pipette.
14. Insert the pipette containing the embryos and blow the embryos into the oviduct. Withdraw the pipette once all the air bubbles have been blown into the oviduct. If the air bubbles are visible within the oviduct then the embryos will now be in the oviduct.
15. Release the micro serrefine clip and carefully replace the reproductive tract into the body and sew up the peritoneum with two surgical knots using suture.
16. If transferring embryos to both sides repeat **steps 9–15** for the other side.
17. Clip the skin with wound clips and remove the female from nose piece.
18. Place into a warmed box and monitor; the female should be up and about within 5 min.

4. Notes

1. The wide columns of the HiSpeed Plasmid Maxi Kit (Qiagen) facilitate large construct purification as they tend not to clog up. However, one should not use the syringes and plungers of the kit for this application as they will shear the DNA.

2. 1 ml containing 324 mg of buprenorphine diluted to 8.1 mg/ml with PBS.
3. Borosillicate glass capillaries (Harvard Apparatus Ltd, Kent, UK):
   - For making embryo-handling pipettes (GC100T-15), hand pull to an inner diameter of 200 μm and flame.
   - For making embryo-holding pipettes (GC100-15), pull to inner diameter of 100 μm by hand and then use the microforge to polish down to 10-μm internal diameter.
   - For making microinjection needles (GC100TF-15), pull on the electrode puller using the appropriate programme.
   - For making embryo transfer pipettes (GC100T-15), pull by hand to inner diameter of 120–150 μm and flame.
4. Manual micromanipulators afford a tighter control and smoother movements in expert hands and they are cheaper. However, they provide a field of movement for the needle that is not horizontal, as tips will move out of focus as they move sideways (along the X axis) in the field. Electronic micromanipulators offer the very useful option to memorise coordinates and send the needle back to a specific position in space at the hit of a button.
5. Air holders require a larger opening at the end of a holding pipette. Conversely, oil lines confer a tighter control of the embryo handling with a smaller opening in the holding pipette. Coating the needle internally with a PVP solution (12%) helps to prevent fragmentation of the oil column in the pipette. A holding pipette should have an opening of 10–20 μm; embryo diameter is 85–100 μm.
6. The Femtojet allows a tight control on the volume of DNA solution injected. It also prevents backflow of media into the needle and will flush through blockages via the 'clean' function. The Femtojet can be used with a foot pedal for injection; this gives more flexibility as it leaves a hand free for ensuring more precise injection as one can hold onto both manipulators at the same time as injecting. It is, however, much more expensive than a simple syringe. Utilisation of a Femtojet slows down the injection process as the apparatus calibrates regularly. Also this is noisy apparatus.
7. The field is divided on the necessity of this step for the preparation of BAC DNA for microinjection. High-quality supercoiled BAC DNA is desirable and is easily injected.
8. Historically, 30 μM spermine and 70 μM spermidine have been added to BAC MIB. We omit these in the BAC MIB that we are currently using without affecting our yields.

9. Powdered hormones are diluted using PBS to the desired IU; this can be variable for the strain of mouse used, ranging from 2.5 to 7.5 IU. Once diluted the hormones are frozen at –20°C on the day of preparation and gently thawed just before use. Replenish stocks after 3 months.

Acknowledgements

We wish to thank the members of the Rigby and Krumlauf laboratories at the National Institute for Medical Research, London, and staff of the Transgenics Service at the Mary Lyon Centre for teaching and sharing the tricks and protocol modifications that are compiled in this chapter.

References

1. Nagy, A. (2003) In: Gertsenstein, M., Vintersten, K., and Behringer, R. (eds.) *Manipulating the Mouse Embryo: A Laboratory Manual (3rd edition)* Cold Spring Harbor Laboratory Press, Cold Spring Harbor, New York.
2. Carvajal, J. J., Cox, D., Summerbell, D., and Rigby, P. W. (2001) A BAC transgenic analysis of the Mrf4/Myf5 locus reveals interdigitated elements that control activation and maintenance of gene expression during muscle development. *Development* **128**, 1857–1868.
3. Hawkins, P., Felton, L. M., van Loo, P., Maconochie, M., Wells, D. J., Dennison, N., Hubrecht, R., and Jennings, M. (2006) Report of the 2005 RSPCA/UFAW Rodent Welfare Group meeting *Lab Animal* **35**, 29–38.

Chapter 9

Gene-Targeting Vectors

J. Simon C. Arthur and Victoria A. McGuire

Summary

Gene targeting in mice has been used extensively to elucidate gene function in vivo. However, for gene targeting to be successful, the targeting vector must be carefully designed. This chapter addresses the rationale behind designing targeting vectors, detailing the essential components, and highlighting specific considerations for different types of vectors, from gene deletions to point mutations and insertions. Examples of vector designs, cloning strategies, and approaches for successful screening of recombinants are described. The use of Cre/LoxP and Flp/frt systems for conditional targeting is described, together with strategies for generating conditional deletions. Methods for generating conditional point mutations are also described and their potential drawbacks discussed.

Key words: Gene-targeting vector, Knockout, Knockin, Cre, Frt

1. Introduction

Gene targeting in mice allows defined modifications, such as deletions, insertions, and point mutations, to be made at specific points in the mouse genome. This is achieved in cells by homologous recombination of a targeting vector that contains the desired change. This is normally carried out in embryonic stem (ES) cells as these cells have high rates of homologous recombination and are able to contribute to the germline when re-introduced into mouse embryos at the morula or blastocyst stage of development *(1–3)*. For gene targeting to be successful, the vector used must have several characteristics, and in this chapter some of the principles behind vector design will be discussed. In addition to containing the desired mutation, the vector must be able to direct recombination at the correct site, it must contain

Elizabeth J. Cartwright (ed.), *Transgenesis Techniques*, Methods in Molecular Biology, vol. 561
DOI 10.1007/978-1-60327-019-9_9, © Humana Press, a part of Springer Science+Business Media, LLC 2009

a marker that allows selection of cells that have incorporated the vector into their genome, and it must modify the genome in such a way that cells can easily be screened for the correct incorporation of the vector.

In order to direct incorporation of the vector to the required site in the genome, targeting vectors will normally contain two 'arms of homology' which correspond to the genomic regions immediately upstream and downstream of the site to be mutated. These arms allow homologous recombination to occur between the vector and the desired site in the genome, resulting in incorporation of the required mutation into the genome **(Fig. 1a)**. Unfortunately homologous recombination of the vector occurs much less frequently than random non-homologous insertion. To overcome this, selection and screening methods are required

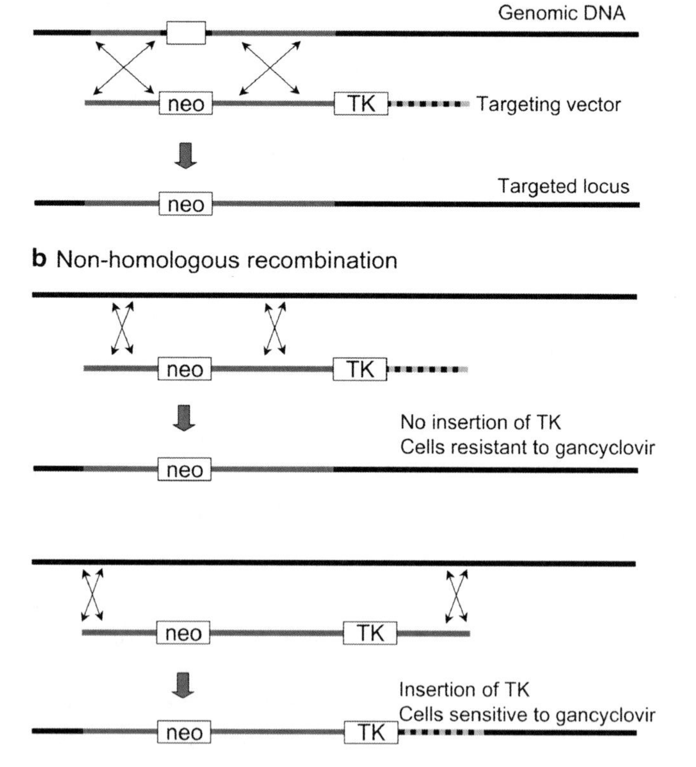

Fig. 1. Recombination in ES cells. (**a**) Gene targeting is achieved in ES cells by homologous recombination of a targeting vector. The targeting vector contains a marker for positive selection (*neo*; neomycin resistance cassette), to select for cells that have incorporated the vector. In addition, a negative selection marker (TK; HSV thymidine kinase) is included. The vector also contains two arms of homology (*grey lines*) which correspond to the genomic sequence 5′ and 3′ to the region to be modified. Homologous recombination occurs between these two arms to incorporate this region of the vector and make the desired change in the genome (in this case deletion). (**b**) In addition to homologous recombination, cells can incorporate all or part of the vector by random insertion. Cells that do not incorporate the *neo* cassette will be selected against by growth in G418. Cells which have randomly inserted the neo cassette from the vector can be divided into two groups: (1) cells where recombination occurred before the TK (*upper panel*) and which therefore do not express the thymidine kinase gene and so are resistant to gancyclovir and (2) cells where recombination occurred after the TK. These cells express thymidine kinase and are therefore killed by gancyclovir. The use of TK for negative selection can therefore remove some, but not all, of the cells where the vector has undergone random insertion; this can result in a two- to fourfold increase in targeting efficiency compared to positive selection alone.

to increase the percentage of ES cell colonies obtained that have undergone homologous recombination and distinguish them from those where random insertion of the vector has occurred.

Two forms of selection, positive and negative, are normally used in gene targeting. Positive selection allows the selection of cells that have incorporated the vector from those which have not. The most commonly used marker for positive selection is a neomycin resistance cassette. This contains the *neo* gene from the bacterial Tn5 transposon which encodes an aminoglycoside 3′-phosphotransferase under the control of a constitutive promoter *(4, 5)*. The *neo* gene confers resistance to G418 (Geneticin), an aminoglycoside antibiotic originally isolated from *Micromonospora rhodorangea*. In both prokaryotic and eukaryotic cells lacking the *neo* gene, G418 blocks polypeptide synthesis by inhibiting the elongation step, thus inhibiting cell growth. Other markers for positive selection, including Hygromycin B, Puromycin, and Blasticidin S resistance genes, have also been described *(6, 7)*. In theory, the effectiveness of these selectable markers can be improved by either promoter or polyA trapping, although for most mouse genes this is not necessary. For promoter trapping, no promoter is included in the *neo* selection cassette, and therefore in order for expression of the *neo* gene to occur it must insert in the correct orientation downstream of an active promoter in the genome. Additionally, to allow expression of the *neo* gene, the *neo* open reading frame must be fused in frame with one of the exons in the targeting vector from the gene to be modified. Alternatively a splice acceptor site (to trap the splicing from the upstream exons), followed by a stop codon, IRES sequence, and then *neo* open reading frame could be used. A drawback of promoter trapping is that it will not work for genes that are not expressed in ES cells. In polyA traps, the *neo* cassette does not contain a polyA sequence but does contain a splice acceptor site at its 3′ end. Expression of the *neo* mRNA therefore requires the insertion of the *neo* upstream of a functional polyA sequence. The advantage of this over promoter trapping is that it is not dependent on the targeted gene being expressed in ES cells. While not commonly used in gene targeting, both have been extensively used in gene trap vectors *(8–11)*.

While these positive selection markers allow the selection of colonies that have stably incorporated the vector, none of them efficiently distinguishes between colonies that have undergone homologous recombination or random insertion of the vector. In order to improve the ratio of homologous recombination to random insertion, negative selection is used. Homologous recombination only incorporates the desired parts of the targeting vector into the genome, while the plasmid backbone is lost. In contrast, random insertion can incorporate all (including the plasmid backbone) or part of the vector (**Fig. 1b**). Inclusion of a negative selection cassette outside the arms of homology allows cells which have randomly incorporated the entire vector to be selected against. The most common method for negative selection makes use of the herpes simplex virus (HSV) thymidine kinase (TK) gene. HSV thymidine kinase is able to phosphorylate gancyclovir, a synthetic analogue of 2′-deoxy-guanosine *(4)*. This results in a deoxyguanosine triphosphate (dGTP) analogue that competitively inhibits the incorporation of dGTP by DNA polymerase and therefore blocks DNA synthesis. Cells which do not express TK are unaffected by gancyclovir, while cells expressing the TK gene, due to random insertion of the targeting vector, are no longer able to replicate due to inhibited DNA synthesis. Other methods of negative selection, including expression of the diphtheria toxin A fragment or GFP, as well as selection markers that combine both positive and negative selection have also been described *(12–14)*.

While the use of negative selection can increase the frequency of correctly targeted clones obtained, it does not remove all the colonies that have randomly incorporated the vector. It is therefore necessary to be able to screen for correctly targeted clones. The most definitive screening method is Southern blotting. For this, a specific restriction site that does not cut in the arms of homology is either introduced or removed adjacent to the mutation in the targeting vector, resulting in a change in size in the band when this enzyme is used for a Southern blot. The detailed design of the targeting vector depends on the nature of the mutation to be made. Most gene-targeted mice are either gene deletions, conditional gene deletions or point mutations; however, insertions and conditional point mutations are also possible. The basic architecture of these vectors is discussed as follows.

2. Deletion Vectors

Gene-targeted deletions (or knockouts) are perhaps the most straightforward gene-targeted mice to make. In an ideal situation, the entire open reading frame of the gene of interest would be deleted, as this would prevent the expression of a fragment of the protein that could either retain partial activity or function as a dominant negative. Despite this, in practice, it is normally only a fragment of the gene that is deleted. This is because mouse genes range from a couple of kilo bases in size to several hundred. In theory gene targeting should be able to make very large deletions; however, in practice the targeting frequency for very large deletions in ES cells is very low. For this reason, most gene knockouts delete between 1 and 4 kb of sequence, and this is insufficient to completely delete many mouse genes. This means that it is usually necessary to select a section of the gene to be deleted. In general it is best to select a region that encodes part of the protein that is essential for its function. In addition, targeting exons at the 5′ end of the gene can help prevent expression of a truncated protein, especially if when these exons are deleted it would result in a frame shift mutation between the remaining exons. Other factors to consider are if alternative splicing isoforms or promoters exist for the gene to be targeted. For instance, deletion of an exon that is only used in some splice forms of the protein will not result in a complete knockout for the protein. In addition, it is also worth checking for overlapping genes and avoiding deletion of the regions common to both genes. Finally it should also be remembered that intronic sequences may have functions independent of the protein encoded by the gene. For instance some microRNAs are produced from the introns of protein-encoding genes, and deletion of these introns may give rise to a phenotype that is due to the loss of the microRNA rather than deletion of the protein.

A schematic of a hypothetical deletion vector is shown in **Fig. 2a**. In this example the gene contains 5 exons and it was decided to delete exons 2 and 3. To make the deletion vector, the regions up- and down-stream of exons 2 and 3 were used as arms of homology. There is no set size for the arms of homology although lengths of 2–6 kb are normal. It is possible that short arms of homology may decrease targeting frequency, so we normally aim to have at least 6–8 kb of sequence between the two arms. One end of the arm is determined by where the deletion is; the other end is normally constrained by the presence of restriction sites that will be required for either cloning or screening. The two arms of homology are cloned either side of the positive selection marker (in this example neomycin). In addition, a negative selection marker (TK) is placed at one end and a unique restriction site at the other. This restriction site is required to linearise the vector before it is transfected into the ES cells. These components are usually assembled in a basic bacterial cloning vector, such as pBluescript.

Several methods can be used to clone the vector. For early mouse knockouts, the genomic clones for the gene of interest were isolated from libraries made from the strain of ES cells to be used, and the arms of homology cloned using pre-existing restriction sites in the genomic sequence. The use of congenic DNA is desirable, as DNA from sources not matched to the ES cells may contain some differences in intronic or intragenic regions that might compromise targeting frequency. Screening libraries is however time consuming, and other options are now possible. Most commonly used ES cell lines are either from 129SvJ or C57BL/6 mice, and both 129SvJ and C57BL/6 BAC

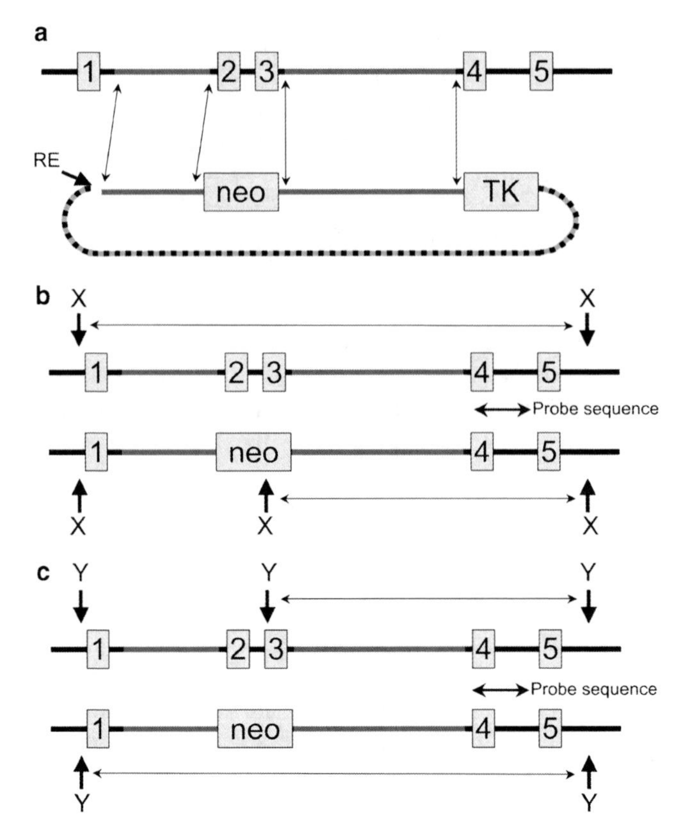

Fig. 2. Deletion vectors. (**a**) Most mouse genes are too large to be easily deleted in one round of targeting and as a result it is necessary to select a group of exons to be deleted. In this example, the hypothetical gene has 5 exons, and exons 2 and 3 were selected for deletion. The targeting vector consists of two arms of homology (*grey lines*) corresponding to the genomic regions 5′ and 3′ to exons 2 and 3 placed either side of a *neo* cassette (for positive selection). At the end of the arms of homology a TK cassette is included (negative selection marker) followed by the plasmid backbone (*dashed line*) to allow replication of the vector in *E. coli.* At the other end of the region of homology a unique restriction site (RE) is included to allow the vector to be linearised before it is transformed into ES cells. This is normally a 7-bp cutter, such as Not I, as these cut infrequently in genomic DNA. (**b**, **c**) Screening strategies must be used to identify correctly targeted ES cells. While initial screens can be done by PCR, screening by Southern blotting is the most definitive method. Ideally, correct integration should be checked at both 5′ and 3′ ends of the vector; however, only examples for screening the 3′ end are shown. For this a probe (*double-headed arrow*) is used that is external to the targeting vector, making the screen specific for the locus to be targeted and preventing detection of false positives due to random insertion of the vector. In the first example, a restriction enzyme (X) which cuts the locus in two positions in the wild-type allele is used: one 3′ to the region used for the probe and one 5′ to the region to be deleted. In the targeted allele a third site is introduced due to incorporation of the *neo* cassette. As a result the targeted allele will give a smaller band in the Southern than the wild-type allele. In the second example an enzyme (Y) that cuts the wild-type locus in three places, including once in the region to be deleted, is used. This site is removed by deletion of exons 2 and 3 and their replacement by the *neo* cassette. This would result in the targeted locus giving a larger band on the Southern blot than the wild-type allele. From a practical point of view, the size of the band from the targeted locus should be less than 12 kb, as bands larger than this are hard to resolve on normal Southern blots. It should also be remembered that, with the exception of targeting of X-linked genes, most ES cell clones are heterozygous for the targeted allele, and therefore both wild-type and targeted bands will be seen on the Southern for the desired ES clone, and the size difference between the bands should be sufficient to allow them to be resolved by Southern blotting.

clones have been mapped onto the mouse genome (http://www.ensembl.org/index.html). As a result, specific BAC clones for the gene of interest can be ordered eliminating the need for library screening. These can be used for subcloning of homology arms by either restriction enzyme digestion or by PCR. The PCR amplification of arms has the advantage that it allows additional short sequences to be included at the end of the primers. This means that both extra restriction sites, for either cloning or screening purposes, and LoxP or Flp sites can easily be incorporated into the PCR product. PCR products should however be sequenced to confirm that PCR-generated errors have not been introduced.

An alternative method makes use of homologous recombination in either bacteria or yeast *(15, 16)*. For this, short sequences either side of the region of homology are cloned into a plasmid vector. The plasmid is then linearised in between these two sequences and transfected, along with the appropriate genomic BAC clone, into a bacterial or yeast strain that supports homologous recombination. As linear DNA will not be replicated, only plasmid that has undergone recombination with the BAC, and been circularised as a result, will be replicated.

To allow screening by Southern blotting, it is necessary to introduce or delete sites for selected restriction enzymes. For instance, in **Fig. 2b**, the restriction enzyme X cuts the allele both 5′ and 3′ to the sequence used in the targeting vector. If another site for the same enzyme is introduced, in or next to the *neo* cassette, then the targeted allele will give a smaller band on a Southern blot than the wild-type allele.

Alternatively, the restriction enzyme Y cuts the allele both 5′ and 3′ to the sequence used in the targeting vector, as well as in the region to be deleted. In the targeted allele, however, the middle site has been lost, and as a result the targeted allele will give a larger band than the wild-type allele (**Fig. 2c**).

Ideally recombination should be checked with both 5′ and 3′ probes to ensure that both ends of the vector have been incorporated correctly. An important aspect is that the probes used for detection in the Southern blot should come from a sequence outside that which is used for the targeting vector. This makes the Southern blot specific for the genomic locus to be targeted and prevents false-positive results. If a probe internal to the vector was used, it is possible that random insertions could occasionally give the same size band on the Southern blot as a correctly targeted clone. Additionally the probe should not lie in a sequence that is present in multiple copies in the genome, as this can appear as a smear on the Southern blot. This can be checked by BLAST searching the probe sequence against the mouse genome. It should also be remembered that, with the exception of X-linked genes, most targeted ES cell clones will be heterozygous for the targeted allele, and thus both wild-type and targeted bands will be seen in the Southern blot. There is also a limit on the practical size of the targeted band as conventional agarose gels will not easily resolve bands in excess of 12 kb. PCR across an arm of homology, from the *neo* to a region outside the targeting vector, is also possible as a screening strategy, but this method can suffer from problems with either non-specific PCR products or poor PCR efficiency from some genomic sequences.

3. Conditional Deletions

The deletion of some genes in mice results in embryonic lethality, and this precludes the study of the function of these genes in the adult animals. It is also sometimes desirable to study the function of a gene in a specific tissue by specifically deleting the gene in this tissue, but not in the rest of the body. These problems can be resolved by conditional gene deletions which make use of the properties of recombinases to recognise specific target sequences that can be introduced into genomic DNA (***(17–19)***, also *see* Chapter 16). The most common system used in mice is the Cre/LoxP system. This makes use of the P1 bacteriophage Cre recombinase that recognises LoxP sites, a 34-bp sequence that comprises two 13-bp inverted repeats flanking an 8-bp spacer region. This 8-bp sequence confers directionality on the LoxP site. If two LoxP sites are present in opposite orientations in a

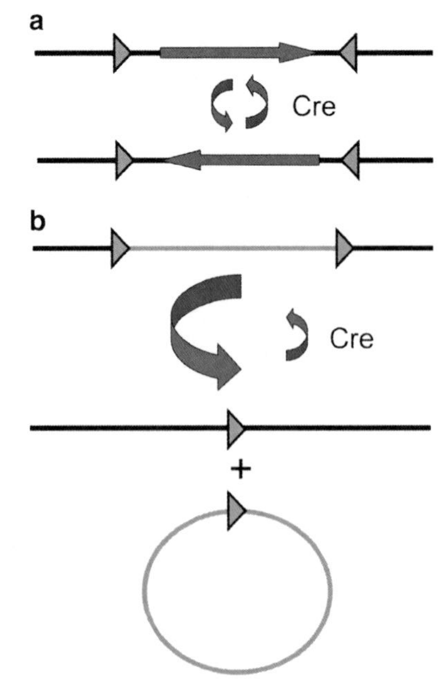

Fig. 3. Cre/LoxP recombination. Cre is a recombinase that recognises LoxP (*grey triangles*) sites in DNA. If, as shown in **a**, the sites are placed in opposite orientations Cre will cause the inversion of the intervening sequence of DNA. Alternatively as shown in **b**, if the LoxP sites are in the same orientation then Cre will catalyse the removal of the intervening sequence, leaving one LoxP site on the original DNA. The intervening sequence (*grey line*) is removed as a circular piece of DNA also containing one LoxP site. In theory Cre can also catalyse the reinsertion of this sequence. However, as the LoxP sites are now on separate DNA molecules, the rate of this reaction is much slower than the rate of removal. As the circular DNA fragment will not be maintained in eukaryotic cells, this effectively prevents reinsertion occurring in vivo.

DNA sequence, Cre will catalyse the inversion of the intervening sequence (**Fig. 3a**). If both LoxP sites are present in the same orientation then Cre will catalyse the removal of the intervening sequence (**Fig. 3b**). It should be noted that the reverse reaction can also occur, although the rate of this reaction is slower as the LoxP sites for this reaction are on separate DNA molecules. As the excised piece of DNA will not be maintained in eukaryotic cells, this effectively makes the rate of the reverse reaction zero in vivo.

For conditional deletions, the gene of interest is modified to introduce LoxP sites in the introns 5′ and 3′ to a selected group of exons that are essential to the function of the gene. These LoxP sites do not interfere with the transcription of the gene. As the transcribed protein has wild-type sequence, the mice with this modified allele should be normal. If however Cre is expressed in the cells of these mice, the exons between the LoxP sites will be deleted, resulting in a knockout allele. Cre expression can be achieved using viruses, although the most common method is to cross the mice to a Cre-expressing transgenic strain. Many Cre-expressing transgenics have been made (as described later in this book), including ones which express Cre under the control of inducible or tissue specific promoters, thereby allowing control of where and when the deletion occurs.

The principles for the design of a conditional deletion vector are similar to that of the deletion vector described earlier. The major difference, however, is that the region that is eventually to be deleted is included in the vector in between the positive selection marker and one of the arms of homology. In addition, LoxP sites are placed either side of the region to be deleted, while a third LoxP site is placed at the far end of the positive selection cassette. The placement of the LoxP sites is important. They should be placed in intronic sequence, ideally at least 200 bp away from the splice sites so as to minimise the possibility that they will interfere with splicing. The placement of LoxP sites in promoter regions can also be problematic, as they could interfere with transcription factor binding to the promoter. Based on the assumption that regulatory elements are more likely to be conserved than other regions in intronic or intragenic DNA, a way to reduce the possibility of interrupting a regulatory sequence in the locus is to place the LoxP in regions not conserved between mouse and human. An example of a conditional deletion vector is shown in **Fig. 4a**.

Initial targeting and screening is carried out in ES cells as for a deletion vector. It should be noted however that the LoxP-flanked region and one arm of homology effectively form one long region of homology with the genomic locus. Homologous recombination can therefore occur before or after the LoxP site, but only recombination in the arm of homology is useful for making

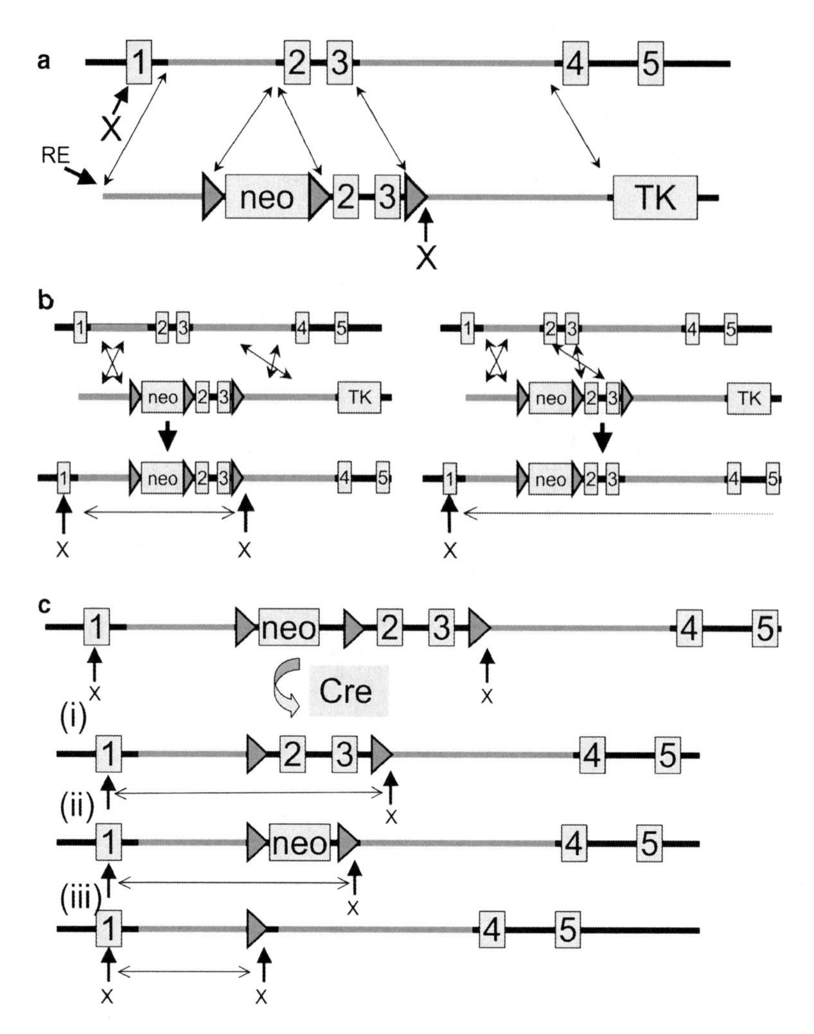

Fig. 4. Conditional vectors. (**a**) Vectors for conditional deletion are similar to those for deletion described in **Fig. 2** except that they also contain the region to be deleted (in this example exons 2 and 3) flanked by LoxP sequences (*grey triangles*). Next to this is placed the positive selection marker (*neo*); in this example it is 5′; however, it could also be placed 3′ to exons 2 and 3. A further LoxP sequence is included so that the *neo* is also flanked by LoxP sites. All the LoxP sites are inserted in intronic sequence away from intron/exon splice junctions. Similar to deletion vectors, a TK cassette is included for negative selection. (**b**) In this example, homologous recombination can occur in the 3′ arm of homology (*left panel*) to give correct incorporation of the vector. Homologous recombination could however occur in the region containing exons 2 and 3 (*right panel*). In this case the 3′ LoxP site would not be incorporated and the resulting allele would not be able to be used for conditional deletion. It is therefore necessary to screen for the insertion of this LoxP site. If the vector has already been shown to be incorporated by homologous recombination at the correct site, this can be done by PCR across the LoxP site. If the PCR is designed to be 200–300 bp, then the presence of the LoxP site will result in a larger fragment than wild-type sequence (note, because most ES cells are heterozygous for the targeted allele, both wild-type and LoxP-containing bands would be amplified). Alternatively, a new restriction site (X) can be included with the LoxP site. Southerns with this enzyme and a probe to this region will therefore give a smaller band in cells where the LoxP site has been correctly incorporated. (**c**) Correct incorporation of the targeting vector results in inclusion of both LoxP sites and *neo* into the allele. As the presence of *neo* can interfere with expression of the allele, it should be removed by transient expression of Cre in the ES cells. This results in three possible options, removal of the *neo* but retention of exons 2 and 3 flanked by LoxP sites (i). This is the desired result as this modified allele can be used for conditional deletion in mice. It is also possible that exons 2 and 3 could be removed but *neo* left (ii), or that both *neo* and exons 2 and 3 could be removed (iii); either of these results in a constitutive deletion of the gene. It is therefore necessary to have a screen that will distinguish the desired allele with LoxP-flanked exons 2 and 3 from both the parent clone and the clones with exons two and three deleted. One example of how this can be done is shown – if a restriction site is included 3′ to the final LoxP site then the sizes obtained by Southern blotting are different for each possible deletion, and for the non-deleted parent clone. For each vector it is necessary to determine the best strategy based on the sizes of the regions involved.

the conditional knockout mice. It is therefore necessary to ensure the screening protocol used confirms that recombination has occurred in the desired region and that the LoxP site has been incorporated. This can be done by the inclusion of a restriction site for Southern screening next to the LoxP site (**Fig. 4b**). Once correctly targeted ES cells have been obtained, it is necessary to remove the positive selection cassette. This is because the promoter and polyadenylation sites, and possible cryptic splice sites in the selection cassette could interfere with the normal expression of the targeted gene. This could result in a hypomorphic allele, which may result in a phenotype even without recombinase-mediated deletion of the gene. The positive selection marker is normally removed by transient, low-level expression of Cre in the ES cells, followed by re-isolation and screening of colonies. Transient expression of Cre results in several possible deletion configurations between the three LoxP sites. The desired result is deletion of the positive selection marker only; however, deletion of the targeted region only, or both targeted region and selectable marker also occurs. When designing the vector it is necessary to allow for Southern strategies to distinguish these possibilities (**Fig. 4c**).

An improvement on the conditional deletion vector is a system that uses a second recombinase system to remove the selectable marker. A yeast recombinase, Flp, has been used successfully for this purpose. The Flp recombinase recognises frt sites, and similar to Cre, it will delete the region between two frt (GAAGTTCCTATTC-TCTAGAAA-GTATAGGAACTTC) sites. An initial drawback of Flp was that as it was a yeast enzyme its optimum temperature was significantly below the body temperature of mice. This has been overcome as mutated versions of Flp with an increased optimum temperature for use in mice have now been described *(20)*. Using this approach, the region to be deleted is flanked by LoxP sites, while frt sites are placed either side of the positive selection marker (**Fig. 5**). This method allows mice to be made after the first round of targeting without the need to remove the selectable maker in ES cell culture. Chimeric mice are then crossed to mice expressing an Flp transgene that is capable of deletion in germline cells. Assuming that germline transmission occurs, this will generate mice heterozygous for the targeted gene and Flp recombinase. The offspring of these mice will have deleted the selectable marker and can then be crossed away from the Flp transgene and onto to Cre-expressing mice as normal. It would also be possible to use the Flp/frt system to make a conditional deletion instead of the Cre/LoxP system; however, this is rarely done because while there are many characterised tissue-specific or regulated Cre recombinase transgenic lines available to induce deletion, there are still relatively few Flp recombinase transgenic lines available.

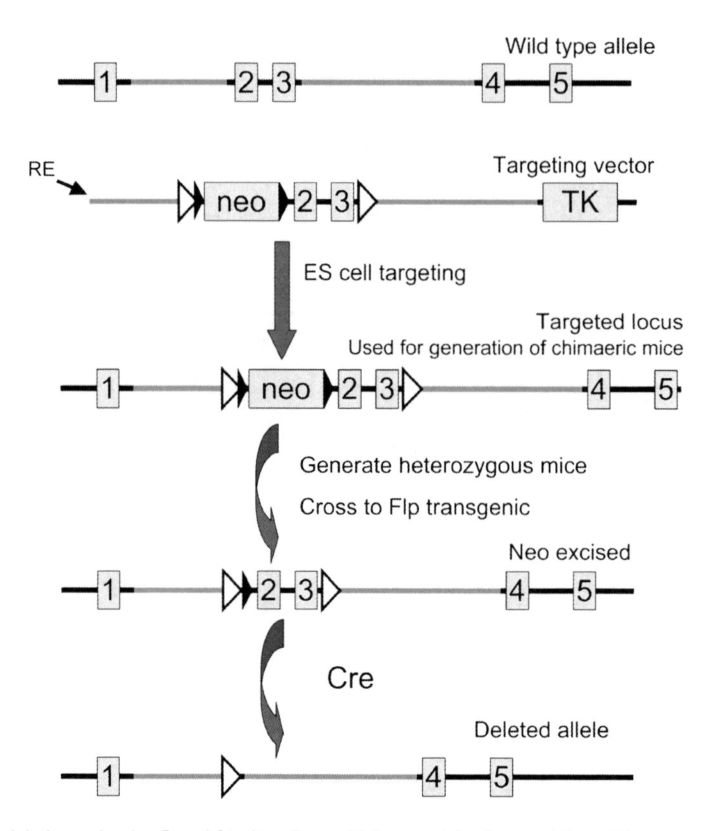

Fig. 5. Conditional deletion using LoxP and frt sites. To avoid the need for Cre excision of the neo cassette, an alternative method to make conditional deletions uses a combination of LoxP sites and frt sites. This vector is similar to the conditional vector described in **Fig. 4**; however, only two LoxP sites (*white triangles*) are used, which flank the region to be deleted (exons 2 and 3 in this example). The neo cassette is flanked by frt sites (*black triangles*). TK is included for negative selection and a unique restriction site (RE) to linearise the vector. This vector is used to target ES cells, resulting in an allele containing the LoxP-flanked exons and *neo* cassette. These cells are used to generate chimeric mice, and chimeras which have the capability to give germline transmission are identified as normal. These chimeras are then crossed to mice homozygous for the expression of an Flp recombinase. The offspring of these mice will undergo flp-mediated deletion of the *neo* cassette, and these mice can then be bred onto the appropriate Cre-expressing mice to make the conditional deletion.

4. Point Mutation (Knockin) and Insertion Vectors

Sometimes instead of deleting a specific gene, it may be desirable to make a point mutation or knockin, or alternatively introduce a tag onto an endogenous protein. For instance it may be preferable to introduce a mutation that destroys catalytic activity rather than completely delete a protein, as this may give less possibility of compensation from a related enzyme.

The principle behind this is very similar to making a conditional deletion; however, instead of the region of interest being flanked by LoxP sequences, the genomic sequence incorporating the desired point mutation or insertion is used in the vector. As with conditional deletions, it is necessary to remove the selectable marker, so this should still be flanked with either LoxP or frt sites. It is also possible to flank the mutated region with recombinase sites, as this would allow the option of deleting the gene in mice – in this case the vector would be identical to a conditional vector with the exception that it uses a mutated rather than wild-type sequence in the vector **(Fig. 6)**. As with conditional vectors, it is possible for recombination to occur between the positive selection marker and the mutation, resulting in the mutation not being incorporated. It is therefore necessary to ensure that the screening method used will distinguish between those clones containing or not containing the mutation. Similar to conditional deletion vectors this can be done by the introduction of a restriction site for screening so that the mutation lies between the new site and the selectable marker. It is possible to make a series of point mutations at the same locus using recombinase-mediated cassette exchange (RMCE); for reviews of this technique *see* (***(21–23)***, also *see* Chapter 16).

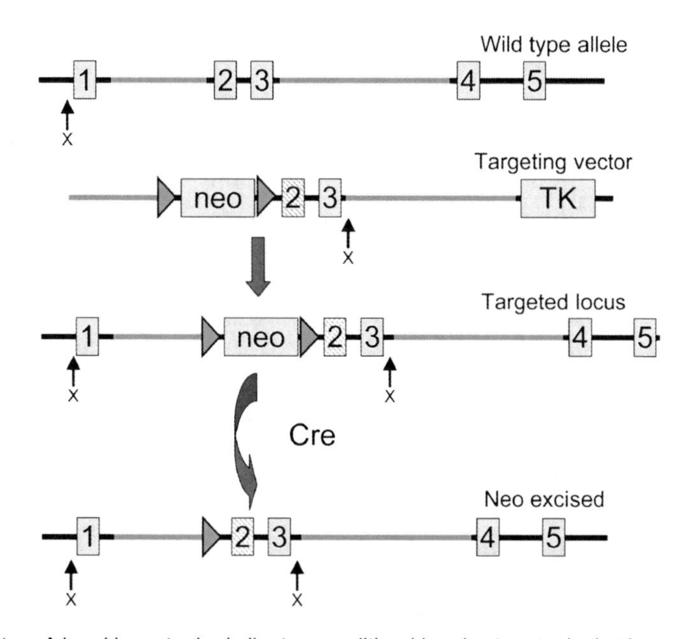

Fig. 6. Knockin vectors. A knockin vector is similar to a conditional knockout vector in that it contains the region to be mutated (equivalent to the region to be deleted in **Fig. 4**, in this example exon 2, with the mutated exon shown as a *hatched rectangle*) in addition to the two arms of homology (*grey lines*) and *neo* and TK selection cassettes. The desired homologous recombination of this vector results in the incorporation of the mutation and *neo* cassette. It is possible that recombination can occur between the *neo* and the mutation resulting in the mutation not being incorporated, and it is therefore necessary to screen for the presence of the mutation. This can be done by introducing a new restriction site (X) so that the mutation lies in between the restriction site and the neo cassette. It is also advisable to confirm the presence of the mutation by either RT-PCR of cDNA or PCR of genomic DNA for the mutated region, followed by sequencing of the PCR product.

5. Conditional Point Mutations

As with gene deletions, point mutations can result in embryonic lethality, and in this event it may be necessary to make a conditional point mutation. Several possible methods, summarised in **Fig. 7**, have been described for this *(24–29)*; however, they all have potential problems.

The first possibility is to cross a conventional point mutation with a conditional deletion of the same gene **(Fig. 7a)**. This results in a mouse in which one copy of the gene is wild type but floxed, while the other is mutated but not floxed. The idea is that expression from the one wild-type gene is sufficient to rescue the effect of the point mutation. When Cre is expressed in these mice, the wild-type floxed allele is removed, leaving only the mutated allele. This, however, makes several assumptions. Firstly, that the point mutation does not exert a dominant negative effect that inhibits the wild-type protein. Secondly, it assumes that transcription from one allele, as would be the case following Cre-mediated removal of the floxed allele, would not cause a reduction in the protein levels for the gene that might result in a phenotype independent of the mutation. Finally, if the gene is imprinted then one allele of the gene will be silenced, resulting in either a wild-type or mutated mouse rather than a conditional mutation.

The other methods aim to convert a wild-type allele into a mutated allele. This can be done either via Cre-mediated inversion or by excision of a minigene.

The Cre-mediated inversion method makes use of the ability of Cre to invert the sequence between two LoxP sites that are placed in opposite directions. In the construct, a wild-type copy of the exon to be targeted is placed in the forward direction, and this is followed by an inverted mutated exon **(Fig. 7b, c)**. Expression of Cre will cause inversion, and the mutated rather than wild-type exon will then be spliced into the mRNA. With two wild-type LoxP sequences however there is nothing to stop the sequences inverting back to their original orientation. To overcome this, mutated LoxP sites can be used **(Table 1)**. One suggestion is to use a combination of a Lox61 site and Lox71 site *(27, 28)*. These contain mutations at opposite ends of the LoxP sequence. While Cre can catalyse inversion between these two LoxP sites, this would result in their conversion to a wild-type sequence and double mutated sequence that can no longer be recognised by Cre **(Fig. 7b)**. An alternative is to use pairs of wild-type and mutated LoxP in the spacer region that cannot undergo recombination with each other at either end of the region to be inverted, referred to as a FLEx switch (*(24, 25)*, also *see* Chapter 16) **(Fig. 7c)**. When Cre-mediated inversion occurs between

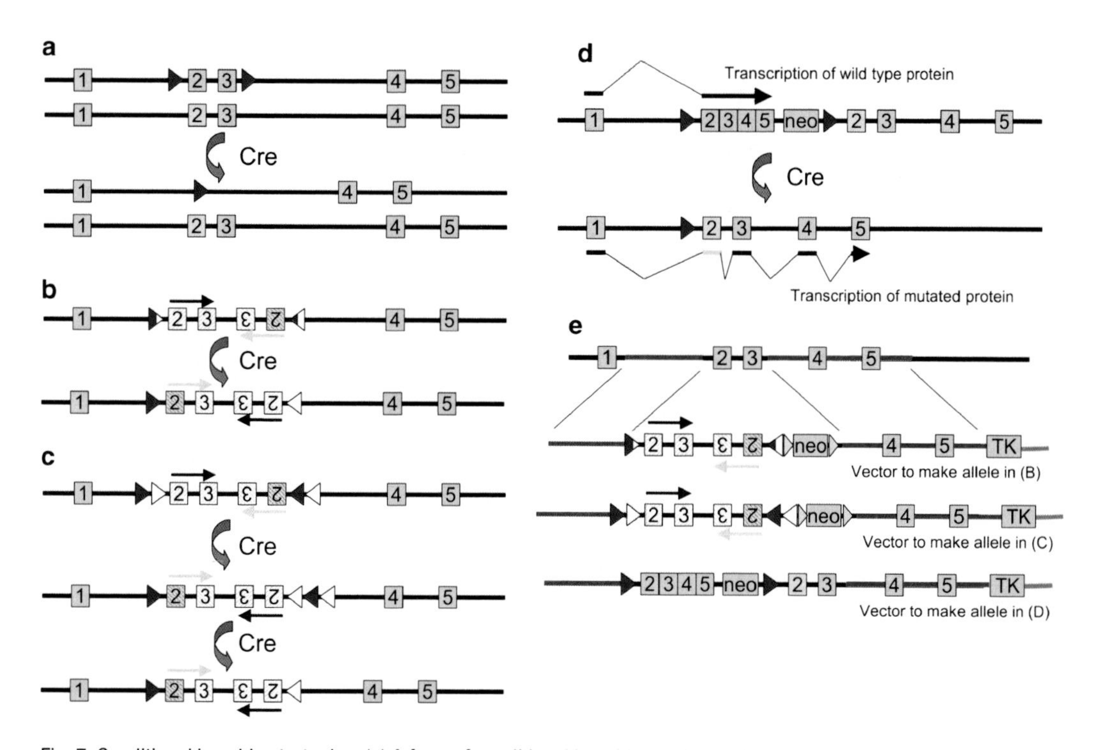

Fig. 7. Conditional knockin strategies. (**a**) A form of conditional knockin can be made by crossing a knockin allele (*yellow* exon) to a wild-type floxed (*blue triangles*) allele. This results in a mouse that will express a mixture of both wild-type and mutated protein. However, in cells which express Cre, the wild-type allele will be deleted resulting in expression of only the mutated protein. This method requires that the wild-type protein be dominant over the mutation and cannot be used for X-linked or imprinted genes. (**b**) Conditional knockin can also be achieved using Cre-mediated inversion. The region to be mutated is flanked by two LoxP sites placed in opposite directions. In between is the genomic region containing a wild-type copy of the exon to be mutated. This is followed by an inverted copy of the same sequence but with the exon mutated (*yellow* exon). Expression of Cre results in inversion so that the mutated exon is then transcribed. To prevent inversion back again, Lox61 and Lox71 sites are used. These contain mutations at opposite ends of the LoxP sequence. While the single mutants can still be recognised by Cre, inversion results in the formation of a double mutant (*white triangle*) that cannot recombine with the wild-type sequence (*blue triangle*). (**c**) Similar to (**b**) except pairs of wild-type (*blue*) and a LoxP site mutated in the spacer region (*white*) are used at each end of the vector. This method requires that the wild-type and mutated LoxP sequence are not able to undergo recombination together in the presence of Cre. Cre will however cause inversion between the two wild-type or two mutated LoxP sites. This results in a series of three LoxP sites, which will undergo deletion to leave one site, which will be of a different type to the site at the other end of the inverted region. (**d**) Conditional deletion can also be achieved by excision of a LoxP-flanked minigene. The minigene contains the wild-type sequence of the exon to be mutated, including the splice acceptor and upstream intronic sequence. This is fused to the cDNA corresponding to the remaining exons including 3′ UTR and polyA sequence. This is followed by the selectable marker. The minigene is placed upstream of the mutated exon (*yellow*) in the vector. The minigene should trap splicing of the upstream exons, which are still transcribed from their endogenous promoter. After Cre-mediated excision, the mutated exon is transcribed. (**e**) Examples of vectors to make the allele are described in **b–d**. The mutated exon is highlighted in *yellow* and the arms of homology in *red*. Note that for **b** and **c** the vector contains a selectable marker (*neo*) that must be removed. This could be done by including an additional LoxP site; however, a more straightforward way would be to place frt sites (*yellow triangles*) either side of the *neo*.

either the two wild-type or two mutant sequences this results in a series of three LoxP sites in the same direction. These would then undergo Cre-mediated deletion to leave a single site (**Fig. 7c**). As a result, the insertion would be flanked by a wild-type LoxP

Table 1
LoxP sequences

LoxP	ATAACTTCGTATAatgtatgcTATACGAAGTTAT
LoxP511	ATAACTTCGTATAatgtatacTATACGAAGTTAT
LOX5171	ATAACTTCGTATAatgtgtacTATACGAAGTTAT
Lox2272	ATAACTTCGTATAaagtatccTATACGAAGTTAT
M2	ATAACTTCGTATAagaaaccaTATACGAAGTTAT
Lox61	ATAACTTCGTATAatgtatgcTATACGAAaggta
Lox71	taccgTTCGTATAatgtatgcTATACGAAGTTAT

Sequences of the wild-type and reported mutations of the LoxP sites are shown. Mutations relative to the wild-type sequence are underlined

site at one end and a mutated one at the other. Cre-mediated inversion would no longer be possible. Both these methods have been described in ES cells; however, it is not clear if the mutated LoxP sites act as efficient enough substrates for Cre in mice to give efficient knockin with many Cre-expressing transgenics in vivo.

The final method makes use of Cre-mediated excision of a minigene *(26, 29)*. In this method, a minigene consisting of the exon to be mutated with the splice acceptor site and upstream intronic sequence at the 5′ end intact, is introduced upstream of the exon to be mutated. The exon is then fused to the remaining 3′ cDNA sequence of the gene of interest, including the 3′ UTR and polyadenylation sequence **(Fig. 7d)**. This minigene, along with the positive selection marker, is flanked by LoxP sites to allow it to be deleted in mice. The theory is that before expression of Cre, transcription from the endogenous promoter occurs, but it is terminated by the minigene polyA signal, and splicing occurs from the upstream exons onto the minigene to give a wild-type mRNA. Following Cre expression, the minigene is removed allowing splicing onto the endogenous 3′ exon, resulting in the mutated exon being used and translated to give a mutated protein. This has been described in mice with some success. The main concern, however, is that when the minigene is present it may not completely block transcription of the downstream exons, giving rise to a mixture of wild-type and mutated protein before Cre expression. The extent to which this occurs seems to depend on the construct used *(26, 29)* and could in theory be reduced further by the inclusion of multiple transcriptional stop sequences and the 3′ end of the minigene.

References

1. van der Weyden, L., Adams, D. J., and Bradley, A. (2002) Tools for targeted manipulation of the mouse genome *Physiol Genomics* **11**, 133–64.
2. Bradley, A. (1991) Modifying the mammalian genome by gene targeting *Curr Opin Biotechnol* **2**, 823–9.
3. Bradley, A. (1993) Site-directed mutagenesis in the mouse *Recent Prog Horm Res* **48**, 237–51.
4. Horie, K., Nishiguchi, S., Maeda, S., and Shimada, K. (1994) Structures of replacement vectors for efficient gene targeting *J Biochem* **115**, 477–85.
5. von Melchner, H., DeGregori, J. V., Rayburn, H., Reddy, S., Friedel, C., and Ruley, H. E. (1992) Selective disruption of genes expressed in totipotent embryonal stem cells *Genes Dev* **6**, 919–27.
6. Watanabe, S., Kai, N., Yasuda, M., Kohmura, N., Sanbo, M., Mishina, M., and Yagi, T. (1995) Stable production of mutant mice from double gene converted ES cells with puromycin and neomycin *Biochem Biophys Res Commun* **213**, 130–7.
7. Tucker, K. L., Wang, Y., Dausman, J., and Jaenisch, R. (1997) A transgenic mouse strain expressing four drug-selectable marker genes *Nucleic Acids Res* **25**, 3745–6.
8. Cecconi, F., and Gruss, P. (2002) From ES cells to mice: the gene trap approach *Methods Mol Biol* **185**, 335–46.
9. Stanford, W. L., Cohn, J. B., and Cordes, S. P. (2001) Gene-trap mutagenesis: past, present and beyond *Nat Rev Genet* **2**, 756–68.
10. Salminen, M., Meyer, B. I., and Gruss, P. (1998) Efficient poly A trap approach allows the capture of genes specifically active in differentiated embryonic stem cells and in mouse embryos *Dev Dyn* **212**, 326–33.
11. Abuin, A., Hansen, G. M., and Zambrowicz, B. (2007) Gene trap mutagenesis *Handb Exp Pharmacol* **178**, 129–47.
12. Araki, K., Araki, M., and Yamamura, K. (2006) Negative selection with the Diphtheria toxin A fragment gene improves frequency of Cre-mediated cassette exchange in ES cells *J Biochem* **140**, 793–8.
13. Karreman, C. (1998) New positive/negative selectable markers for mammalian cells on the basis of Blasticidin deaminase-thymidine kinase fusions *Nucleic Acids Res* **26**, 2508–10.
14. Chen, Y. T., and Bradley, A. (2000) A new positive/negative selectable marker, puDeltatk, for use in embryonic stem cells *Genesis* **28**, 31–5.
15. Muyrers, J. P., Zhang, Y., and Stewart, A. F. (2001) Techniques: recombinogenic engineering – new options for cloning and manipulating DNA *Trends Biochem Sci* **26**, 325–31.
16. Testa, G., Vintersten, K., Zhang, Y., Benes, V., Muyrers, J. P., and Stewart, A. F. (2004) BAC engineering for the generation of ES cell-targeting constructs and mouse transgenes *Methods Mol Biol* **256**, 123–39.
17. Nagy, A. (2000) Cre recombinase: the universal reagent for genome tailoring *Genesis* **26**, 99–109.
18. Sauer, B. (1998) Inducible gene targeting in mice using the Cre/lox system *Methods* **14**, 381–92.
19. Branda, C. S., and Dymecki, S. M. (2004) Talking about a revolution: the impact of site-specific recombinases on genetic analyses in mice *Dev Cell* **6**, 7–28.
20. Buchholz, F., Angrand, P. O., and Stewart, A. F. (1998) Improved properties of FLP recombinase evolved by cycling mutagenesis *Nat Biotechnol* **16**, 657–62.
21. Wirth, D., Gama-Norton, L., Riemer, P., Sandhu, U., Schucht, R., and Hauser, H. (2007) Road to precision: recombinase-based targeting technologies for genome engineering *Curr Opin Biotechnol* **18**, 411–9.
22. Baer, A., and Bode, J. (2001) Coping with kinetic and thermodynamic barriers: RMCE, an efficient strategy for the targeted integration of transgenes *Curr Opin Biotechnol* **12**, 473–80.
23. Seibler, J., Schubeler, D., Fiering, S., Groudine, M., and Bode, J. (1998) DNA cassette exchange in ES cells mediated by Flp recombinase: an efficient strategy for repeated modification of tagged loci by marker-free constructs *Biochemistry* **37**, 6229–34.
24. Schnutgen, F., Doerflinger, N., Calleja, C., Wendling, O., Chambon, P., and Ghyselinck, N. B. (2003) A directional strategy for monitoring Cre-mediated recombination at the cellular level in the mouse *Nat Biotechnol* **21**, 562–5.
25. Schnutgen, F., and Ghyselinck, N. B. (2007) Adopting the good reFLEXes when generating conditional alterations in the mouse genome *Transgenic Res* **16**, 405–13.
26. Bayascas, J. R., Sakamoto, K., Armit, L., Arthur, J. S., and Alessi, D. R. (2006) Evaluation of approaches to generation of tissue-specific knock-in mice *J Biol Chem* **281**, 28772–81.
27. Oberdoerffer, P., Otipoby, K. L., Maruyama, M., and Rajewsky, K. (2003) Unidirectional

Cre-mediated genetic inversion in mice using the mutant loxP pair lox66/lox71 *Nucleic Acids Res* **31**, e140.

28. Zhang, Z., and Lutz, B. (2002) Cre recombinase-mediated inversion using lox66 and lox71: method to introduce conditional point mutations into the CREB-binding protein *Nucleic Acids Res* **30**, e90.
29. Skvorak, K., Vissel, B., and Homanics, G. E. (2006) Production of conditional point mutant knockin mice *Genesis* **44**, 345–53.

Chapter 10

Gene Trap: Knockout on the Fast Lane

Melanie Ullrich and Kai Schuh

Summary

Gene trapping is a powerful tool to ablate gene function and to analyze in vivo promoter activity of the trapped gene in parallel. The gene trap strategy is not as commonly used as the conventional gene-targeting strategy, although it offers appealing options. Nowadays, a wide collection of embryonic stem cell clones, with a huge variety of trapped genes, have been identified and are available through the members of the International Gene Trap Consortium (IGTC). This chapter focuses on BLAST searches for the appropriate stem cell clones, the confirmation of vector insertion by RT-PCR or X-Gal staining, and the characterization of the exact insertion site to develop a PCR-based genotyping strategy. Furthermore, protocols to follow the activity of the commonly used β-galactosidase reporter are given.

Key words: Gene trap, Gene disruption, Promoter activity, Embryonic stem cells, Mouse genome, β-Galactosidase, β-Geo, X-gal staining

1. Introduction

Although the human and mouse genomes have recently been sequenced and published *(1–3)*, adequate expression patterns and functional information are available for only 10–15% of the genes. An attractive experimental strategy to achieve both gene ablation and functional analysis of murine genes is gene trapping *(4–7)*. With this technique, a loss of function mutation by random insertion of the gene trap vector and expression profiling of an artificial reporter gene under control of the endogenous promoter is obtained simultaneously. While initial mutagenesis approaches in mice, like X-ray or chemical mutagenesis, caused either complex chromosomal rearrangements affecting multiple genes or generated point mutations without landmarks for

Elizabeth J. Cartwright (ed.), *Transgenesis Techniques*, Methods in Molecular Biology, vol. 561
DOI 10.1007/978-1-60327-019-9_10, © Humana Press, a part of Springer Science+Business Media, LLC 2009

cloning *(8–10)*, gene trapping is the first technique taking a middle path between these random mutations and the defined mutations caused by gene targeting. As compared to gene targeting, gene trapping is often faster because of the broad availability of ES cell clones with identified trapped genes, making the cloning of a knockout vector and subsequent homologous recombination in ES cells unnecessary.

The first step after your decision to generate a knockout mouse by gene trapping is to search online databases, for example,

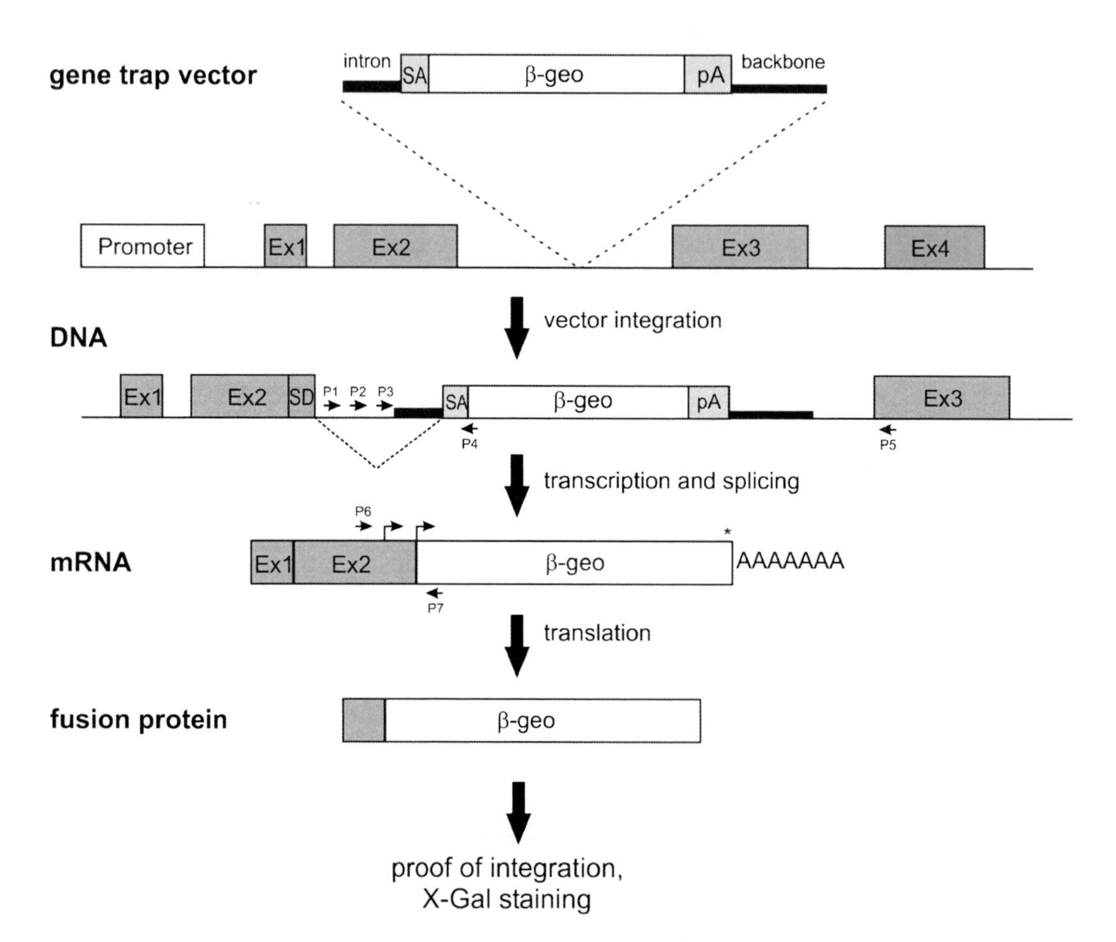

Fig. 1. Scheme of a gene trap approach. The gene trap vector consists of typical elements like an upstream intron, containing a splice branch site, a splice acceptor site (SA), a reporter and a resistance gene, in this case a fusion gene of lacZ and neoR, called β-geo, a polyadenylation signal (pA), and part of the vector backbone, necessary for propagation of the vector. The construct provides, after integration in the sense orientation into an intron of the target gene, the splice acceptor for the next upstream located exon, which itself carries the splice donor site (SD). After splicing and processing of the RNA transcript, the resulting mRNA comprises the conventional features needed for correct translation, e.g. the original start codon of the target gene and/or an internal start codon of the β-geo gene (*bent arrows*), an in-frame stop codon (*asterisk*), and the poly A tail. Translation of this mRNA results in a fusion protein of at least the β-galactosidase, the neomycin resistance gene product aminoglycoside phosphotransferase and, depending on the integration site, maybe part of the target protein. The primers located in the intron of the target gene (P1–3) and a reverse primer located in the gene trap vector (P4) are used to estimate the exact integration site of the vector. Together, with a primer for the wild-type allele localized downstream of the gene trap vector (P5), they are also useful for genotyping of ES cells and transgenic mice. Primers 6 and 7 (P6, P7) are used to run the RT-PCR reactions.

the International Gene Trap Consortium, for an ES cell line with your gene of interest trapped. Using the provided information about gene sequence it is possible to predict the chromosomal localization, the exon/intron organization of the trapped gene, and roughly the insertion site of the vector. Before starting a knockout mouse project, it is also important to confirm the gene trap vector insertion by RT-PCR with RNA isolated from ES cells. Additionally, most gene trap vectors contain a reporter gene, like LacZ (*see* **Fig. 1**), which allows X-Gal stainings of cultivated ES cells, confirming correct insertion and splicing of the gene trap vector in the ES cell clone. A 'primer walking' strategy can be used to define the exact insertion locus and to design adequate primers for a genotyping PCR with genomic DNA. After generating knockout mice with the trapped ES cell clone the activity of the endogenous promoter can be studied by assessing the β-Galactosidase reporter gene expression in embryos, organs, and tissue sections *(11, 12)*.

2. Materials

2.1. Data Mining

1. A computer with fast internet access and an internet browser installed. Pop-up windows should be allowed transiently to make browsing and BLAST searches easier.
2. A sequence file of the cDNA of interest in FASTA or plain text (*.txt) format; if possible from mouse.
3. Enough hard disk space and/or a printer to save and print out results.

2.2. Verification of Gene Trap Vector Insertion by RT-PCR

1. 6-well plate (BD Falcon).
2. ES cell culture medium: 1× D-MEM high glucose (Dulbecco's modified Eagle Medium, Invitrogen) supplemented with 0.1 M non-essential amino acids (100× stock, Invitrogen), 1 mM sodium pyruvate (100× stock, Invitrogen), 2 mM l-glutamine (100× stock, Invitrogen), 15% fetal bovine serum (Invitrogen), penicillin/streptomycin mix (final concentration 50 μg/mL each, Invitrogen), 1 nM β-mercaptoethanol (Sigma), LIF (leukaemia inhibitory factor, 1,000 units/mL, Chemicon).
3. PBS, pH 7.4 (Phosphate-buffered saline, Invitrogen).
4. RNase-free reaction tubes.
5. TRIZOL (Invitrogen); Chloroform; Isopropanol.
6. RNase-free water: 0.1% DEPC water. Dissolve diethylpyrocarbonate (DEPC) in distilled water and stir overnight.

Autoclave the water to inactivate the DEPC; otherwise, it might inhibit the PCR reaction.

7. 75% Ethanol, prepared with DEPC water.
8. Verso 1-Step ReddyMix Kit (Abgene).
9. Primers for the trapped allele.
10. 1% standard agarose in 1× TAE buffer: 40 mM Tris–acetate, 1 mM EDTA. 50× TAE stock solution: 242 g Tris base, 57.1 mL of glacial acetic acid, 100 mL of 0.5 M EDTA, pH 8.0, ad 1000 mL H_2O. The same buffer is used as running buffer.
11. 1% Ethidium bromide solution (Sigma).
12. DNA Ladder (Fermentas).

2.3. Verification of Gene Trap Vector Insertion by X-Gal Staining of ES Cells

1. LabTek one-well glass chamber slides (Nunc); microscope cover slips (24 × 60 mm).
2. 0.1% gelatin: Dissolve in PBS and autoclave.
3. ES cell culture medium: (*see* **Subheading 2.2**, **step 2**).
4. PBS, pH 7.4 (Phosphate-buffered saline, Invitrogen).
5. 0.1 M Phosphate buffer: 27 mM NaH_2PO_4, 73 mM Na_2HPO_4.
6. Fix buffer: Supplement 0.1 M phosphate buffer with 5 mM EGTA, pH 8, 0, 2 mM $MgCl_2$, and 0.2% glutaraldehyde. This buffer can be stored at 4°C for up to 4 months (*see* **Note 1**).
7. Wash buffer: Supplement 0.1 M phosphate buffer with 2 mM $MgCl_2$; store at 4°C.
8. X-Gal stock solution (50 mg/mL, sufficient for 500 mL staining solution): Add 10 mL dimethylformamide to a 500-mg bottle of X-Gal and store at –20°C in the dark.
9. Staining buffer: Supplement 0.1 M phosphate buffer with 2 mM $MgCl_2$, 5 mM $K_4Fe(CN)_6$, and 5 mM $K_3Fe(CN)_6$. Store in the dark at 4°C. Before use, add X-Gal to a final concentration of 1 mg/mL and filter the solution with a fluted filter (*see* **Note 2**).
10. Mounting medium: Add 2.4 g Mowiol 4-88 (Roth) to 6 g glycerol and mix. Add 6 mL distilled water and mix overnight on a magnetic stirrer at room temperature. Add 0.2 M Tris–HCl, pH 8.5 and heat to 50°C for 10 min while stirring. Store 0.5-mL aliquots at –20°C.

2.4. Simple and Efficient DNA Preparation and PCR: Genotyping in Half a Day

1. REDExtract-N-Amp Tissue PCR Kit (Sigma).
2. Primers for wild-type and knockout alleles.
3. Standard agarose gel equipment (*see* **2.2**, **step 10**).

2.5. Expression Profiling in Embryos, Organs, and Sections by X-Gal Stainings

Essentially, the X-Gal staining procedure for entire organs, embryos, and for tissue sections is very similar to the procedure used for embryonic stem cells. Additionally, you need

1. The detergents deoxycholate, sodium salt; Nonidet P40 (Roche).
2. 2-methylbutane.
3. OCT-compound.
4. Nuclear fast red solution: 5% aluminium sulphate in distilled water. Add 0.1 g of Nuclear Fast Red to 100 mL and slowly heat until boiling. Cool the staining solution and pass through a fluted filter.
5. SuperFrost Plus microscope slides (Menzel); microscope cover slips (24 × 60 mm).
6. Mounting Media: Glycerin (Nuclear Fast Red counterstaining); Mowiol (X-Gal staining, *see* **2.3, step 10**).

3. Methods

3.1. Data Mining

3.1.1. Identifying an Embryonic Stem Cell Line with the Desired Trapped Gene

The following shows the procedure for the Sanger Institute Gene Trap Resource (SIGTR), which is quite similar to the procedure of other members of the International Gene Trap Consortium (IGTC).

1. Go to the site of the Sanger Institute Gene Trap Resource: http://www.sanger.ac.uk/PostGenomics/genetrap/
2. Choose 'Data Access' or go directly to the BLAST site: http://www.sanger.ac.uk/cgi-bin/blast/submitblast/genetrap
3. Copy and paste your cDNA sequence as FASTA or plain text in the indicated query box or locate the file on your hard disk (this option will work only with FASTA or plain text files!), choose the database (e.g. SIGTR or IGTC) and start BLAST search.
4. Retrieve your BLAST results by clicking the 'retrieve' button or by following the given hyperlink (BLAST search may take some minutes).
5. BLAST results will be given as a graphical view as well as sequence alignments of your query sequence with matching sequences in the database.
6. Click on the links with the highest scores or scroll down the page to view the sequence alignments. When using a mouse cDNA as query, the sequence identity should be 100% over a long stretch of the sequence. This is important in cases with more than one possible isoform, to avoid false identification of a gene trap of another protein family member.

7. Once you are sure that one or more clones have the trap in the desired gene, go to the IGTC site (http://www.genetrap.org/dataaccess/index.html) and search for additional information on the chosen clone. Here you will also find information about gene trap vectors, and again a sequence alignment image, giving a first idea, in which intron the trap was inserted.
8. Choose the clone(s) with the gene trap vector inserted most upstream in the gene, to avoid later expression of a fusion protein consisting of a large portion of your protein and the β-geo fusion protein. Such an artificial protein may act in a dominant way on an organism, resulting in a complex phenotype in heterozygous animals, as well. Furthermore, it will be quite difficult to differentiate between a dominant acting factor and haploinsufficiency.
9. Follow the instructions on the IGTC website to contact the distributor and to obtain your clone(s) of interest. This procedure varies a little bit, depending on the provider.

3.1.2. Exact Determination of Intronic Insertion Locus of the Gene Trap Vector

1. After doing the BLAST search, when gathering additional information on the IGTC site, you will also find the sequence tag information, i.e. the sequence used to identify the trapped gene. Copy and paste this sequence into the BLAST window of the Ensembl mouse database (http://www.ensembl.org/Mus_musculus/blastview) and start the search.
2. After clicking the 'view' button, the results will be displayed. In the 'Alignment Locations vs. Karyotype' frame, the alignment with highest score is boxed. Clicking on this will open a pull-down menu with the options 'Genomic Sequence' and 'Contig View'. The first option will give you the genomic sequence of the trapped gene with the exon sequences highlighted. The gene trap is located after the most downstream exon displayed. The second option guides you to the contig view, which shows again the exon/intron structure of the trapped gene in the region of the sequence tag. Clicking on one exon will open the option 'Gene View', giving you all the needed information of this gene to design primers for a genotyping PCR strategy.

3.1.3. Development of a PCR Strategy for Later Genotyping

1. To this point it will be clear after which exon the trap is inserted; however, the exact insertion locus in the downstream intron needs to be identified. Based on the genomic information saved previously, choose forward primers located in this intron with a distance of 1–2 kb, covering the complete intron, and run PCRs with a reverse primer located in the gene trap vector (Primers 1–3 and 4 in **Fig. 1**).

2. With a forward primer located closely to the insertion point, you will get a PCR product. Sequence comparison of this PCR product with the gene sequence and with the gene trap vector sequence will show the exact insertion point and, additionally, if part of the intron upstream of the vector splice acceptor was cut off by exonucleases.
3. Use the best working primer combination to identify the trapped allele in the mouse line derived from the ES cell line. To identify the wild-type allele choose a primer located in the exon downstream of the insertion (*see* **Note 3** and Primer 5 in **Fig. 1**).

3.2. Verification of Gene Trap Vector Insertion by RT-PCR

In order to confirm gene trap vector insertion and proper splicing to the exon upstream of the insertion point, it is recommended to perform an RT-PCR to detect the fusion transcript.

1. Design primers for RT-PCR by placing the forward primer in the next upstream exon, and the reverse primer in the gene trap vector sequence (Primers 6 and 7 in **Fig. 1**).
2. Grow feeder-independent ES cells to confluence in a well of a 6-well plate (9.6 cm^2).
3. Aspirate off the ES cell medium and wash cells once with PBS.
4. Add 1 mL TRIZOL (per 10 cm^2 culture area) directly to the well, lyse the cells by passing the lysate several times through a pipette, and transfer the lysate into a 1.5-mL reaction tube.
5. Incubate the homogenized samples for 5 min at room temperature for complete dissociation of nucleoprotein complexes.
6. Add 0.2 mL chloroform per 1 mL of TRIZOL and cap the tube securely.
7. Shake the tube vigorously for 15 s and incubate it for 2–3 min at room temperature.
8. Centrifuge the sample at ≤12,000 × *g* for 15 min.
9. Transfer the colourless upper aqueous phase to a new tube (*see* **Note 4**).
10. Add 0.5 mL isopropyl alcohol per 1 mL of TRIZOL used for the initial homogenization and precipitate the RNA by inverting the tube several times carefully. Incubate sample for 10 min at room temperature.
11. Centrifuge the sample at ≤12,000 × *g* for 10 min at 4°C and remove the supernatant. The RNA precipitate appears colourless to white at the bottom of the tube (*see* **Note 5**).
12. Wash the RNA pellet once with at least 1 mL 75% ethanol per 1 mL TRIZOL. Mix by vortexing, centrifuge at ≤7,500 × *g* for 5 min at 4°C, and remove the supernatant.

13. Air dry or vacuum dry the pellet for 5-10 min (*see* **Note 6**).
14. Dissolve RNA in RNase-free water by passing the solution a few times through a pipette tip, incubate the sample for 10 min at 55-60°C, and proceed to the RT–PCR.
15. Set up the following reaction mix with 50 μL final volume for one sample (*see* **Note 7**):

Component	Volume	Final concentration
1-Step PCR Reddy Mix (2×)	25 μL	1×
Forward primer (10 μM)	1 μL	200 nM
Reverse primer (10 μM)	1 μL	200 nM
RT Enhancer	2.5 μL	
Verso Enzyme Mix	1 μL	
RNA	1-5 μL	1 μg
RNase-free water	fill up to 50 μL	
Total volume	50 μL	

The 1-Step PCR Reddy Mix includes a reaction buffer allowing both reverse transcription and PCR amplification in the same reaction, the DNA Polymerase Thermoprime Plus, dNTPs, and a loading dye for agarose gel electrophoresis. The Verso Enzyme Mix contains a reverse transcriptase together with an RNase inhibitor. The RT Enhancer removes contaminating DNA, thereby eliminating the need for DNaseI treatment.

16. Perform one-Step RT-PCR with the following cycling program.

Step	Temperature	Time	Number of cycles
cDNA synthesis	50°C	15 min	1
Verso inactivation	95°C	2 min	1
Denaturation	95°C	20 s	
Annealing	50–60°C	30 s	35–45
Extension	72°C	1 min	
Final extension	72°C	5 min	1

17. Prepare a 0.5–2% agarose gel (depending on the PCR product size).

18. Load 10 μL of the DNA ladder and the PCR products and run the gel until the visible front of the loading dye reaches the end of the gel (*see* **Note 8**).
19. To visualize PCR products, put the gel on a transilluminator and compare the sizes of the PCR products with the DNA ladder.

3.3. Verification of Gene Trap Vector Insertion by X-Gal Staining of ES Cells

Integration and expression of the reporter gene β-galactosidase can easily be checked by staining of the ES cells with its substrate X-Gal, which is converted into an indigo dye.

1. Coat one well of a chamber slide with 0.1% gelatin for 1 h.
2. Aspirate off the residual gelatin solution, split approximately 2×10^6 ES cells into the well (9.4 cm^2), and incubate for 24 h at 37°C in a humidified 5% CO_2 incubator.
3. Aspirate off the ES cell culture medium and wash the cells with 2 mL PBS.
4. Fix the ES cell monolayer with 2 mL fix buffer at room temperature for 15 min.
5. Wash the cells twice with 2 mL wash buffer at room temperature for 5 min each.
6. Add 2 mL staining buffer containing 1 mg/mL X-Gal and incubate cells overnight at 37°C in a humidified chamber.
7. Aspirate off the staining buffer and add wash buffer (*see* **Note 9**).
8. ES cells containing the gene trap vector will stain blue and can be identified by bright field microscopy.
9. To mount the slides, carefully remove the gasket (with a razor blade etc.); add 50 μL Mowiol and the cover slip (*see* **Note 10**). The samples can be viewed after 16 h.

3.4. Simple and Efficient DNA Preparation and PCR: Genotyping in Half a Day

A very fast and easy way to genotype the offspring is to isolate genomic DNA from the tail tip (or ear snip) and do a PCR with wild-type and knockout-specific primers. Here we use the REDExtract-N-Amp Tissue PCR kit from Sigma, because this kit allows genotyping in just a few hours.

1. Disinfect scissors and forceps prior to use and between different samples.
2. Take a mouse tail tip biopsy (approximately 0.5 cm) and place it in a 1.5-mL reaction tube with the cut end down. If necessary, store biopsies at –20°C.
3. Add 100 μL extraction and 25 μL tissue preparation solution; mix by vortexing (*see* **Note 11**).
4. Ensure that tails tips are submersed and incubate samples for 12 min at 55°C (*see* **Note 12**).

5. Boil samples for 3 min at 95°C. It is normal that the tails tips are not completely digested.
6. Add 100 μL neutralization solution and mix by vortexing.
7. Continue immediately with genotyping PCR or store the samples at 4°C. For long-term storage discard the undigested tissue and store samples at 4°C for up to 6 months.
8. Set up the following reaction mix with a 20-μL final volume for each sample:

Component	Volume	Final concentration
REDExtract-N-Amp PCR Reaction Mix (2×)	10 μL	1×
Forward primer (10 μM)	0.8 μL	400 nM
Reverse primer wild-type allele (10 μM)	0.8 μL	400 nM
Reverse primer knockout allele (10 μM)	0.8 μL	400 nM
Tissue extract	4 μL	
Sterile water	fill up to 20 μL	
Final volume	20 μL	

The REDExtract-N-Amp PCR Reaction Mix contains buffer, salts, dNTPs, and Taq polymerase, JumpStart Taq antibody for hot start, and REDTaq loading dye. Wild-type and knockout PCR can be performed in one tube (*see* **Note 3**).

9. Perform genotyping PCR with the following cycling program:

Step	Temperature	Time	Number of cycles
Initial denaturation	94°C	3 min	1
Denaturation	94°C	30 s to 1 min	
Annealing	45–68°C	30 s to 1 min	30–35
Extension	72°C	1 to 2 min	
Final extension	72°C	10 min	1

10. To visualize the PCR products, run them on a 0.5–2% agarose gel, put the gel on a transilluminator, and compare the sizes of the PCR products with the DNA ladder.

3.5. Expression Profiling in Embryos, Organs, and Sections by X-Gal Stainings

Many gene trap vectors contain the lacZ gene or the artificial β-geo gene, enabling activity studies of the endogenous promoter normally controlling the ablated gene. To visualize expression profiles, X-Gal staining of embryos, organs, and tissue sections can be performed (*see* **Fig. 2**).

1. Kill a pregnant female mouse (for recovery of embryos) or any other heterozygous or knockout mice (for organs).
2. Pin the mouse on its back, cut along the ventral midline from the groin to the chin, and make an incision from the start of the first incision to the knee joint on both sides of the animal. Pin aside the skin, cut the abdominal wall similar to the skin, and put aside. Locate your organs of interest and keep them in PBS. For embryo preparation, find the uterus, take it out, place it into a Petri dish filled with PBS, and carefully release the embryos by cutting between the decidual swellings (*see* **Note 13**). For staining go to **step 7**.
3. For embryo or tissue sections, either place the tissue into a plastic tube containing 2-methylbutane or embed it in OCT compound and orientate the tissue with a spatula. Then, quick freeze the specimen with liquid nitrogen (*see* **Note 14**).
4. Place OCT compound on to the pre-cooled (–20°C) mounting block of the cryostat, add and orientate your tissue while the embedding medium is liquid, and wait until it is frozen.
5. Fix the block on the holder and fine adjust the angle between your tissue and the knife.
6. Set the section thickness from 50 to 100 μm, move the knife towards your tissue, and trim away the embedding medium covering your sample. When you reach the tissue or the desired area of the tissue, adjust the section thickness to 5–10 μm and collect the sections.
7. Collect the sections on room temperature Superfrost Plus slides by approaching the section still lying on the surface of the knife. When the slide just touches the section, it will melt to the glass. Do not press the slide to the section. Several sections can be collected on one slide.
8. Store the slides at room temperature and allow to dry.
9. Perform X-Gal staining of tissues or tissue sections as indicated (*see* **Note 15**):

	Embryos up to 8.5 days	Embryos up to 11.5 days	Organs	Tissue sections
Fixation	15 min	30 min	15–60 min	15 min
Wash	3 × 5 min	3 × 15 min	3 × 15–30 min	2 × 5 min

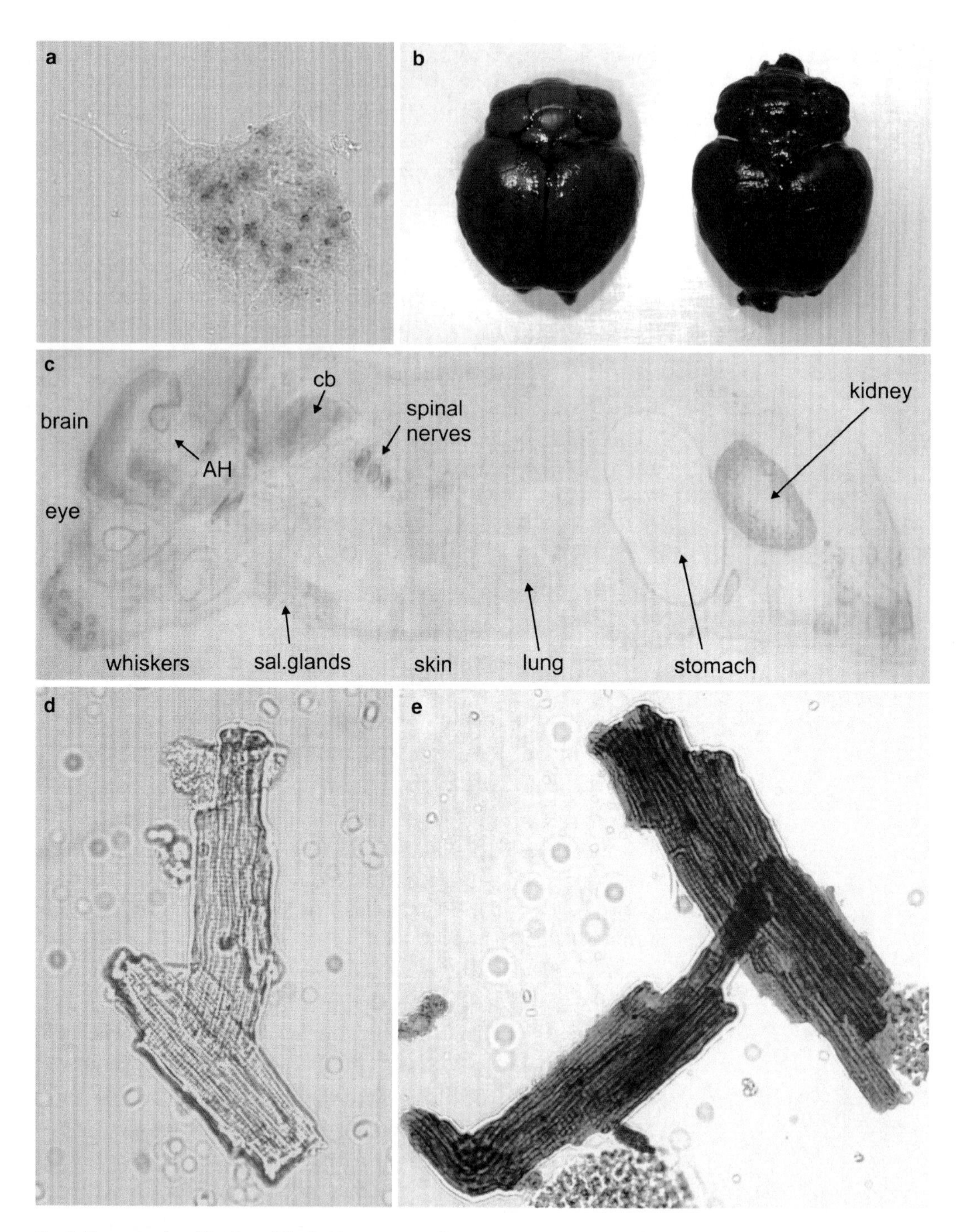

Fig. 2. Examples for utilization of the lacZ reporter gene expression under control of the endogenous promoter. (**a**) X-Gal staining of an ES cell clone growing in a feeder cell-free medium showing blue staining, indicative of correct gene trap vector integration. (**b**) Brains of a wild-type mouse (*left*) and a mouse with a trapped gene showing strong promoter activity (*right*) in the brain. (**c**) Cross-section and X-Gal staining of a newborn mouse demonstrating variable promoter activity in different parts of the body. (**d**) Control staining of isolated cardiac myocytes of wild-type mice in comparison to cardiac myocytes of mice with a trapped gene displaying high expression in this cell type, indicative of strong promoter activity in this tissue (**e**).

The volumes and incubation times of fix-, wash-, and staining buffers required depend on the size of embryos or tissues. Add sufficient buffer to cover the specimen. Stain tissues or sections for 1–24 h at 37°C in a humidified chamber; the speed and intensity of staining is dependent on the promoter activity of the trapped gene. For incubation of slides with sections use histological staining trays that allow easy handling of 20 slides at the same time.

10. Discard the staining buffer and add wash buffer (*see* **Note 9**).
11. To counterstain the tissue sections with Nuclear Fast Red, transfer the slides to distilled water; otherwise, continue with **step 14** (*see* **Note 16**).
12. Add Nuclear Fast Red solution to the tray and incubate for 1–10 min. The staining solution can be reused several times.
13. Wash the sections several times with distilled water until the water remains clean.
14. Mount the slides with glycerine, cover, and seal with nail polish. The sections can be viewed after the nail polish hardened (*see* **Notes 10** and **17**).

4. Notes

1. To avoid degradation of glutaraldehyde in the fix buffer, add this reagent prior to use. The buffer without glutaraldehyde can be stored at 4°C. Caution: Glutaraldehyde is toxic.
2. Filtration of the staining solution minimizes the formation of crystals. Mix staining buffer and X-Gal stock solution at least 1 h before you start staining and filter prior to use.
3. PCR products should not be too long (up to 1,000 bp). To run two reactions in parallel in one tube, the corresponding wild-type product should differ by at least 100 bp from the gene trap product.
4. Following centrifugation, the mixture separates into a pink phenol–chloroform phase at the bottom of the tube, a white interface, and a colourless upper phase. RNA remains exclusively in the aqueous phase and its volume is about 60% of the TRIZOL volume used for homogenization. Be careful to transfer just the aqueous phase to avoid DNA or protein contamination of the RNA. The RNA extraction is based on the method of Chomczynski and Sacchi *(13)* and can also be used to isolate proteins and DNA from the same sample.
5. The RNA precipitate is often poorly visible. Therefore, remember the orientation, in which you put the tube into

the centrifuge because the pellet should be at the outer side. Avoid touching this region with the pipette while taking off the supernatant.

6. Do not dry the RNA pellet in a SpeedVac as this will greatly decrease its solubility.
7. Performing a one-step RT-PCR reaction is much easier and faster than performing the cDNA synthesis and the DNA amplification in two separate reactions. The protocol given here works very well in most cases and optimization of primer and RNA concentrations are not necessary. Similar kits to the one described are available from other suppliers.
8. The voltage to be used depends on the distance of the electrodes. Typically you run the gel with a maximum of 5V/cm until you can easily distinguish the different DNA fragment sizes.
9. Do not incubate the specimen in the wash buffer for longer than 24 h, because unspecific staining of cells or tissues not expressing β-Galactosidase may occur. For long-term storage of samples add fix buffer, wrap in parafilm, and store at 4°C. Slides can also be mounted as indicated in **Subheadings 3.3** and **3.5**.
10. Avoid air bubbles in the mounting medium by adding it drop by drop along the longer edge of the slide, apply the cover slip on the same edge, and let it fall on the slide. Cover slips can be sealed with nail polish.
11. If you want to perform several extractions, premix sufficient volumes of extraction and tissue preparation solution in a ratio of 4:1 up to 2 h before use.
12. In an amendment to the protocol given by Sigma, we incubate the samples at 55°C instead of room temperature for 12 min instead of 10 min, because this enhances the DNA yield.
13. For expression profiling in embryos older than 12 days p.c. dissect embryos or do cryostat sections. Staining is limited by the size of the specimen due to impaired penetration of the substrate or endogenous β-galactosidase activity. The same is true for whole organs.
14. Compact tissues, like heart, kidney, or spleen, can be frozen without embedding in OCT compound and fixed directly at the mounting block of the cryostat. Longish and hollow tissues like aorta, intestine, or bladder need prior embedding.
15. To permeabilize embryos and organs, include 0.01% deoxycholate and 0.02% Nonidet P-40 in both wash and staining buffer and agitate gently during incubation.
16. Nuclear Fast Red, as indicated in the name of the dye, stains the cell nuclei in red and the cytoplasm in pink. This staining

method allows an approximate histological analysis of the tissues and is easier and faster than haematoxylin eosin staining.

17. Do not use Mowiol or any other aqueous mounting medium because this will wash out the staining of Nuclear Fast Red. If you just performed X-Gal staining, you can use Mowiol.

Acknowledgments

The authors would like to thank Tobias Fischer and Peter M. Benz for providing organs of mice with different genes trapped to demonstrate X-Gal stainings, and Karin Bundschu for providing X-Gal stained tissue sections of mice with a trapped *spred2* gene.

References

1. Lander, E. S., Linton, L. M., Birren, B., Nusbaum, C., Zody, M. C., Baldwin, J., Devon, K., Dewar, K., Doyle, M., FitzHugh, W. et al. (2001). Initial sequencing and analysis of the human genome. *Nature*, **409**, 860–921
2. Venter, J. C., Adams, M. D., Myers, E. W., Li, P. W., Mural, R. J., Sutton, G. G., Smith, H. O., Yandell, M., Evans, C. A., Holt, R. A.et al. (2001). The sequence of the human genome. *Science*, **291**, 1304–1351
3. Waterston, R. H., Lindblad-Toh, K., Birney, E., Rogers, J., Abril, J. F., Agarwal, P., Agarwala, R., Ainscough, R., Alexandersson, M., An, P. et al. (2002). Initial sequencing and comparative analysis of the mouse genome. *Nature*, **420**, 520–562
4. Friedrich, G. and Soriano, P. (1991). Promoter traps in embryonic stem cells: a genetic screen to identify and mutate developmental genes in mice. *Genes Dev*, **5**, 1513–1523
5. Gossler, A., Joyner, A. L., Rossant, J. and Skarnes, W. C. (1989). Mouse embryonic stem cells and reporter constructs to detect developmentally regulated genes. *Science*, **244**, 463–465
6. Skarnes, W. C., Auerbach, B. A. and Joyner, A. L. (1992). A gene trap approach in mouse embryonic stem cells: the lacZ reported is activated by splicing, reflects endogenous gene expression, and is mutagenic in mice. *Genes Dev*, **6**, 903–918
7. Wurst, W., Rossant, J., Prideaux, V., Kownacka, M., Joyner, A., Hill, D. P., Guillemot, F., Gasca, S., Cado, D., Auerbach, A. et al. (1995). A large-scale gene-trap screen for insertional mutations in developmentally regulated genes in mice. *Genetics*, **139**, 889–899
8. Van Buul, P. P. and Leonard, A. (1974) Translocations in mouse spermatogonia after exposure to unequally fractionated doses of x-rays. *Mutat Res*, **25**, 361–365
9. Russell, L. B., Hunsicker, P. R., Cacheiro, N. L., Bangham, J. W., Russell, W. L. and Shelby, M. D. (1989). Chlorambucil effectively induces deletion mutations in mouse germ cells. *Proc Natl Acad Sci U S A*, **86**, 3704–3708
10. Russell, W. L., Kelly, E. M., Hunsicker, P. R., Bangham, J. W., Maddux, S. C. and Phipps, E. L. (1979). Specific-locus test shows ethylnitrosourea to be the most potent mutagen in the mouse. *Proc Natl Acad Sci U S A*, **76**, 5818–5819
11. Bundschu, K., Gattenlohner, S., Knobeloch, K. P., Walter, U. and Schuh, K. (2006). Tissue-specific Spred-2 promoter activity characterized by a gene trap approach. *Gene Expr Patterns*, **6**, 247–255
12. Bundschu, K., Knobeloch, K. P., Ullrich, M., Schinke, T., Amling, M., Engelhardt, C. M., Renne, T., Walter, U. and Schuh, K. (2005). Gene disruption of Spred-2 causes dwarfism. *J Biol Chem*, **280**, 28572–28580
13. Chomczynski, P. and Sacchi, N. (1987). Single-step method of RNA isolation by acid guanidinium thiocyanate-phenol-chloroform extraction. *Anal Biochem*, **162**, 156–159

Chapter 11

Culture of Murine Embryonic Stem Cells

Ivana Barbaric and T. Neil Dear

Summary

Appropriate culture of murine embryonic stem cells is critical to enabling introduced mutations to be passed through the germ line. ES cells must be carefully cultured to ensure that pluripotency is maintained. The feeder cells and foetal calf serum used to culture the cells can significantly influence germ line transmission potential. Additionally, ES cells can be karyotypically unstable, so determining the karyotype of ES cell lines will increase your confidence in the ability of the manipulated cells to contribute to the germ line in chimaeric mice.

Key words: Karyotype, LIF, MEF, Feeder cells, Freezing, Thawing

1. Introduction

Martin Evans and Matthew Kaufman, and Gail Martin first achieved the establishment of pluripotent embryonic stem cells in 1981 *(1, 2)*. Critical to the ability to isolate these cells was the use of feeder cells. It has been subsequently demonstrated that the soluble factor Leukemia Inhibitory Factor (LIF) is required to maintain the cells in an undifferentiated state *(3, 4)* (for a review of the history of embryonic stem cell research see *(5)*. The choice of ES cell line to use in gene-targeting experiments is critical to success. Most gene-targeting experiments have used ES cell lines derived from the 129 strains of mice. This is because ES cells from these strains are easy to culture, maintain their germ line potential after prolonged culture, and have high rates of germ line transmission. Some suitable ES cell lines are freely provided to academic collaborators, e.g. the R1 line from Andras Nagy *(6)* or the AB2.2 line from Alan Bradley *(7)*. As C57BL/6J is the

Elizabeth J. Cartwright (ed.), *Transgenesis Techniques*, Methods in Molecular Biology, vol. 561
DOI 10.1007/978-1-60327-019-9_11, © Humana Press, a part of Springer Science+Business Media, LLC 2009

most widely used mouse strain, there is a growing trend to try and use ES cells derived from C57BL/6 mice, some recent lines of which are proving to be highly germ line competent *(8)*.

The basic methods required for use of ES cells ready for generating gene-targeted mice are (1) culture of ES cells, (2) preparation of feeder cells, (3) batch testing foetal calf serum, (4) karyotyping ES cell lines, and (5) testing ES cell lines for presence of infectious agents. All these techniques are described in this chapter.

2. Materials

It is recommended that ES cells be cultured in a separate facility from other cells. This reduces the risk from infectious agents (e.g. *Mycoplasma*) or contamination with other cell lines. Throughout the document the term 'water' refers to water that has a high resistivity, typically around 18 MΩ-cm. All reagents should be of AnalaR™ quality.

2.1. Preparation of Mouse Embryonic Fibroblasts

1. 0.1% (w/v) gelatin solution: Warm a solution of 2% (w/v) porcine gelatin (Sigma) to 37°C, remove 25 mL, and add to 475 mL of sterile PBS. Mix by inversion. Store at 4°C.
2. Feeder cell medium: Dulbecco's modified Eagle's medium (D-MEM), high glucose, without pyruvate, containing GlutaMAX I™ (Invitrogen) (or, alternatively, 2 mM glutamine) supplemented with 10% (v/v) foetal calf serum (FCS) (*see* **Note 1**).
3. Trypsin/EDTA solution: 2.5 g/L porcine trypsin, 0.2 g/L EDTA tetrasodium salt in Hanks' Balanced Salt Solution (Sigma).
4. PBS: 1.54 mM KH_2PO_4, 2.71 mM Na_2HPO_4, 155.17 mM NaCl, pH 7.2. We purchase ready prepared from Invitrogen.
5. Freezing medium: 10% (v/v) sterile DMSO (Sigma) in standard FCS. Make fresh on day of use.
6. Mitomycin C medium: Dissolve one 2 mg vial mitomycin C (Sigma) in 2 mL PBS. Mix with 196 mL feeder cell medium and sterilize by filtration using a 0.22-μm filter.
7. Dissection equipment: Fine forceps (Dumont No. 5) and fine surgical scissors.
8. Tissue culture flasks (150 cm^2) and cryogenic vials, e.g. CryoTube™ (Nunc).
9. Freezing container: Nalgene Mr Frosty (Sigma). Fill with isopropanol as directed.

2.2. ES Cell Culture

1. Feeder cell medium: *See* **Subheading 2.1, item 2**.
2. ES cell medium: KnockOut™ Dulbecco's modified Eagle's medium (D-MEM) (Invitrogen) supplemented with 2 mM glutamine, 1× non-essential amino acids (Invitrogen), 10 μM β-mercaptoethanol, 15% foetal calf serum (FCS) (tested for ability to support ES cell growth), 1,000 U/mL leukaemia inhibitory factor (LIF) (Millipore) (optional if using feeders).
3. 50 mM β-mercaptoethanol: Add 70 μL β-mercaptoethanol to 20 mL PBS and filter sterilize using a 0.22-μm filter (Millipore, Watford, UK). Store at 4°C for no longer than 1 month. This should be diluted 1 in 5,000 when preparing ES cell medium as earlier.
4. PBS: 1.54 mM KH_2PO_4, 2.71 mM Na_2HPO_4, 155.17 mM NaCl, pH 7.2.
5. Trypsin-EDTA solution: 2.5 g/L porcine trypsin, 0.2 g/L EDTA tetrasodium salt in Hanks' balanced salt solution (Sigma).
6. Cell freezing medium: 10% sterile DMSO (Sigma) in ES cell-batch tested FCS. Make fresh on day of use.

2.3. Batch-Testing Foetal Calf Serum

1. ES cell medium: As prepared in **Subheading 2.2, item 2** but *without LIF*.
2. Six-well plates (Nunc).
3. Methylene blue hydrate solution (2%): Mix 2 g methylene blue hydrate (Sigma) in 100 mL PBS. Filter through Whatman No. 1 paper and store at 4°C.
4. BCIP/NBT solution: We use the ready prepared 5-bromo-4-chloro-3-indolyl phosphate (BCIP/Nitroblue Tetrazolium (NBT) liquid substrate system, Sigma). For the original description of the reagent, see Blake et al. *(9)*.
5. Acetone/methanol fixative: Mix acetone:methanol 1:1 on day of use.

2.4. Karyotyping ES Cells

1. 0.075 M KCl: Dissolve 5.7 g KCl in water and make up to 1 L. Sterilize by autoclaving.
2. Colcemid®: Prepare a 10-μg/mL solution in PBS. A pre-prepared solution (KaryoMAX® Colcemid) is available from Invitrogen. Shortly before use prepare ES cell medium containing a 1/1,000 dilution of the Colcemid stock (final concentration = 10 ng/mL)
3. Fixative: 3:1 methanol:acetic acid. Prepare fresh by adding 12.5 mL acetic acid to 37.5 mL methanol.

4. Glass Pasteur pipettes, glass slides with frosted ends and cover slips (22 × 40 × 0.15 mm), and a drying rack for the slides (Fisher Scientific).
5. Giemsa solution: We use Karyomax Giemsa (Gurr's Improved R66, Invitrogen).
6. Gurr's buffer, pH 6.8: The convenient way is to purchase buffer tablets (e.g. from Invitrogen). These are simply mixed with water.
7. Mounting medium for coverslips, e.g. Clarion Mounting medium (Sigma).

2.5. Testing ES Cells for Presence of Mouse Pathogens

1. PCR primers and control plasmids: Controls are listed in Ayral et al. *(10)* and can be obtained from the authors.
2. DNA lysis buffer: 50 mM Tris–HCl of pH 8.5, 1 mM EDTA of pH 8.0, 0.5% Tween 20. Sterile filter and store at 4°C. Add proteinase K to a final concentration of 300 μg/mL on the day of use.
3. Reagents for cDNA synthesis: MMLV reverse transcriptase and 10× buffer, random nonamers, RNase inhibitor (New England Biolabs).
4. Plasmids should be amplified in 15 μL volumes using 'GoTaq' Taq DNA polymerase (Promega) in 1× manufacturer's reaction buffer, 1.5 mM $MgCl_2$, 200 μM each dNTP, 1.5 pmoles each of forward and reverse primers, and 0.2 units Taq polymerase. Prepare the reaction mix in 12 μL and then add 3 μL DNA or water to bring to the final volume.

3. Methods

Before commencing culture of ES cells, feeder cells need to be prepared. Culturing of ES cells on feeders is not mandatory as some groups culture ES cells in the absence of feeders using medium supplemented with LIF. The feeder cells to be used will depend on the ES cell line. Two kinds of feeder cells are common - mouse embryonic fibroblasts (MEFs) and STO cells. The isolation of MEFs is detailed. There are also two methods commonly used to block cell proliferation – treatment with mitomycin C or γ-irradiation. The latter is easier and there is no toxic waste to be disposed, but requires access to a radiation source.

3.1. Preparing Feeder Cells

3.1.1. Preparation of Mouse Embryonic Fibroblasts

1. Mate a female mouse (*see* **Note 2**) and check early the next morning for presence of a copulation plug. After 14 days, kill the pregnant females either by cervical dislocation or some other humane method. The embryos will be approximately 14.5-days old. Typically, we harvest

embryos from one or two females and process a total of 10–20 embryos.

2. On day of harvest, warm trypsin/EDTA and feeder culture medium to 37°C, wash dissection equipment in 70% (v/v) ethanol, and prepare three 100-mm Petri dishes each containing 10 mL ice-cold PBS.
3. With the culled mouse on its back, wipe down the abdominal skin and fur with 70% (v/v) ethanol.
4. Make an incision down the midsection using scissors, thereby exposing the uterine horns. Observe how the embryos are located on each side of the uterus and in the amniotic sacs.
5. Pull the embryos and uterine horns away from the abdomen and carefully detach them from the animal, placing them in a dish with PBS.
6. With fine-tipped scissors, make a longitudinal cut down the entire uterine horn, thus releasing the embryos into the PBS.
7. Carefully detach embryos from the amniotic sac and placenta and place in a fresh dish of PBS.
8. Kill the embryos as rapidly as possible. To do this, hold each embryo with forceps and decapitate it using the fine scissors. Repeat until all embryos have been killed in this manner.
9. Carefully pull out the liver as well as any other internal organs from each embryo (this can be done under a dissecting microscope) and place the carcasses in the third dish of fresh PBS, trying to wash away as many red blood cells as possible.
10. Pool all the carcasses in a fresh Petri dish without PBS and gently mince using a scalpel blade.
11. Add 20 mL trypsin/EDTA to the plate, pipette up and down 8–10 times to disperse the tissue, then transfer to the sterile conical flask containing glass beads, and stir in an incubator at 37°C for 30 min.
12. Remove the medium from the conical flask to a sterile 50-mL sterile tube. Wash to recover remaining cells in the flask by adding 20 mL feeder culture medium to the flask, swirl, and remove the medium and pool (*see* **Note 3**).
13. Count the viable cells and plate at a density of 5×10^6 cells per 150 cm^2 tissue culture flask, each in a total of 20 mL feeder culture medium. Incubate overnight at 37°C in 5% CO_2.

3.1.2. Freezing of MEFs

1. Allow the plates to grow until confluent. This should only take a day or two.
2. Wash the flask once with 20 mL PBS.

3. Add 2 mL trypsin/EDTA solution and incubate for 4 min at 37°C.
4. Tap the flask to remove semi-adherent cells. Add 5 mL feeder cell medium and remove all cells from the flask by pipetting up medium and washing over the flask surface 4–5 times.
5. Pool the cells from all flasks in a 50-mL tube. Centrifuge at 500 × *g* for 5 min at 4°C.
6. Label cryotubes with 'MEF P1' (P1 indicates "passage 1") and the date.
7. Resuspend the pellet in 0.5 mL freezing medium per flask of cells harvested. For example, if there are ten flasks, resuspend the cells in 5 mL freezing medium.
8. Pipette up and down 8–10 times to disperse the pellet into single cells.
9. Aliquot 0.5 mL into 1-mL cryotubes.
10. Place cryotubes in the freezing container filled with isopropanol and place at –80°C overnight.
11. Next day, transfer the tubes into liquid nitrogen.
12. Routinely, some of the MEFs should be checked for resistance to the cytotoxic drugs being used. For example, for MEFs carrying the Neo^R gene, plate a sufficient number of MEFs onto each well of a 6-well plate to achieve 50% confluency (approximately 2×10^5 cells). Treat the wells 1–6 with an increasing concentration of G418 as follows – 50, 100, 200, 300, 400 μg/mL. Leave for 5 days in a 37°C, 5% CO_2 incubator, inspecting the cells daily. G418-resistant MEFs should be able to tolerate the highest dose of G418. They should continue to expand until the well is confluent.

3.1.3. Growth Arrest of Feeder Cells

Passaging MEFs

1. Warm the medium, PBS, and trypsin to 37°C before commencing.
2. Determine the location of the cells in the LN_2 tank and remove. Wear gloves and face shield or glasses.
3. Loosen the cap slightly and thaw vial of cells quickly in a 37°C water bath.
4. Transfer the contents of the cryotube to a sterile tube. Add 1 mL of pre-warmed medium dropwise to the tube while gently mixing.
5. Transfer the contents to a tube containing 14 mL of pre-warmed medium and gently mix.
6. Pipette 5 mL into each of three 150-cm^2 flasks.
7. Add a further 15 mL of feeder cell medium to each flask. Swirl the plate to mix.

8. Incubate at 37°C in 5% CO_2.
9. On about day 3, the cells should be confluent. If not, incubate further until confluent.
10. Warm MEF medium, PBS, and trypsin to 37°C before commencing.
11. Wash flask with 20 mL PBS and aspirate.
12. Add 2 mL trypsin/EDTA solution and swirl the flask to mix. Incubate for 5 min at 37°C.
13. Tap flasks a few times to dislodge cells. Check cells to ensure none is still attached. If still attached, tap a few times and return to the incubator for a further 5 min.
14. Add a total of 14 mL feeder cell medium to the three flasks to stop the action of the trypsin.
15. Pool all the cells and pipette the cell suspension up and down 8–10 times to make a single-cell suspension.
16. Dispense 2 mL into each of ten 150-cm^2 flasks (includes the three used flasks and seven new ones).
17. Add 18 mL feeder cell medium to each flask, swirl to mix, and return to the incubator.
18. When plates are approaching confluency, which should be in about 3 days, proceed to mitomycin C treatment or γ-irradiation (*see* **Note 4**).

Mitomycin C Treatment

1. Prepare mitomycin C-containing medium (*see* **Note 5**).
2. Aspirate the medium and add mitomycin C-containing medium.
3. Incubate for 2–4 h at 37°C.
4. Aspirate the medium, wash with 20 mL PBS and add 2 mL trypsin/EDTA solution per flask.
5. Incubate for 5 min at 37°C.
6. Add ~5 mL medium to one flask and wash down well to collect cells. Transfer the medium to another flask and wash down. Continue transferring and washing down each flask.
7. Once all the flasks are washed and cells harvested pool the contents in a single sterile tube.
8. Wash down all flasks a second time using ~10 mL medium to ensure all cells are harvested.
9. Once all ten flasks are washed and cells harvested pool the cells and centrifuge at 500 × *g* for 5 min.
10. Aspirate the medium with a pipette and transfer to the waste bottle.
11. Resuspend the pellet in a total of 20 mL PBS. Count the cells at this stage.

12. Centrifuge at 500 × *g* for 5 min, aspirate PBS and wash again with a further 20 mL PBS. This second wash is important to remove any traces of mitomycin C.
13. Resuspend in an appropriate concentration of freezing medium such that cells are in 0.5 mL cell freezing medium per cryotube. Feeders are plated at a density of about 5×10^4 cells/cm^2. As an example, 2.75×10^6 cells are required to coat a 100-mm dish, so freezing 5.5×10^6 cells in one cryotube would be enough for 2 × 100 mm dishes. The number of feeders required for each size of well/dish is shown in **Table 1**.

γ-Irradiation

1. Harvest the cells as in **Subheading "Mitomycin C Treatment" steps 4–12**.
2. Resuspend the pellet in 5 mL feeder cell medium.
3. Irradiate with 25–40 Gy. The exact dose will need to be determined empirically.
4. Centrifuge at 500 × *g* for 5 min, aspirate medium, and resuspend in an appropriate concentration of cell freezing medium (*see* **Subheading "Mitomycin C Treatment", step 13**).

3.1.4. Plating Growth-Arrested Feeder Cells

1. Thaw a vial of growth-arrested feeder cells rapidly in a 37°C water bath.

Table 1
Plating feeder cells

Dish/flask size	Culture area per plate (cm^2)	Culture area per well (cm^2)	Medium volume per well/dish/flask (mL)	Dishes/wells per ampoule feeder cells[a]
100-mm dish	55	–	10	2
60-mm dish	21	–	4	5
35-mm dish	8	–	2	12
6-well	57	9.5	2	10 wells
24-well	45.9	1.9	1	52 wells
96-well	31.7	0.33	0.15	303 wells
75-cm^2 flask	75	–	15	1.3
25-cm^2 flask	25	–	5	4

[a]From a vial of 5×10^6 MEFs plated at a density of 5×10^4/cm^2

2. Add about 500 μL feeder medium to the cryotube. Mix gently to thaw completely. Remove contents of vial to a 50-mL tube containing 5 mL feeder cell medium.
3. Add an appropriate amount of feeder cell medium for plating (*see* **Table 1**).
4. Pipette medium containing cells onto plates and place in a 37°C, 5% CO_2 incubator.
5. Allow feeders to attach for at least 4 h or overnight before using. Check the plates to ensure they are at the appropriate density (**Fig. 1**). If there are too few feeder cells, there is no need to start over again; just thaw some more cells and add to the dish/flask. Allow the new cells to attach and then check again for coverage.
6. The plated cells can be kept for approximately 1 week with occasional medium changes (every 3 days), but if not used within this time, the cells should be discarded.

3.2. Culture of ES Cells

1. ES cells grow rapidly. Consequently, they need to be passaged frequently. There is significant variation between ES cell lines though, with doubling times varying from 11 to 24 h *(11)*. To prevent differentiation of ES cells, a source of LIF is required. This can be either provided by the feeder cells or added exogenously to the medium. We do both to optimize our chances of maintaining the cells in a pluripotent state. It is essential that you monitor your cells carefully, on at least a daily basis (*see* **Table 2**). Overgrowth will result in differentiation and a terminal loss of pluripotency. Adopt rigorous cell culture standards when culturing ES cells. Culture your ES cells in a dedicated biohazard hood and incubator and use separate aliquots of medium, PBS, and trypsin for each cell line. Work only on one ES cell line at a time in the biohazard hood.

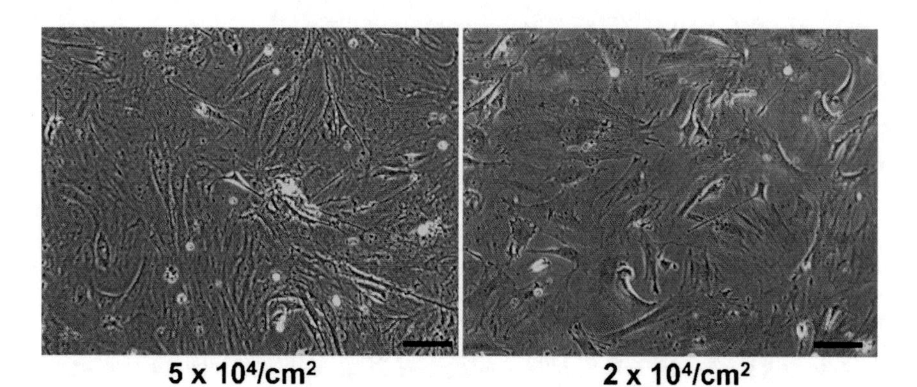

Fig. 1. Examples of MEF feeder cells plated at two different densities. The plate at $2 \times 10^4/cm^2$ is probably too sparse for some ES cell lines as there are quite large gaps where no feeders are attached. Using $5 \times 10^4/cm^2$ results in a suitable density. Scale bar = 0.1 mm.

Table 2
Troubleshooting ES cell culture

Problem	Possible cause	Solution
ES cells did not grow after plating	ES cells are dead	
ES colonies are differentiated, or colonies detach from plate	FCS is poor	FCS must be batch tested
	Feeder cells are of poor quality	Check your feeders against control feeders, or prepare a new batch of feeder cells
	Too few feeder cells	See relevant section of troubleshooting guide
	ES cells too dense	Ensure cells do not grow larger than colonies shown in **Fig. 3d**
ES colonies are small	FCS is poor	FCS must be batch tested
	Medium is poor	Ensure medium contains fresh glutamine. Check quality of individual components (including number of times frozen and thawed, time stored at 4°C). Supplementing with 20% MEF-conditioned medium may help
ES cells did not grow after thawing	Freezing procedure flawed	Was cell freezing medium fresh? Check cooling rate of cells during freezing
	Cells did not survive in liquid nitrogen	Is liquid nitrogen tank regularly filled with LN_2?
	Thawing procedure aggressive	Thaw cells rapidly at 37°C but add pre-warmed ES cell medium dropwise to cells in cryotube. Test whether centrifugation of thawed cells before plating is better or worse than plating directly in diluted cell freezing medium
Feeder cells have overgrown the ES cell culture	Feeder cells have not been growth arrested appropriately and have continued dividing, thus overtaking the ES cell culture	Prepare new feeder cells and increase mitomycin C treatment time or γ-irradiation dose
Feeder cells are too sparse when plated	Seeding density too low	Increase the seeding density by carrying out empirical tests on concentration required for 100% coverage
	Some cells did not survive freezing	Check for dead cells after overnight culture of thawed cells. If many dead cells evident, check DMSO used for freezing and check rate of cooling was not too rapid
MEF yield from embryos poor	Treatment to disperse cells insufficient	Increase incubation time in trypsin
Primary MEFs do not grow	Prolonged treatment with trypsin has killed cells	Decrease incubation time in trypsin

This will avoid the possibility of contamination with other cells or infectious agents such as *Mycoplasma*. A typical time line for the preparation of MEFs and culture of sufficient ES cells required for a gene-targeting experiment is summarized in **Fig. 2**.

3.2.1. Thawing ES Cells

1. Plate feeder cells at least 4 h before culturing ES cells. Alternatively, if feeders are not being used, treat plates with 0.1% gelatin (*see* **Note 6**), leave for several hours to overnight, then aspirate gelatin, and leave dishes with the lids off under the hood for 15 min to dry.
2. Prepare ES cell medium with appropriate foetal calf serum (FCS) (*see* **Note 7**) and, if appropriate, LIF (*see* **Note 8**).
3. Thaw a vial of ES cells quickly in a 37°C water bath.
4. Slowly add 500 μL ES cell medium to the vial. Transfer to a 15 mL tube and slowly add a further 4 mL ES cell medium.
5. Spin down at 500 × *g* for 5 min. Aspirate medium, and resuspend the pellet in an appropriate volume of ES cell medium, pipetting up and down several times to disperse any clumps.
6. Aspirate medium from the feeder cell plates.

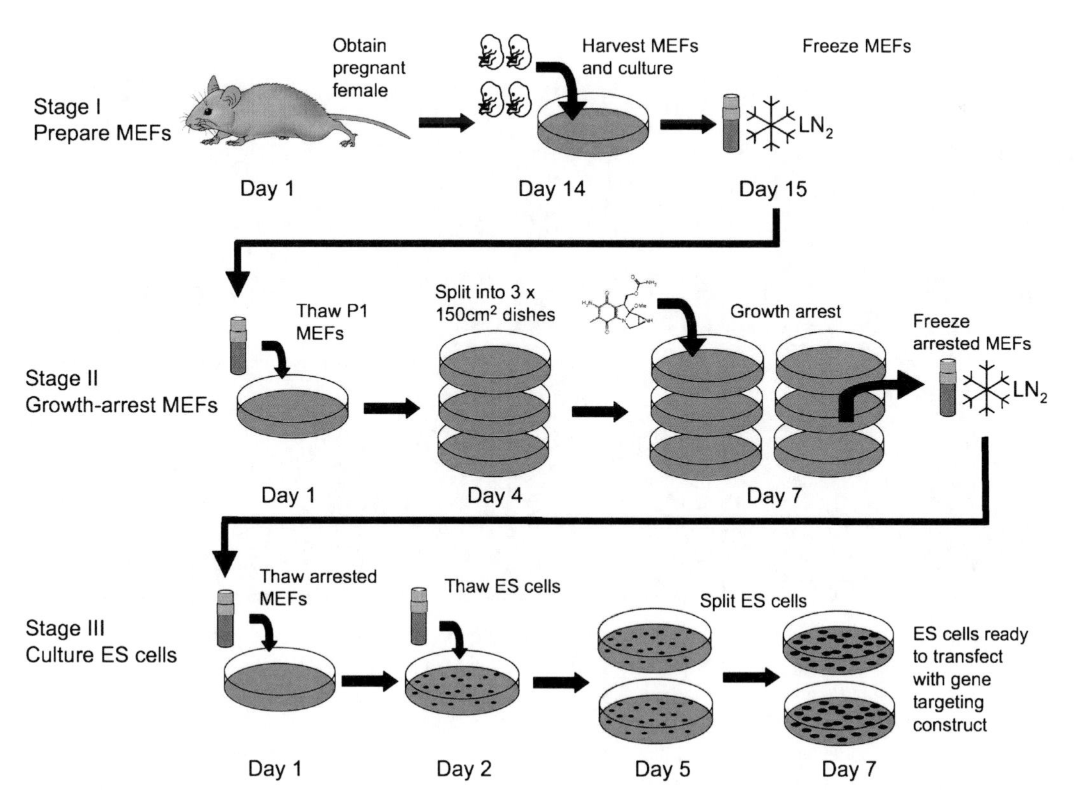

Fig. 2. A typical timeline for production of ES cells ready for transfection with a gene-targeting vector.

7. Gently pipette ES cells on to the dishes/flasks at an appropriate density (we suggest 1–2 × 10^5/cm^2) and place cells in a 37°C, 5% CO_2 incubator.

3.2.2. Culturing and Passaging ES Cells

1. Check the plates and change medium daily. If the cells are at high density, medium may need to be changed twice daily. Ensure your cells do not overgrow. Examples of ES cell characteristics at different densities are shown in **Fig. 3**.
2. When cells are ready to be passaged, prepare feeder dishes or gelatinise dishes in preparation.
3. Change the ES cell medium 4 h before passaging.
4. Aspirate medium and wash cells with an equal volume of PBS (*see* **Note 9**).
5. Add appropriate amount of trypsin/EDTA solution (e.g. 1 mL per 100 mm dish). Incubate for 4–5 min at 37°C.
6. Pipette up and down before adding medium to help disperse the ES cells. Add 5 mL ES cell medium to stop the action of the trypsin.
7. Pipette the cell suspension up and down 8-10 times to make a single-cell suspension (*see* **Note 10**).

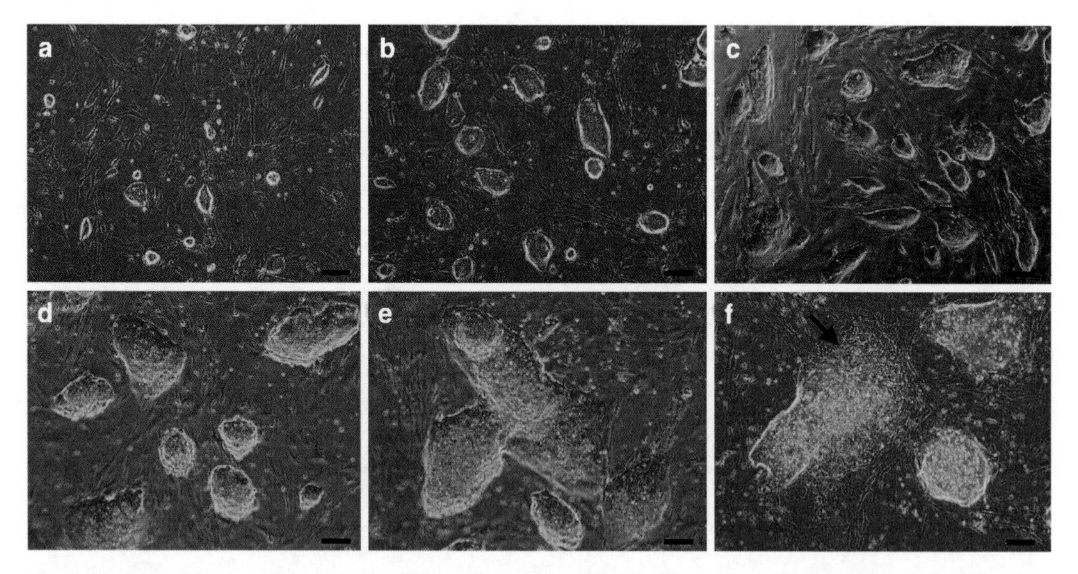

Fig. 3. Examples of the typical appearance of ES cell colonies at different colony sizes. (**a**) 1 day after plating ES cells at an appropriate seeding density. (**b**) 2 days. (**c**) 3 days. (**d**) 4 days – this is the maximum size the colonies should achieve before passaging them. Note the smooth-edged appearance to the colonies. The cells are tightly packed together and there is no suggestion of more than one cell type. (**e**) Overgrown colonies – the colonies are beginning to merge into one another. There is some differentiation at this stage. (**f**) Differentiating colonies – note that cells have begun to grow out from the colony on one side (indicated by arrow) and these have a different morphology, more like fibroblasts than pluripotent ES cells. Scale bar = 0.1 mm.

8. Pipette the cells into a sterile tube and pellet cells by centrifugation at 500 × *g* for 5 min.
9. Aspirate supernatant and resuspend cells in an appropriate amount of ES cell medium. Typically, passage ES cells at 1:6.
10. If freezing cells, count with a haemocytometer, then resuspend in cell freezing medium at an appropriate concentration. We suggest freezing between 2×10^6 and 10^7 cells per 0.5 mL freezing medium/cryotube. Record the cell line name, date, concentration and importantly, the passage number of the cells.

3.3. Batch Testing Foetal Calf Serum

It is important to use FCS that supports the growth of ES cells and maintains them in an undifferentiated state. Batch testing only needs to be done every 3–4 years in small ES cell labs. It can be argued that true batch testing should also involve generation of chimaeric mice from the ES cells and testing for germ line inheritance as there are anecdotal accounts that batches of serum that performed well in cell culture tests still did not lead to germ line transmission of ES-cell derived genomic material. However, this would be an arduous process and unrealistic as suppliers would not reserve a batch of serum long enough to test this way. In most cases, batch testing as outlined later has been successful in choosing an FCS that supported gene targeting and the germ line transmission of the targeted ES cells. This protocol tests ten new batches as well as the current batch of FCS for comparison. A typical example of the results of batch testing is shown in **Fig. 4**.

3.3.1. Preparation

1. Before commencing, obtain aliquots of serum for batch testing (*see* **Note 11**).
2. Thaw a vial of ES cells onto a 100-mm dish previously coated with growth-arrested feeder cells (*see* **Note 12**).
3. Plate feeder cells (*see* **Note 13**) onto 6-well plates. One 6-well plate is required for each batch of FCS to be tested and one for the current batch as a control. Mark the lid and base of each 6-well plate with a number that corresponds to a batch of FCS being tested.
4. Allow feeder cells to attach overnight.

3.3.2. Plating ES Cells

1. Harvest exponentially growing ES cells and count.
2. Mix 7×10^4 ES cells per 120 mL ES cell medium *without serum or LIF*. This gives approximately 10^3 ES cells per 1.7 mL medium. For each FCS batch to be tested, a total of 10.2 mL of this medium is required, so for 11 batches (ten to be tested and the control), 120 mL cells + medium should suffice.

3. Plate 1.7 mL into each well of the plates.
4. For each dish, use a different batch of FCS. Add 0.3 mL to wells 1–4, 0.19 mL FCS to well 5, and 0.73 mL to well 6. This gives a final FCS concentration of 15% in wells 1–4, 10% in well 5, and 30% in well 6.
5. Change the medium daily, replacing with the same concentration of FCS.
6. Culture for 4–5 days but no longer. Begin scoring before the colonies become overgrown.

3.3.3. Scoring ES Cells before Staining

Examine ES colonies under the microscope. Note how the colonies appear and whether there is any example of any of the following (*see* **Fig. 4**):

- An indiscrete border to the colony and the appearance of 'cobblestone'-like cells within the colony. This suggests the presence of endoderm and indicates the cells are differentiating.
- Fibroblast-like cells spreading out from the border into the surrounding feeder cells.
- Loosely packed ES cells instead of tight, densely packed cells.
- A dark centre to the colonies. This indicates dying cells.

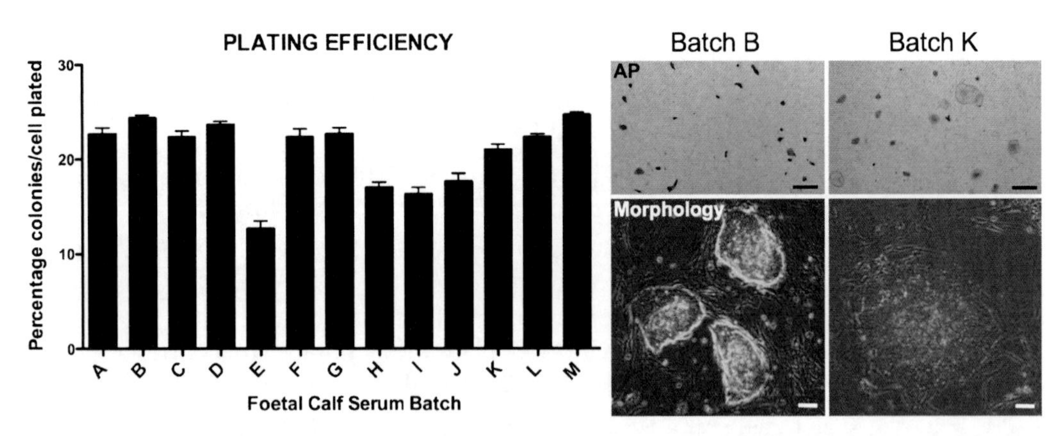

Fig. 4. An example of some results of batch testing foetal calf serum. On the left, counting of the number of colonies after staining with methylene blue allows plating efficiency for the new batches (labelled A–L) to be compared to our current batch (M). Batch E performed particularly badly in this test. On the right are shown examples of ES colonies grown in serum batches B and K. Batch K is particularly poor at maintaining ES cells in an undifferentiated state despite reasonable plating efficiency results. Alkaline phosphatase staining (AP) demonstrates how growth of ES cells in serum batch K has resulted in mainly differentiated colonies while serum batch B has produced mainly non-differentiated (darkly staining colonies). Methylene blue staining allows colonies to be counted and also reflects differences in colony size – note how colonies grown in batch K are much larger than B due to their differentiated nature. Typical examples of morphology with each batch are shown. Batch B was the serum batch finally purchased. Scale bar = 1 mm (AP) and 0.1 mm (Morphology).

- Any other evidence of differentiation, e.g. occasionally 'beating' cells (i.e. resembling contracting heart muscle) has been observed with certain FCS batches.

3.3.4. Staining ES Cells

1. Remove medium from well 1 only and wash with 2 mL PBS.
2. Remove PBS and fix cells in well 1 for 10 min in 2 mL acetone/methanol fixative.
3. Wash the well with 2 mL PBS.
4. Flood the well with 2 mL BCIP/NBT and incubate for 25 min at 37°C.
5. Remove medium from wells 2 to 6 and wash each of these wells with 2 mL PBS.
6. Aspirate PBS from wells 2 to 6 and replace with 2 mL 2% methylene blue.
7. After 5 min, remove the stain in all six wells by simply inverting the plate.
8. Wash the plate in water by dipping in a sink filled with tap water.
9. Leave to dry overnight.

3.3.5. Scoring

1. Differentiation: Combine your previous analysis of the colonies with the intensity of staining for alkaline phosphatase activity (well 1). Undifferentiated stem cells stain strongly, and staining is reduced in more differentiated cells. The number of strongly stained colonies can be quantified and compared between plates (*see* **Fig. 4**).
2. Plating efficiency: Estimate the plating efficiency at 15% by taking the average of the counts in wells 2, 3, and 4, assuming that 1,000 ES cells were plated in each well. A good serum should have a plating efficiency of between 5 and 20%. Also calculate the plating efficiency from well 5 (10% FCS) and compare the reduction in plating efficiency.
3. Toxicity: Compare the plating efficiency at 15% with that of 30%. Any marked reduction at 30% indicates toxicity of the serum. Also note colony size on the 30% plate. Smaller colonies are also an indicator of toxicity.

3.3.6. Phase 2 Testing

1. Select the best two FCS batches and culture ES cells in a single well of a 6-well plate for a minimum of two passages in standard ES cell medium containing FCS and feeders. Seed the same number of cells for each batch.
2. Count the number of cells at each subsequent passage and examine the cells daily for evidence of differentiation. Choose the batch with the best growth characteristics and with the least signs of differentiated ES cells.

3. Place an order for this batch either to be delivered on a monthly basis or as a single bulk purchase.

3.4. Basic Karyotyping of ES Cell Lines

The value of karyotyping of targeted ES cell lines is debated. For targeted ES cell clones, some labs do not perform karyotyping but proceed directly to the blastocyst injection stage – either the cells go germ line or they do not. However, major karyotypic abnormalities (loss of chromosomes or large translocations) in ES cell lines appear to be common. In one study, 7 out of 9 mouse ES cell lines studied exhibited aneuploid rates of between 18 and 60% *(12)*. It has been reported that when more than 50% of the ES cells were aneuploid, germ line transmission was not obtained *(13)*. Moreover, karyotypic abnormalities might affect the phenotype of the mice generated. Trisomy of chromosome 8 seems to be a common abnormality *(14)* perhaps implying that it confers a growth advantage on these cells. Thus, basic karyotyping of clones that are to be used for blastocyst injection or morula aggregation increases efficiency and thereby reduces overall costs and numbers of mice used. If new ES cell lines are being derived or received from elsewhere that will be used for numerous gene-targeting projects, we strongly suggest obtaining the detailed karyotype of the cells by a method such as G-banding or SKY (Spectral Karyotyping)/M-FISH (multi-colour fluorescence in situ hybridization). However, these are not trivial methods and significant experience and patience is needed to identify individual mouse chromosomes and subtle translocations/deletions. It is best to pass your ES cell line on to experts for this sort of analysis. At least be certain of a euploid chromosome number (i.e. 40) before proceeding with use of a new ES cell line for gene-targeting experiments.

3.4.1. Preparation of Metaphase Spreads

1. Plate ES cells onto a 100-mm dish that has been previously treated with 0.1% (w/v) gelatin. Neither a feeder layer nor LIF is required.
2. Grow until approximately 60% confluent.
3. Replace medium with 10 mL pre-warmed ES cell culture medium containing 10 ng/mL colcemid (1/1,000 dilution of stock solution).
4. Incubate for 2–3 h at 37°C (*see* **Note 14**).
5. Trypsinize the cells as normal and wash in 10 mL PBS.
6. Remove the PBS and resuspend the cells dropwise in 2 mL 0.075 M KCl (the solution should be pre-warmed to 37°C). After adding a few drops, gently flick the pellet to assist resuspension. Then continue dropwise addition while gently shaking the tube to mix.

7. Add a further 8 mL of pre-warmed 0.075 M KCl (no need to add dropwise now) and invert the tube to mix.
8. Incubate for six min in a water bath at 37°C.
9. Centrifuge at 1,100 × *g* for 5 min.
10. Remove the KCl and gently resuspend the pellet in 10 mL fixative. Again for the first 2 mL, do this dropwise and flick the pellet to assist in resuspension. Do this slowly and flick after the addition of each few drops to keep the cells evenly suspended and prevent any clumping.
11. Centrifuge at 1,100 × *g* for 5 min.
12. Repeat **steps 10** and **11** twice more.
13. Remove most of the fixative, leaving about 0.5 mL. Gently resuspend the cells by flicking the pellet.
14. Take up the cells in a glass Pasteur pipette and drop them from a height of 12–18 in. onto a clean glass slide (*see* **Note 15**).
15. Place in a drying rack and air dry for at least 1 h. Store at room temperature.

3.4.2. Giemsa Staining of Chromosomes and Counting

1. Prepare 50 mL of staining solution by mixing 2 mL Giemsa stain stock solution with 48 mL Gurr's buffer and pour into a Coplin jar.
2. Place slides in the jar and incubate for 4 min.
3. Remove slides using forceps and transfer to a Coplin jar containing tap water.
4. Place under running tap water until clear.
5. Rinse the Coplin jar several times by filling with distilled water.
6. Air dry.
7. Add two drops of mounting medium to the slide and place a coverslip gently on top. Do not press down but allow the mounting medium to slowly fill the coverslip area.

3.4.3. Analysis

1. Examine a minimum of 30 good metaphase spreads. Ensure the chromosomes are well spread out and not too compacted. A good example is shown in **Fig. 5a**.
2. Count the number of chromosomes and also note any gross abnormalities, e.g. large translocations should be obvious.
3. A reasonable ES cell line should possess a euploid ($2n = 40$) count in at least 70% of cells examined. Examples of typical results are shown in **Fig. 5b**. There is a published table of confidence limits for detecting chromosomal mosaicism *(15)* that can be of assistance when interpreting results.

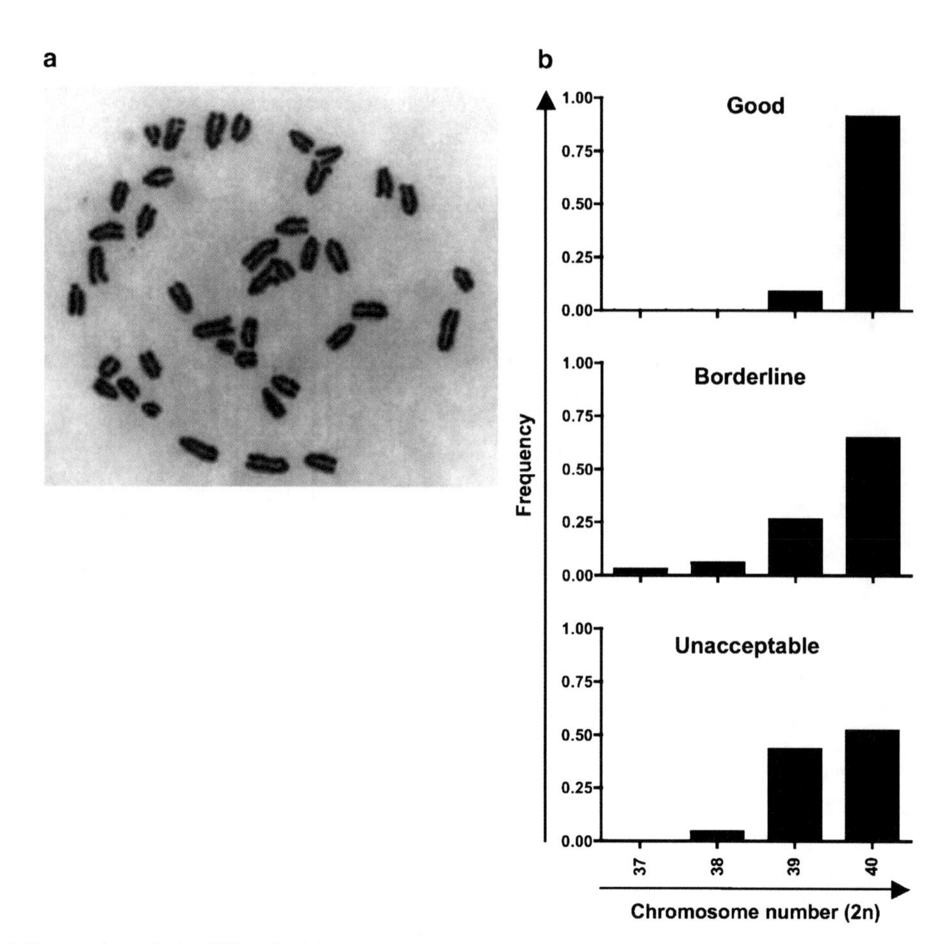

Fig. 5. Karyotypic analysis of ES cells. (**a**) A typical example of a suitable metaphase spread obtained from an ES cell line to use for counting chromosomes. (**b**) Examples of chromosome counts for targeted ES cell lines. All three targeted clones were obtained from the same parent cell line – PC3 *(20)*. The unacceptable cell line is probably a chimaera of two different ES cell types – it was subsequently revealed that the $2n = 39$ cells lacked the Y chromosome.

As an example, if 30 identical $2n = 40$ karyotypes are obtained, then one can be 95% confident that an abnormal karyotype present in 10% of the cells would have been detected.

3.5. Testing ES Cells for Presence of Mouse Pathogens

ES cells end up contributing to mice, so testing ES cells for pathogens is critical to avoid infection of mouse colonies. Infection of ES cell lines is probably uncommon *(16)* though some ES cell lines have been reported to carry *Mycoplasma* and MHV *(17)*. Traditionally, testing of such samples for infectious agents used the Mouse Antibody Production (MAP) test, which involved inoculating a sample of cells into mice and examining the plasma for presence of antibodies directed against the infectious agent several weeks later. PCR-based assays are growing in popularity as a cheap, efficient, and reliable alternative. We describe PCR-based assays to detect the following viruses: Mouse hepatitis virus (MHV), Minute virus of mice (MVM), Sendai virus (SV), Pneumonia virus of mice (PVM), and Theiler's murine encephalomyelitis virus (TMEV), as well as the main murine *Mycoplasma* pathogens - *M. pulmonis*, *M. arthritidis*, *M. neurolyticum*, and *M. collis(10, 18)*. As the viruses are all RNA viruses, cDNA must first be synthesized for use in the PCR assay.

3.5.1. Preparation of Genomic DNA for Mycoplasma Assay

1. Harvest up to 10^6 ES cells and place in a microfuge tube. The exact number of cells is not important, e.g. one well of a 6-well plate would be ample.
2. Pellet by centrifugation at 1,000 × *g* for 5 min.
3. Remove supernatant and add 150 μL of lysis buffer.
4. Incubate overnight at 55°C (with gentle shaking if possible).
5. Vortex briefly next morning.
6. Heat inactivate the proteinase K by incubation at 100°C for 12 min (*see* **Note 16**).
7. Add 600 μL sterile water and vortex briefly to mix.
8. Store at 4°C.

3.5.2. Preparation of cDNA for Virus Assay

1. Isolate total cellular RNA from 10^7 ES cells using a suitable method (*see* **Note 17**). Store RNA dissolved in RNAse-free water at –20°C until ready to synthesize cDNA.
2. Combine 5 μg RNA, 0.2 μg random hexamers (or nonamers) and make up to 14 μL with RNase-free water.
3. Incubate at 65°C for 5 min and chill on ice.
4. Add 2 μL 5× RT buffer, 2 μL 10 mM dNTPs, 1 μL RNase inhibitor, and 1 μL (200U) MMLV reverse transcriptase.
5. Incubate at 42°C for 1 h.
6. Heat inactivate for 5 min at 95°C and chill on ice. Store at –20°C.

3.5.3. PCR Assay

1. For each ES cell line to be assayed, there are eight primer pairs to use in separate reactions (**Table 3**).
2. For each pathogen primer pair, four tubes, each containing 12 μL PCR reaction mixture, are required:
   (a) The positive control plasmid alone.
   (b) The test sample.
   (c) The test sample and the positive control sample together.
   (d) A negative control with the primer pair only and no other exogenous DNA added.
3. Add 3 μL cDNA for each virus PCR or 3 μL genomic DNA for the *Mycoplasma* PCR to tubes (b) and (c). Add 1 ng plasmid DNA to tubes (a) and (c). Add 3 μL ultrapure water to tube (d).

Table 3
Primer pairs used for amplification of target sequences

Target infectious agent or control	Forward primer (5′-3′)	Size of amplicon (bp)	Positive control plasmid[a]
MHV	CACGAGCCGTAGCATGTTTA CGCATACACGCAATTGAAAA	100	pCR2.1-MHV
MVM	AAACTAAGCGCGGCAGAAT TCCAGTCCTCTGGTGAGGTT	99	pCR2.1-MVM
PVM	GGCCCAGTCTAGTGTCAAGG CCCCACCTCAGCATAACTTG	99	pCR2.1-PV
SV	GGGTTCCAGGATCTTTGGTT GGGATGGTGCTGATATCCTG	94	pCR2.1-SV
TMEV	CCCTACGGACCTTCTTTGTG GAGCGGTACGTCAGTCCAGT	100	pCR4-TMEV
Mycoplasma[b]	ACACCATGGGAGCTGGTAAT CCCACGTTCTCGTAGGGATA	145	pCR4-MycoR
D1Mit137	CCAGAAAATAGAATGAGCACAGG CCAGAAAATAGAATGAGCACAGG	187	N/A
Actb	ACCAACTGGGACGACATGGAGAAGA TACGACCAGAGGCATACAGGGACA	190	N/A

[a]As described *(10)*
[b]These primers are able to amplify those *Mycoplasma* species that are the main rodent pathogens

4. Control PCRs are required to test for genomic DNA and cDNA amplification using the D1Mit137 and *Actb* primer pairs, respectively (*see* **Note 18**). For these, only two reaction tubes are required – one containing 3 μL genomic DNA or cDNA, and one containing ultrapure water as a negative control.
5. PCR conditions for all primer pairs are 95°C for 2 min followed by 40 cycles of 95°C for 30 s, 55°C for 30 s, and 72°C for 30 s.
6. Products are separated by electrophoresis in 3% (w/v) TBE agarose gels alongside an appropriate DNA marker.

4. Notes

1. Some labs culture their feeder cells using the same batch of FCS they use for ES cells. In our experience, this has not been necessary and we use a cheap source of standard FCS for culture of feeders.
2. Choose a strain of mouse that will be most economical for all your requirements. For gene-targeting experiments, it is more efficient to isolate MEFs from a mouse line that yields MEFs resistant to commonly used cytotoxic drugs. These MEFs can then be used for culturing wild-type ES cells and gene targeted lines. In contrast, if you isolate MEFs from a wild-type mouse strain, they cannot be used for selection of drug-resistant ES cells, so you will need at least two lines of MEFs. We use the Tg(DR4)1Jae/J line developed by Rudolf Jaenisch's lab *(19)* and available from the Jackson Laboratories (Strain No. 003208). This is resistant to hygromycin, G418, puromycin, and 6-thioguanine. The same applies if you are using STO cells as feeders; it is better to use an STO cell line that is resistant to G418, such as the cell line SNL (available from the Wellcome Trust Sanger Institute, Hinxton, UK).
3. If there are a large number of embryos to harvest, the solution may be very viscous at this stage. This is due to lysed cells releasing their DNA. For this reason, do not centrifuge the cells before plating as the DNA and the cell pellet become intertwined somehow and it is hard to remove the supernatant without removing the pellet accidentally. When the medium is changed on the following day, any genomic DNA clumps will be removed. If you need to resuspend the cells in fresh medium before plating, then take up the solution in a 20-mL syringe and push out through an 18-guage needle. Repeat four or five times. This tends to shear the genomic DNA and assist the pelleting of cells by centrifugation.
4. Although we have only passaged the primary MEFs two times before arresting growth, it is possible to passage the MEFs a further two or three times (splitting the cells 1:4 at each passage) and still obtain feeders that will maintain ES cells appropriately. However, by about passage 6, the cells begin to senesce and are no longer suitable. We prefer to use cells at earlier passages to avoid any possibility that the cells will senesce during expansion.
5. Mitomycin C is toxic. Wear gloves throughout and handle with care. Waste must be disposed of in accordance with local guidelines.

6. ES cells will not attach to tissue culture dishes unless they have been pre-coated with something to assist binding. Gelatin is a cheap coating agent and, thus, widely used. Coating plates with gelatin is not a pre-requisite if the ES cells are being plated on to a layer of feeder cells. In our experience, fibroblast cell lines typically used as feeders for ES cells attach well in the absence of gelatinization. Nevertheless, some labs prefer to coat plates with gelatin even when feeders are being used.
7. There is no need to heat-inactivate FCS used for ES cell culture. The FCS must be suitable for growth of ES cells and must, therefore, either be tested (*see* **Subheading 3.3**) or purchased already tested and certified for use with ES cells (e.g. from Invitrogen). An alternative is to use a serum-free replacement such as Knockout SR™ (Invitrogen) or ESGRO Complete™ Clonal Grade Medium (Millipore), as it is more consistent from batch to batch than FCS and supports ES cell growth well. Note that alternatives to batch testing your own FCS are expensive. We have also noticed that feeder cells produce more extracellular matrix when cultured in some types of serum replacement medium and, as a result, extra care is required to ensure that trypsinization disperses any clumps of feeder cells.
8. LIF is absolutely required if ES cells are grown without feeders. Although we include it in the presence of feeders as an added prevention against differentiation, some labs do not. An alternative is to use a feeder cell line that secretes LIF, such as the cell line SNL (available from Allan Bradley at the Wellcome Trust Sanger Institute).
9. If using serum-free replacement medium instead of FCS, there is no need to wash cells with PBS before trypsinization as there are no trypsin inhibitors present.
10. Be sure you have a single-cell suspension because clumps of ES cells are more likely to form large colonies that will differentiate. Check the plates under the microscope to ensure you have a single-cell suspension before centrifugation.
11. Most companies will supply a small aliquot (50–100 mL) free of charge for batch testing. Ask them to hold enough of that batch for an agreed period of time until the testing is complete and you have had a chance to analyse the results. This requires 6–8 weeks as some suppliers are slower at delivering batches than others. We obtain 2 or 3 batches of FCS from four different suppliers for each round of testing. Some suppliers will even identify batches for testing that have had good results when tested by others for ES cell growth potential, which increases your chances of finding a good batch significantly. The amount to reserve will obviously depend

on your requirements, but even a small gene-targeting lab would purchase at least 10 L.

12. Different ES cell lines may not necessarily respond in the same way to a particular batch of serum, so, ideally, you should test each ES cell line you wish to use. However, this is often impractical. We test all samples on the most common ES cell line in use in our lab first, then select the best two or three batches, and retest on a second ES cell line.
13. Some labs test their FCS on ES cells in the absence of feeders. However, if you intend using feeders in the culture of your ES cells, it is preferable to test the FCS with feeders.
14. The treatment time with colcemid can vary from as short as 15 min up to several hours. Colcemid acts by preventing formation of the mitotic spindle, thereby arresting cells in metaphase. The longer the treatment time in colcemid, the more cells that will be in metaphase; however, the chromosomes will also be more contracted making characterization of individual chromosomes more difficult. Even the ability to identify large translocations after simple Giemsa staining might be compromised if the chromosomes are too contracted. A 2-hour incubation is a good starting point but the time may have to be varied.
15. This step is important. The cells need to be dropped from a height to allow the cell membrane to burst and to assist in the spreading of the chromosomes. Practice it a few times as the biggest problem is your drops missing the slide and ending up on the lab floor.
16. The tube lids can open at this stage if not well secured.
17. We use a commercial kit such as the RNeasy Mini Kit (Qiagen). Commercial kits are also available for the cDNA synthesis step.
18. The alternative is to use both the control and the pathogen primer pairs in the same reaction tube. The procedure and expected results are outlined in the original publication *(10)*. However, co-amplification can vary greatly with the type of *Taq* DNA polymerase and reaction conditions used, so more time is needed at the outset to define the exact reaction conditions for co-amplification. However, it is worth the effort.

Acknowledgements

We thank Monica Neilan for assistance with ES cell culture and Adam Steinberg and the Biochemistry Medialab of the University of Wisconsin for providing the mouse image.

References

1. Martin, G.R. (1981). Isolation of a pluripotent cell line from early mouse embryos cultured in medium conditioned by teratocarcinoma stem cells. *Proc. Natl. Acad. Sci. U.S.A.* **78**, 7634–7638
2. Evans, M.J. and Kaufman, M.H. (1981). Establishment in culture of pluripotential cells from mouse embryos. *Nature* **292**, 154–156
3. Williams, R.L., Hilton, D.J., Pease, S., Willson, T.A, Stewart, C.L., Gearing, D.P., Wagner, E.F., Metcalf, D., Nicola, N.A., and Gough, N.M. (1988). Myeloid leukaemia inhibitory factor maintains the developmental potential of embryonic stem cells. *Nature* **336**, 684–687
4. Smith, A.G., Heath, J.K., Donaldson, D.D., Wong, G.G., Moreau, J., Stahl, M., and Rogers, D. (1988). Inhibition of pluripotential embryonic stem cell differentiation by purified polypeptides. *Nature* **336**, 688–690
5. Solter, D. (2006). From teratocarcinomas to embryonic stem cells and beyond: a history of embryonic stem cell research. *Nat. Rev. Genet.* **7**, 319–327
6. Nagy, A. (2008). URL: www.mshri.on.ca/nagy/r1.htm retrieved on 28th February 2008
7. Wellcome Trust Sanger Institute. (2008). URL: www.sanger.ac.uk/Software/archives/mouse-help.shtml retrieved on 28th February 2008
8. Keskintepe, L., Norris, K., Pacholczyk, G., Dederscheck, S.M., and Eroglu, A. (2007). Derivation and comparison of C57BL/6 embryonic stem cells to a widely used 129 embryonic stem cell line. *Transgenic Res.* **16**, 751–758
9. Blake, M.S., Johnston, K.H., Russell-Jones, G.J., and Gotschlich, E.C. (1984). A rapid, sensitive method for detection of alkaline phosphatase-conjugated anti-antibody on Western blots. *Anal. Biochem.* **136**, 175–179
10. Ayral, A., Clarkson, S., Cheeseman, M., Wells, S., and Dear, T. (2006). A panel of optimized primers and positive control DNAs for PCR detection of common biological contaminants in mouse cell lines and tissue samples. *Lab Anim.* **35**, 31–36
11. Udy, G.B., Parkes, B.D., and Wells, D.N. (1997). ES cell cycle rates affect gene targeting frequencies. *Exp. Cell Res.* **231**, 296–301
12. Balbach, S.T., Jauch, A., Böhm-Steuer, B., Cavaleri, F.M., Han, Y.M., and Boiani M. (2007). Chromosome stability differs in cloned mouse embryos and derivative ES cells. *Dev. Biol.* **308,** 309–321
13. Longo, L., Bygrave, A., Grosveld, F.G., and Pandolfi, P.P. (1997). The chromosome make-up of mouse embryonic stem cells is predictive of somatic and germ cell chimaerism. *Transgenic Res.* **6**, 321–328
14. Liu, X., Wu, H., Loring, J., Hormuzdi, S., Disteche, C.M., Bornstein, P., and Jaenisch R. (1997). Trisomy eight in ES cells is a common potential problem in gene targeting and interferes with germ line transmission. *Dev. Dyn.* **209**, 85–91
15. Hook, E.B. (1977). Exclusion of chromosomal mosaicism: Tables of 90%, 95%, and 99% confidence limits and comments on use. *Am. J. Hum. Genet.* **29**, 94–97
16. Nicklas, W., and Weiss. J. (2000). Survey of embryonic stem cells for murine infective agents. *Comp. Med.* **50**, 410–411
17. Kyuwa, S. (1997). Replication of murine coronaviruses in mouse embryonic stem cell lines in vitro. *Exp. Anim.* **46**, 311–313
18. Lindsey, J., Boorman, G., Collins, M., Hsu, C.-K., Van Hoosier., Jr.,G., and Wagner, J. (1991) Infectious Diseases in Mice and Rats. National Academy Press, Washington. DC. pp. 42–48
19. Tucker, K.L., Wang, Y., Dausman, J., and Jaenisch, R. (1997). A transgenic mouse strain expressing four drug-selectable marker genes. *Nucleic Acids Res.* **25**, 3745–3746
20. O'Gorman, S., Dagenais, N.A., Qian, M., and Marchuk, Y. (1997). Protamine-Cre recombinase transgenes efficiently recombine target sequences in the male germ line of mice, but not in embryonic stem cells. *Proc. Natl. Acad. Sci. U.S.A.* **94**, 14602–14607

Chapter 12

Targeting Embryonic Stem Cells

Roland H. Friedel

Summary

Gene targeting of mouse embryonic stem cells facilitates the deliberate engineering of the mouse genome. The gene targeting vector is introduced into cells by electroporation, and cell clones carrying the targeting vector are obtained by drug selection. Clones can be screened effectively for correct recombination events by either Southern blot analysis with external and internal probes, or by long-range PCR. Positive clones are subsequently expanded and prepared for the generation of chimeric mice.

Key words: Gene Targeting, ES cells, Southern blot analysis, Long-range PCR

1. Introduction

Gene targeting in mouse embryonic stem (ES) cells is a key technology for the genetic analysis of mammalian biology. The homologous recombination between a DNA targeting vector and the target locus allows for the designed modification of the genome. This technology has been developed in the 1980s by the pioneering work of the Evans, Smithies, and Capecchi laboratories *(1–3)*, and has since been successfully applied by thousands of laboratories worldwide.

This chapter provides protocols for the procedures of electroporating ES cells with a targeting vector, the identification of correctly targeted ES cell clones by either Southern blot analysis or long-range PCR, and the preparation of a positive clone for the generation of chimeric mice. The workflow of the experiments is shown in **Fig. 1**.

A correct gene targeting event meets the following three criteria: First, homologous recombination has occurred between

Elizabeth J. Cartwright (ed.), *Transgenesis Techniques*, Methods in Molecular Biology, vol. 561
DOI 10.1007/978-1-60327-019-9_12, © Humana Press, a part of Springer Science+Business Media, LLC 2009

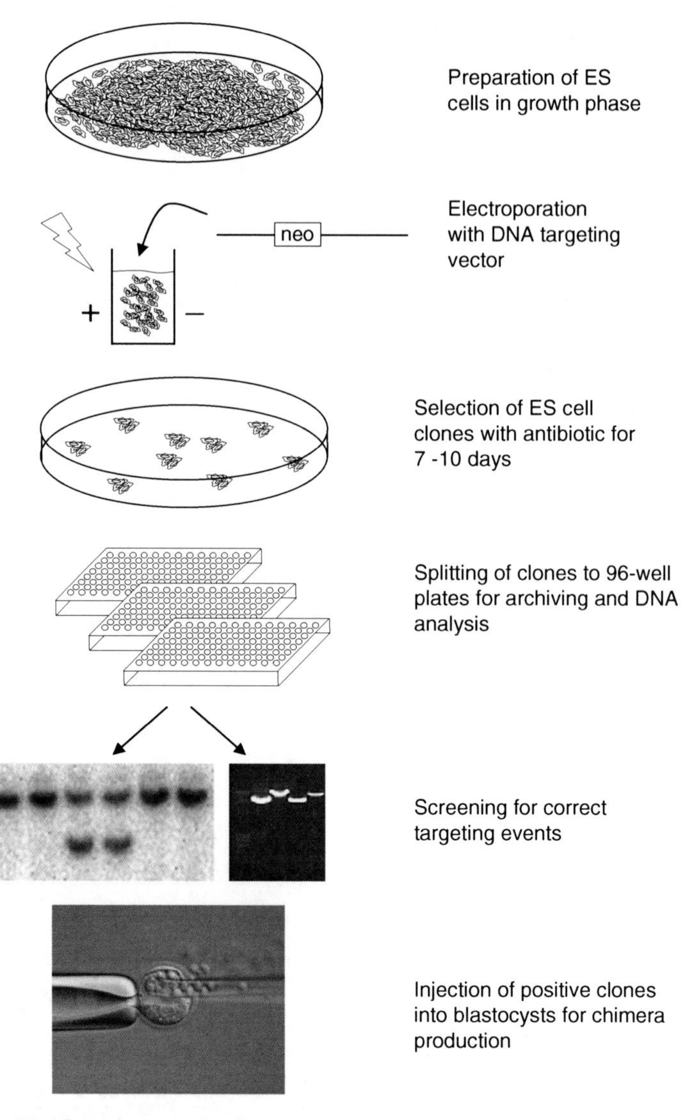

Fig. 1. Workflow of a gene targeting experiment.

the 5′ homology arm of the targeting vector and the target locus. This can be either tested by Southern blot analysis with a probe that lies outside the homology arm ("5′ external probe" **Fig. 2a**) or by long-range PCR with a primer outside the homology arm ("5′ external primer" **Fig. 2b**). Second, homologous recombination has occurred between the 3′ homology arm of the targeting vector and the target locus. This can again be tested with a Southern probe ("3′ external probe" **Fig. 2a**) or PCR primer ("3′ external primer" **Fig. 2b**). Third, no integration of the targeting vector into the mouse genome, besides the target locus, has occurred. This can be tested by Southern blot analysis with

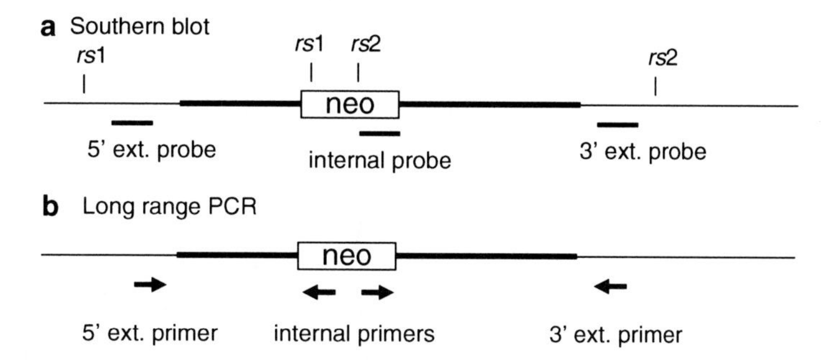

Fig. 2. Strategies for the screening of targeted ES cell clones. (**a**) For Southern blot analysis with an external probe (ext. probe), restriction sites are chosen that allow a correctly targeted allele to be distinguished from a wild-type allele. The internal probe will detect illegitimate integrations of the vector. The homology arms of the vector are indicated by a *bold line*. rs1, rs2: restriction sites. (**b**) For long-range PCR, external primers (ext. primer) are used in combination with internal primers to detect correct recombinants.

an "internal probe" **(Fig. 2a)**. In general, it is advisable to test for all the three criteria. However, depending on the specific experimental requirements, it may be sufficient to screen only for correct 5′ and/or 3′ recombination.

Both Southern blot analysis and long-range PCR are equally well suited for the detection of the correct recombination of the homology arms (*see* **Note 1**). An investigator should decide which method to use depending on the equipment and technical expertise available. Testing for single insertion of the targeting vector, however, can only be done by Southern blot analysis with an internal probe, which can detect illegitimate integrations of the vector. It is possible to combine Southern blot and PCR analysis for the screening of clones. For example, a first screen for the correct 5′ recombination can be performed by Southern blot analysis of clones from 96-well plates. Then, positive clones are expanded and tested for correct 3′ recombination by long-range PCR and for single integration by an internal Southern probe.

2. Materials

2.1. ES Cell Culture

1. ES cell media and trypsin solution according to the requirements of cell line used.
2. Phosphate-buffered saline (PBS).
3. Cell culture plasticware (e.g. Nunc, Costar): 10 cm dishes, 96-well flat-bottom plates, 96-well V-bottom plates.

4. Electroporation apparatus, e.g. "Gene Pulser Xcell" (Biorad).
5. Electroporation cuvette, 0.4 mm gap width (Biorad).
6. Geneticin (G418). Prepare a 100 mg/mL stock solution from the G418 powder (Invitrogen) in PBS according to the active concentration of the batch (indicated on each vial) (*see* **Note 2**). Filter through 0.2 μm syringe filter. Store aliquots at –20°C. Keep working stock at 4°C for up to 1 month.
7. 2× ES cell freezing medium: ES cell medium containing 20% DMSO.

2.2. Southern Blot Analysis

1. Sarcosyl–DNA lysis buffer: 10 mM Tris–HCl, pH 7.5, 10 mM EDTA, 10 mM NaCl, 0.5% sarcosyl (Lauroylsarcosine, Sigma). Store at room temperature.
2. Proteinase K (Roche): Prepare 10 mg/mL stock solution and store aliquots at –20°C.
3. Salt–ethanol solution: Add 150 μL of 5 M NaCl to 10 mL of ice-cold absolute ethanol.
4. 70% ethanol: Store at room temperature.
5. Spermidine (Sigma): Prepare 0.1 M stock solution and store aliquots at –20°C.
6. RNaseA (Roche): Prepare 10 mg/mL stock solution. Boil RNaseA stock solution for 10 min to remove any contaminating DNase activity. Store aliquots at –20°C.
7. Restriction digest mix for one well of a 96-well plate: 1× restriction buffer, 1 mM spermidine, 100 μg/mL BSA, 50 μg/mL RNase A, 10 units restriction enzyme; add H_2O to 30 μL.
8. 6× DNA loading buffer: 15% ficoll in TE, 0.02% orange-G (Sigma).
9. Hybond-XL nylon membrane (GE Healthcare).
10. Whatman 3 M filter paper.
11. Denaturation buffer: 1.5 M NaCl, 0.5 M NaOH, store at room temperature.
12. Neutralization buffer: 0.5 M NaCl, 0.1 M Tris–HCl, pH 7.5, store at room temperature.
13. 20× SSC buffer: 3 M NaCl, 0.3 M —sodium citrate, store at room temperature.
14. UV Cross-linker (Stratagene).
15. alpha-^{32}P-dCTP (10 μCi/μL).
16. High prime probe labelling kit (Roche).
17. Mini spin prep purification column (Roche).

18. Fish sperm DNA (Roche), prepare 10 mg/mL solution, store at –20°C.
19. Quickhyb hybridization buffer (Stratagene).

2.3. Long-Range PCR

1. PCR machine with 384-well block, e.g. "DNA engine" (Biorad).
2. SDS–DNA lysis buffer: 10 mM Tris-HCl, pH7.5, 10 mM EDTA, 10 mM NaCl, 0.5% SDS. Store at room temperature.
3. Proteinase K (Roche). Prepare 10 mg/mL stock solution and store aliquots at –20°C.
4. RNaseA (Roche). Prepare 20 mg/mL stock solution. Boil RNaseA stock solution for 10 min to remove any contaminating DNase activity and store aliquots at –20°C.
5. Salt–ethanol solution: Add 150 μL of 5 M NaCl to 10 mL of ice-cold absolute ethanol.
6. PCR reagent: AccuPrime High Fidelity Taq DNA Polymerase (Invitrogen).
7. 5× Cresol Red loading buffer: Mix 19 mL of a 0.1% Cresol Red (Sigma) stock solution with 30 g sucrose, add H_2O to 50 mL. Dissolve at 65°C and store aliquots at –20°C.
8. 384-well PCR plates and sealing foil (Abgene).

3. Methods

The protocols for ES cell culture before and after electroporation are described here only briefly. The specific recipes for the ES cell medium, the trypsin solution, and the trypsinization time differ among various ES cell lines. Please refer for details to the Chapter 11 in this book or to protocols for the specific ES cell line.

The 5′ and 3′ external probes for the Southern blot should be about 1 kb in length, and designed to detect bands of different sizes for the wild type and the targeted alleles (**Fig. 2a**). The internal probe is typically a fragment of the neo cassette or another vector-specific element. To facilitate the experimental design, a DNA sequence editing software should be used to generate sequence files of the wild-type locus and of the predicted targeted locus, which can then be used to choose appropriate restriction enzymes and to predict the size of the resulting bands.

It is important for the success of the Southern analysis that the external probe does not contain repetitive elements of the mouse genome (which would hybridize to multiple fragments). The first control to avoid repetitive elements is to align the probe sequence against the mouse genome sequence and only a single high-quality hit should be obtained (this can be done, for example, by using the "BLAT" function at the website genome.ucsc.edu). The second control is to perform a test Southern on wild-type DNA. Only a single band of the expected size should be obtained.

The external Southern probes can be easily generated by PCR from a preparation of genomic DNA of wild-type ES cells. The internal probe is generated either by PCR or by restriction digest of the targeting vector plasmid. The DNA fragments used for probe labelling should be purified by excision from an agarose gel.

For long-range PCR analysis, two external primers are designed that lie outside the homology regions and two internal primers that are located inside the targeting cassette **(Fig. 2b)**. The primers should be 24-mers, with a G/C content of ≥50%. Align primer sequences with the mouse genome sequence to avoid repetitive DNA elements. The external primers can also be validated with a PCR on wild-type DNA. The size of the PCR products can vary between 1.5 and 8 kb, depending on the length of the homology arms of the targeting vector. We routinely genotype ES cell clones with PCR products of about 6 kb.

3.1. ES Cell Culture Before and After Electroporation

3.1.1. Electroporation of Mouse ES Cells

1. The targeting vector DNA should be linearized by restriction digest the day before electroporation. For a standard targeting experiment with a promoter-driven resistance cassette (e.g. PGK-neo), 10 μg DNA should be prepared (*see* **Note 3**). Precipitate the digest and resuspend DNA in 50 μL PBS. Allow for 1 h of resuspension of DNA at room temperature and store at 4°C overnight.
2. Expand ES cells for electroporation into two 10 cm dishes. The cells should have been split at least twice after initial thawing of the frozen vial. Optimal results are obtained when cells have reached almost confluence with dividing cells still present (subconfluence) at the time of electroporation.
3. On the day of electroporation, change media on cells in the morning, 2–4 h before electroporation.
4. Trypsinize cells, spin down, resuspend in 10 mL PBS, and count. Spin down 1×10^7 cells and resuspend with PBS to a total volume of 750 μL. Mix with 50 μL linearized targeting vector DNA.
5. Set parameters of electroporation apparatus to 800 V, 10 μF and exponential decay mode (*see* **Note 4**). Transfer 800 μL of cell suspension into an electroporation cuvette and electroporate. The resulting discharge time constant should be in the range of 0.1–0.3 ms.

6. Let cells rest at room temperature for 10 min.
7. Distribute cells on five prepared 10 cm dishes with regular medium. To get plates with optimal clone density, plate one dish with 5×10^6 cells, one dish with 2×10^6 cells, and three dishes with 1×10^6 cells (*see* **Note 5**).

3.1.2. Selection, Picking, and Expansion of ES Cell Clones

1. The day after electroporation, switch to media with selective antibiotic. For targeting with PGK-neo cassette, use 200 μg/mL G418 in the media (*see* **Note 6**). Change media every day. Selection time depends on the cell line, and is typically between 7 and 10 days. Clones should be picked when they reach the size of a standard micropipette tip. Extended selection times may lead to differentiation of ES clones.
2. For picking of clones, wash cells on selection plate once with PBS and cover them again with PBS. Pick clones with a Pipetman P20 and a volume of 10 μL by gently scratching them from the bottom of the plate. Transfer picked clones into a 96-well V-bottom plate. For a typical gene targeting experiment, at least three 96-well plates should be picked.
3. Add 30 μL of Trypsin to each well. Incubate for 10 min at 37°C. Add 150 μL of medium (+G418) to each well, triturate cells thoroughly to generate a single cell suspension. Transfer suspension to a prepared 96-well flat-bottom plate.
4. Change medium (+G418) on 96-well plates every day, 150 μL per well.
5. After 2–4 days, when the majority of wells have reached subconfluence (do not wait for all clones, you may loose those that grow rapidly), split cells to three 96-well plates (copies A and B for freezing cells, copy C for further splitting, to be used for DNA analysis).
6. Freeze copies A and B when the majority of wells have reached subconfluence. For freezing, wash cells with PBS, trypsinize cells with 50 μL trypsin and add 50 μL medium and 100 μL 2× freezing medium. Cover wells with 30 μL mineral oil. Freeze in Styrofoam box in –80°C freezer and store for up to 3 months.
7. Split copy C to three copies (C1, C2, and C3) for DNA analysis. Grow copies C1-C3 for 2–3 extra days. The majority of wells should be very dense; turning the media yellow. Wash cells with PBS. Store plates at –20°C or process directly for Southern blot analysis or PCR.

3.2. Screening of Clones from 96-Well Plate by Southern Blot Analysis

1. For DNA lysis of cells in a 96-well plate, mix sarcosyl-DNA lysis buffer with Proteinase K stock solution to a final concentration of 1 mg/mL Proteinase K. Add 50 μL of the mix to each well. Incubate the plate in a humidified box at 60°C overnight (*see* **Note 7**).

2. The next day, for ethanol precipitation, mix ice-cold salt–ethanol solution thoroughly. Add 100 μL of salt-ethanol solution to each well. Let the plate sit at room temperature for 1 h, do not shake, vortex or spin the plate.
3. After DNA has precipitated, threads of DNA are visible on the bottom of the wells. Slowly invert the plate on paper towels to remove the salt–ethanol solution, the DNA precipitate will stick to the plastic. Drain excess liquid by blotting the plate on paper towels. Do not spin the plate. It is important to wash the plate three times with room temperature 70% ethanol to remove excess salt that was used for precipitation. Remove ethanol each time by slowly inverting the plate on a paper towel. After the final wash, blot the plate on paper towels and leave the plate inverted to dry. Plates can be stored at –20°C before further processing.
4. Prepare restriction digest mix, which contains in a volume of 30 μL: 1× restriction buffer (as recommended for enzyme), 1 mM spermidine, 100 μg/mL BSA, 50 μg/mL RNase A, 10 units restriction enzyme. Add 30 μL of the mix per well and incubate overnight in a humidified box at 37°C.
5. Prepare a 0.7% agarose gel with TAE buffer that can fit at least 100 samples, for example, by using two 50-tooth combs. To each well of the 96-well plate, add 6 μL of 6× DNA loading buffer. Load the digests on to the agarose gel; include a DNA size marker for each row. Run the gel to separate DNA fragments for 2–4 h at 4 V/cm (*see* **Note 8**).
6. Take a photo of the gel on a transilluminator. It is important to also take a photo of the DNA size marker with a ruler at the side of the gel so that the sizes of Southern blot bands can later be calculated by comparing this photo with the position of the bands on the gel.
7. Shake gel in denaturation buffer for 1 h (*see* **Note 9**). Be generous with the volume of the denaturation buffer to ensure complete denaturation.
8. Cut one sheet of Hybond-XL membrane and two sheets of Whatman paper to the size of the agarose gel. A tray with larger dimensions than the gel should be used as reservoir for the transfer. Fill the tray with denaturation buffer to a height of about 1 cm. Place the inverted gel tray of the agarose gel, as a transfer base, into the reservoir tray and cover it with a sheet of Whatman paper that connects to the denaturation buffer. Wet the paper with denaturation buffer and roll out air bubbles under the paper with a Pasteur pipette.
9. Place the denatured agarose gel upside-down onto the transfer base (*see* **Note 10**). Wet the Hybond-XL membrane and the two sheets of Whatman paper with water. Place successively over the

gel by using a pipette to roll out air bubbles. Add paper towels to a height of about 15 cm over the gel stack. Cover the stack of paper towels with a glass plate, put a weight of about 500 g on top (e.g. a catalog book). Seal uncovered sides of the base with plastic wrap to avoid sit shortcuts between the base and the paper towels. Transfer overnight at room temperature.

10. On the following day, disassemble the blot and discard the paper towels. While lifting the membrane from the gel, mark on the membrane the DNA side and the position of the combs with a pencil. Rinse the membrane for 10 min in neutralization buffer. Then rinse twice for 10 min in 2× SSC and dry the membrane on Whatman paper. Cross-link the membrane in a UV cross-linker. The membrane can be wrapped in plastic wrap and stored at −20°C until hybridization.
11. Prepare the radioactively labelled probe with a "High Prime" labelling Kit (*see* **Note 11**). For one labelling reaction, use 25 ng of purified DNA fragment and dilute it with H_2O to a total volume of 11 μL. Boil for 5 min to denature the DNA and chill on ice. Add 4 μL High prime mix and 5 μL ^{32}P-dCTP. Incubate the reaction for 10 min at 37°C. Add 30 μL H_2O, and purify the reaction product with a Sephadex-based mini Quick Spin column. This will yield about 50 μL of purified probe. Successful labelling of the probe can be controlled by measuring it's radioactivity with a scintillation counter.
12. Prehybridize the membrane in a roller bottle with 10–20 mL Quickhyb for 1 h at 65°C (*see* **Note 12**). The amount of hybridization buffer has to be adjusted according to the size of the membrane to allow sufficient coverage of the membrane by the buffer.
13. For a hybridization volume of 10–20 mL, usually one-half of the labelled probe should be added. This will yield a probe concentration of about $0.1–1 \times 10^6$ cpm/mL. Before adding the probe, boil it for 10 min in 500 μL of fish sperm DNA (10 mg/mL), then chill on ice. Add the denatured probe to Hyb buffer and hybridize at 65°C overnight.
14. The next day, wash the membrane once for 20 min in 2× SSC, 0.1% SDS at 65°C, and three times for 20 min in 0.2× SSC, 0.1% SDS at 65°C. Remove the membrane from the roller bottle and briefly dry on Whatman paper. Cover the membrane with plastic wrap and expose on a phosphoimager plate for 3–4 h. Alternatively, expose the membrane on X-ray film at −80°C overnight, develop the film and judge further exposition time by the signal strength obtained from overnight exposure.

3.3. Screening of Clones from 96-Well Plate by Long-Range PCR

1. If plates have been stored after the PBS wash at −20°C, they can be used directly for DNA lysis. If plates had not been frozen, freeze cells now for 20 min at −80°C to facilitate cell lysis.
2. For DNA lysis, mix SDS–DNA lysis buffer with Proteinase K to a final concentration of 1 mg/mL Proteinase K. Add 50 μL of the mix to each well. Incubate the plate in a humidified box at 55°C for at least 3 h or overnight (*see* **Note 7**).
3. Let the plate cool to room temperature and briefly spin down to prevent droplets spilling over to other wells. Add 5 μL of the RNaseA solution per well and incubate in a humidified box for 30 min at 37°C.
4. Briefly spin down the plate and add 150 μL of ice-cold salt–ethanol solution to each well for DNA precipitation. Let it sit at room temperature for 1 h, do not shake or vortex the plate.
5. After DNA has precipitated, threads of DNA are visible on the bottom of the wells. Invert the plate slowly on paper towels to drain the salt–ethanol solution, the DNA precipitate will stick to the plastic. Remove excess liquid by blotting plate on paper towels. Do not spin the plate. It is important to wash the plate three times with room temperature with 70%ethanol to remove the excess salt that was used for precipitation. Remove ethanol each time by slowly inverting the plate on a paper towel. After the final wash, blot the plate on paper towels and leave inverted to dry.
6. Add 50 μL TE buffer per well to resuspend the DNA. Incubate in a humidified box at 55°C overnight (*see* **Note 13**). After incubation, briefly spin down the plate to avoid spill over of droplets to other wells. Cover plates tightly with sealing foil to avoid evaporation, and store at 4°C.
7. For long-range PCR in a 384-well format, for each well prepare a reaction volume of 11 μL (*see* **Note 14**): 5.9 μL H_2O, 1.1 μL AccuPrime buffer II, 0.5 μL of forward primer (10 μM), 0.5 μL of reverse primer (10 μM), 2 μL of 5× Cresol Red loading buffer, 0.05 μL AccuPrime enzyme (*see* **Note 15**), and 1 μL of DNA preparation.
8. Run PCR with the following thermal protocol (*see* **Note 16**): Initial denaturation for 3 min at 93°C. Eight cycles with 15 s at 93°C, 30 s at 68°C (decrease by 1°C each cycle), and 6 min at 68°C (*see* **Note 17**). 30 cycles with 15 s at 93°C, 30 s at 60°C, and 6 min at 68°C (increase time by 20 s each cycle). Final extension for 7 min at 68°C, then hold at 10°C.

9. Analyze PCR products on a 0.8% agarose gel. Since Cresol Red loading buffer is already present in the PCR reaction, PCR products can be loaded directly on gel.

3.4. Expansion of Selected ES Cell Clones

1. If available, thaw several candidate clones from one 96-well plate, because the remaining clones on the plate will show reduced viability after repeated freezing and thawing.
2. Place the frozen 96-well plate in a 37°C incubator and thaw for several minutes. As soon as the cells have thawed, immediately take out the plate.
3. Transfer the entire contents of a well into a 24-well plate containing 2 mL medium (*see* **Note 18**). This serves to dilute the DMSO of the freezing media. At this point, it is no longer necessary to add selective antibiotic to the media.
4. Change media every day. After 3–5 days, when cells have reached subconfluence, transfer cells to a 6-well plate.
5. When cells have reached subconfluence, split 1:4 into one well of a 6-well plate, and the remaining cells into one 10 cm dish. The cells on the 6-well plate can be used for blastocyst injection 1 or 2 days after splitting, depending on the growth rate of the cells. The cells of the 10 cm plate can be frozen 1–3 days later into three cryo vials for archiving.
6. On injection/aggregation day, trypsinize the cells of the subconfluent 6-well plate and triturated thoroughly with 2 mL media by pipetting up and down 20 times. Transfer cell suspension into a 15 mL conical tube on ice. Add Hepes, pH 7.4, to a concentration of 10 mM to the cells to buffer the pH. Cells can now be transferred on ice to a transgenic facility for blastocyst injection or morula aggregation.

4. Notes

1. For gene targeting using modified BACs as targeting vectors *(4)*, Southern blot and long-range PCR cannot be used because of the large size of the homology arms. In this case, quantitative real-time PCR has to be performed with primers that are specific for the wild-type allele ("allele counting"). A targeted clone contains only one wild-type allele, and will therefore generate only 50% of the PCR product of a wild-type clone *(4)*.
2. We have observed inconsistent results with ready-made solutions of G418. It is recommended to prepare the G418 solution freshly from powder.

3. For promoterless targeting constructs, which generally yield fewer clones, electroporate 50 μg of DNA *(5)*.
4. These electroporation parameters work well for E14Tg2a ES cells and JM8 ES cells. Parameters may have to be adjusted for other ES cell lines.
5. For promoterless targeting constructs, two times 5 × 10^6 cells should be plated on 10 cm dishes.
6. The 200 μg/mL G418 concentration is for ES cells targeted with a PGK-neo selection cassette. For promoterless targeting, use 125 μg/mL G418 for E14 cells and 100 μg/mL for JM8 cells.
7. To generate a humidified box, paper towels are soaked generously with water and placed inside a plastic, sealable container. It is not necessary to seal the 96-well plates inside the humidified box. However, the box itself should be sealed tightly. If necessary, seal the box by wrapping with plastic foil.
8. Gel running times are variable and depend on the requirements of the experiment. The gel should run long enough to separate the different fragment sizes that are expected by the Southern blot analysis.
9. This chapter describes the alkali transfer method for Southern blotting. An alternative method is the neutral transfer protocol. For neutral transfer, rinse gel after denaturation briefly with H_2O and wash twice for 30 min with neutralization buffer. Use 20× SSC as the transfer buffer. The recommended membrane type for neutral transfer is Hybond-N+ (GE Healthcare).
10. Placing the gel in an upside-down orientation on the base helps to avoid air bubbles between the gel and membrane. Large gels can be moved by holding it between two old X-ray films. Very large gels can also be cut into pieces and assembled again on the transfer base.
11. We have obtained consistent results with ^{32}P labelling kits from Roche Applied Science. Labelling kits and Sephadex-based purification columns are also available from other companies.
12. Commercial hybridization buffers, like Quickhyb from Stratagene, can yield about double signal intensity when compared to standard buffers. However, satisfying results can also be obtained by using a Church hybridization buffer: 0.5 M Na_2HPO_4, 0.5 M NaH_2PO_4, 10 mg/mL BSA, 7% SDS, 1 mM EDTA, 100 μg/mL salmon sperm DNA.
13. It is very important for the success of the PCR that the DNA has been resuspended completely. This step should not be abbreviated.

14. For PCR reactions in 96-well format, adjust reaction volume to 25 μL.
15. We have obtained good results for long-range PCR from genomic DNA with the AccuPrime High Fidelity enzyme (Invitrogen) and the Expand High Fidelity enzyme (Roche Applied Sciences).
16. This PCR program is optimized for using 384-well PCR plates (Abgene) on "DNA Engine" PCR machines (MJ Research/Biorad). Temperatures, times, and cycle numbers may have to be adjusted for other PCR machines.
17. The extension time depends on the size of the expected PCR product. Calculate as initial extension 1 min/kb product. In the above protocol, 6 min have been calculated for an expected PCR product of 6 kb.
18. Thawing of a single well of a 96-well plate to a 24-well plate works well for E14 ES cells. For ES cell lines with less robust growth, thaw plate, spin down cells, aspirate carefully the media supernatant, resuspend with 150 μL media, and transfer to a new 96-well plate.

Acknowledgements

The author would like to thank Susanne Bourier, Rene Schadowski, and Claudia Seisenberger for the critical reading of the manuscript. This work was supported by the European Union FP6 framework program: EUCOMM, European Conditional Mouse Mutagenesis Program and the German Federal Ministry of Education & Research (BMBF) in the framework of the National Genome Research Network (NGFN), Grant 01GR0430.

References

1. Bradley, A., Evans, M., Kaufman, M. H., Robertson, E. (1984). Formation of germ-line chimaeras from embryo-derived teratocarcinoma cell lines. *Nature* **309**, 255–256
2. Doetschman, T., Gregg, R. G., Maeda, N., Hooper, M. L., Melton, D. W., Thompson, S., Smithies, O. (1987). Targetted correction of a mutant HPRT gene in mouse embryonic stem cells. *Nature* **330**, 576–578
3. Thomas, K.R. and M.R. Capecchi (1987). Site-directed mutagenesis by gene targeting in mouse embryo-derived stem cells. *Cell* **51**, 503–512
4. Valenzuela, D. M., Murphy, A. J., Frendewey, D., Gale, N. W., Economides, A. N., Auerbach, W., et al. (2003). High-throughput engineering of the mouse genome coupled with high-resolution expression analysis. *Nat. Biotechnol.* **21**, 652–659
5. Friedel, R. H., Plump, A., Lu, X., Spilker, K., Jolicoeur, C., Wong, K., et al., (2005). Gene targeting using a promoterless gene trap vector ("targeted trapping") is an efficient method to mutate a large fraction of genes. *Proc. Natl. Acad. Sci. U.S.A.* **102**, 13188–13193

Chapter 13

Generation of Chimeras by Microinjection

Anne Plück and Christian Klasen

Summary

Since the technique of introducing a targeted mutation ('gene targeting') into the mouse genome was published almost 20 years ago (Cell 51:503–512, 1987), the number of mouse mutants (mouse models) is increasing, especially after the advent of the full mouse genomic sequence in 2002 and the human genomic sequences in 2003 that reveals more and more large stretches of similarity between the two species at the genomic level. This chapter describes the tools and the experimental route of targeted manipulation by microinjection in the mouse using targeted embryonic stem cells (ES cells).

The techniques have become standardized over recent years (Nature 309:255–256, 1984; *Practical Approach*. IRL Press, Oxford, 254 pp, 1987; Science 240:1468–1475, 1988; *Practical Approach*. IRL Press, Oxford, New York, 1993; *Transgenic Animal Technology: A Laboratory Handbook*, 2nd edition. Academic Press, San Deigo, 2002; *Manipulating the Mouse Embryo – A Laboratory Manual*, 3rd edition. Cold Spring Harbor Laboratory Press, Cold Spring Harbor, NY, 2003) and basically two methods have been used to generate chimeric mice that transmit the mutation of interest via the ES cell genome to the offspring:

Microinjection of ES cells into blastocyst or morula stage embryos (this chapter) or aggregation of ES cells with morula stage embryos (*see* Chapter 14).Microinjection of ES cells into the blastocoel (cavity) of the blastocyst stage embryo and also morula injections using micropipettes driven by micromanipulators require sophisticated manual skills and an expensive phase contrast inverted microscope. Although most commonly used, it is quite expensive to establish this technique in a laboratory, in particular, if piezo- or laser- supported routes come into play. Although the establishment of germ-line potent ES cells was first published in 1981 (Proc Natl Acad Sci U S A 78:7634–7636, 1981; Nature 292:154–156, 1981), up to now it has not been possible to establish germ-line transmitting ES cells from any other mammalian species, not even from rat which is closely related, nor was it possible to introduce targeted mutations by different means to the germ-line of mammals. After 20 years, the mouse is still the only mammalian species where mutations can be introduced in a targeted manner and therefore it is very important to many fields in biology, like immunology, neurobiology, and developmental biology to study gene function and disease. Through means of introducing even point mutations to single relevant molecules of a signal transduction pathway in order to study regulation of cellular and physiological processes in complex organisms in a tissue specific or inducible manner (conditional gene targeting, (Cell 73:1155–1164, 1993; Science 265:103–106, 1994)), more recently the field has expanded exponentially.

Elizabeth J. Cartwright (ed.), *Transgenesis Techniques*, Methods in Molecular Biology, vol. 561
DOI 10.1007/978-1-60327-019-9_13,

© Humana Press, a part of Springer Science+Business Media, LLC 2009

Key words: Gene targeting, ES cells for microinjection, ES cell transfer techniques, Microinjection, Piezo-supported microinjection, ES mice, Laser-mediated microinjection, Blastocyst-stage embryos, Morula-stage embryos, Embryo donor mice, Stud males, Germline transmission

1. Introduction

Since the late eighties, the generation of embryonic stem cell-mediated genetically modified mice has become an important tool to generate mouse mutants that carry replacement mutations or insertions in a precise chromosomal position (locus of interest) – so-called targeted mutations ('gene targeting') via homologous recombination in ES cells. These ES cells are used to generate chimeric animals, which subsequently transmit the recombinant genotype to a proportion of their offspring.

This is in contrast to the transgenic approach using vectors or BACs for microinjection of two-cell stage preimplantation embryos where the mutation is also stably integrated in the genome but the insertion takes place randomly. In the latter case, it is unknown where the target locus is, how it is modified by the insertion and how this would affect the subsequent phenotype.

Several developments led to the final success of the gene targeting technique and led to the first knockout mouse published by Thomas and Capecchi in 1990 *(14)*.

1. In vitro culture of preimplantation embryos.
2. Generation of chimeras by microinjection using blastocyst stage embryos *(10)*.
3. Development of germline potent embryonic stem cells *(11, 12)*.
4. In vitro homologous recombination of the human β-globin gene in erythroleucemia cells *(13)*.
5. Gene targeting of embryonic stem cells *(1)*.

The ultimate success then was achieved by the combination of **steps 2** and **5**.

Later, the targeting methods were further refined thereby making conditional gene targeting possible, where the mutation occurs in a tissue-specific manner *(8, 9)* and/or it can be induced to occur later in life of the individual organism *(15–19)*.

Since the application of these findings led to great success in many fields of biological sciences, this work was honoured by the Nobel prize committee in Molecular Medicine in 2007 (http://nobelprize.org/nobel_prizes/medicine/laureates/2007/index.html).

Generation of chimeras by microinjection using blastocyst stage embryos was developed in 1968 *(10)* and has developed to become the conventional microinjection technique which is still widely applied to date. This chapter will describe the generation of chimeras by microinjection of ES cells into blastocyst stage embryos. In addition, we will highlight refined applications and new tools on the market; we will briefly address the generation of one-step knockout mice by piezo-supported microinjection of tetraploid blastocysts and laser-mediated injection.

2. Materials

2.1. Animals

The following animals are needed to generate chimeras by ES cell transfer into blastocysts:

1. Embryo donors for micromanipulation and stud males.
2. Foster mothers/vasectomized or sterile males.
3. Wild-type female mice for mating with chimeric males.

2.2. Embryo Culture

1. Humidified incubator with 5% CO_2 and 37°C dedicated for embryos only.
2. M2 medium, used for handling of embryos outside the incubator: NaCl 5.533 g/L, KCl 0.356 g/L, $CaCl_2 \cdot 2H_2O$ 0.252 g/L, KH_2PO_4 0.162 g/L, $MgSO_4 \cdot 7H_2O$ 0.293 g/L, $NaHCO_3$ 0.349 g/L, HEPES 4.969 g/L, sodium lactate 2.61 g/L, sodium pyruvate 0.036 g/L, glucose 1 g/L, BSA 4 g/L, Penicillin G 0.06 g/L, streptomycin 0.05 g/L, phenol red 0.01 g/L, water (for embryo transfer, Sigma). Store at 4°C for a maximum of 2 weeks.
3. M16 medium, used for culture of embryos in 5% CO_2: NaCl, 5.533 g/L, KCl 0.356 g/L, $CaCl_2 \cdot 2H_2O$ 0.252 g/L, KH_2PO_4 0.162 g/L, $MgSO_4 \cdot 7H_2O$ 0.293 g/L, sodium lactate 2.61 g/L, sodium pyruvate 0.036 g/L, glucose 1 g/L, BSA 4 g/L, penicillin G 0.06 g/L, streptomycin 0.05 g/L, phenol red 0.01 g/L, water (for embryo transfer, Sigma). Store at 4°C for a maximum of 2 weeks.

   You will find it more convenient to prepare concentrated stock solutions (e.g. 10x concentrated) and freeze aliquots at –20°C. Thaw the volume you need each week (e.g. 10 ml of M16 and 100 ml of M2).
4. Embryo-handling pipette made from glass capillaries (Blaubrand Micropipettes).
5. Mouth pipette (e.g. Aspirator tube assemblies, Sigma).
6. Embryo-tested mineral oil (Sigma), stored in the dark.
7. Micro drop culture dishes (Falcon).
8. Centre-well organ culture dish (Falcon).

2.3. Cell Culture

1. Humidified incubator with 5% CO_2 and 37°C dedicated for ES cells only.
2. Embryonic stem cells (ES cells).
3. Mouse embryonic fibroblasts (MEF cells) (inactivated).
4. ES cell medium: 400 ml Dulbecco's MEM, with high glucose, 75 ml heat-inactivated foetal bovine serum (batch tested for ES cells), 5 ml 100× NEAA (non-essential amino acids, 5 ml 100× sodium pyruvate, 5 ml 100× L-glutamine, 5 ml penicillin-streptomycin, 5 ml β-mercaptoethanol solution (50 ml DMEM + 35 μl β-mercaptoethanol), 1,000 units/ml ESGRO mLIF (Millipore). Mix by swirling the bottle and store at 4°C for a maximum of 4 weeks.
5. MEF medium: 450 ml Dulbecco's MEM with high glucose, 50 ml heat-inactivated fetal bovine serum, 2.5 ml 100× L-glutamine, 2.5 ml penicillin–streptomycin, 2.5 ml β-mercaptoethanol solution (50 ml DMEM + 35 μL β-mercaptoethanol). Mix by swirling the bottle and store at 4°C for a maximum of 4 weeks.
6. 0.05% Trypsin–EDTA.
7. Sterile PBS (without Ca^{2+} and Mg^{2+}).
8. 1 M HEPES buffer.
9. 2-Mercaptoethanol.
10. Mitomycin C.
11. 70% Ethanol.
12. 0.1% gelatin solution.

2.4. Microinjection

1. Inverted microscope with either phase contrast or differential interference contrast (DIC) optics.
2. Two micromanipulators (from Eppendorf, Leica, or Narishige).
3. Cell Tram Air for control of the holding capillary (Eppendorf).
4. Cell Tram Vario for control of the injection needle (Eppendorf).
5. Microinjection chamber (e.g. EMBL, www.embl-heidelberg.de/LocalInfo/mechanical-engineering/projects.html or Biomedical Instruments www.pipettes.de).
6. Injection capillary (from BioMedical Instruments or TransferTips (ES), Eppendorf).
7. Holding capillary (VacuTips, Eppendorf, or from BioMedical Instruments).
8. Embryo-tested mineral oil (Sigma), stored in the dark.

9. Blastocyst stage embryos.
10. ES cells prepared for injection.
11. ES cell medium.
12. Mouth pipette assembly (Sigma).
13. HEPES-buffered ES medium: 200 μL 1 M HEPES in 10 ml ES cell medium.
14. Centre-well organ culture dish (Falcon).

3. Methods

3.1. Animals

The following animals are needed to generate chimeras by ES cell transfer into blastocyst or morula-stage embryos:

1. Embryo donors for micromanipulation and stud males: Mate wild-type females (C57BL/6 are most commonly used) with wild-type stud males (C57BL/6 are most commonly used) in order to produce host embryos that will receive ES cells by microinjection. Stud males must be individually caged, but they can be used for several rounds of mating with new groups of female mice; the number of uses will depend upon either performance or age (use up to 6 months of age). Place females in a cage with a resident male, not vice versa. Embryos could either be produced by normal mating (1:2 ratio male:females) or using superovulated juvenile female mice for mating (1:1 ratio). The method of choice depends on the strain of mice and the available protocols in the corresponding animal facility. Many ES cell lines presently available and in use are derived from the 129 mouse strain (129/Sv). The 129 mice are homozygous wild-type A/A at the agouti locus. To recognize chimeras by coat colour, it is recommended to inject the 129 derived ES cells into a blastocyst from an albino or homozygous non-agouti (a/a) strain. The C57BL/6 mouse strain is the strain of choice as host embryos if 129-derived ES cells are used. The ES cells seem to have a developmental advantage and the coat colour is easily distinguishable. Unfortunately, C57BL/6 mice do not give good yield of blastocysts from natural matings (on average 3–5/ animal) and plugged females do not always contain embryos. If the ES cells are from a non-agouti (e.g. C57BL/6) strain (Bruce 4, *(20)*), then the host blastocyst should be from an albino or agouti strain (e.g. BALB/c, CB/20). More recently, embryonic stem cells from hybrid strains (129sv × C57Bl/6) have been favoured (V6.5, *(21)*)

since they keep their germline capabilities even after several rounds of transfection *(22)*. However, the coat colour chimerism will no longer be fully representative. It is usual to inject such ES cells into C57BL/6 host embryos since these mice are most commonly available and to test chimerism by PCR of tail biopsies.

2. Foster mothers and vasectomized or sterile males (*see* Chapter 15):
   Fertile wild-type female mice are mated to vasectomized (or sterile) males (normal mating 2:1 ratio) in order to produce suitable females in which the manipulated embryos are reimplanted. Successful mating is verified by the presence of a vaginal plug; these females are called pseudopregnant foster mothers.

   Any surplus plugged, pseudopregnant foster females can be recycled into the foster mating pool after 2 weeks.The genotype of these females is irrelevant to the experiment. These females should be good mothers as they must carry the re-implanted embryos to term and nurse the newborns to weaning age. They should also be happy to allow inspection of the newborn pups and accept pups from another mother. Most commonly, outbred stock CD1 females or alternatively F1 C57BL/6 × BALB/c females fulfil all these needs and can be recommended.

   Usually, the genetic background of the vasectomised or naturally sterile males is also not critical as these males should not produce any offspring but they should be good breeding performers. Since outbred stock males grow fat very soon (and then do not perform well anymore), we recommend the use of F1 males.

3. Wild-type female mice for mating with chimeric males.
   For the establishment of a mutated mouse line, the resulting chimeric animals (most often males; as most ES cells have an XY karyotype there should be a shift of the sex ratio towards male mice among the chimeras born *(3)*) must be mated to the same strain as the host embryos in order to visualise coat colour transmission.

   It is good practice to cross to wild-type mice for two further generations before crossing to homozygosity in order to get rid of other non-intended mutations that occur during in vitro culturing of ES cells and transfections. In addition, one should always produce germline chimeras from at least two independent clones, to be sure that the phenotype is directly related to the intended mutation.

   Depending on the individual experience concerning a specific combination of ES cell genotype and host embryo, it may be worth test breeding all male chimeras with 50% or more agouti colour for germ-line transmission or even breed female chimeras for certain ES cell lines (i.e. IB10). For some

ES cell lines, 80% chimeras give better germline performance than their 100% chimeric littermates since they may be sterile or do not even mate at all for other unknown reasons.

3.2. Embryo Culture

1. Prepare a micro drop culture dish of M16 overlaid with mineral oil, pre-incubate at 37°C, 5% CO_2 for at least 1 h.
2. In the morning of day 3.5 dpc, sacrifice the female embryo donor mice humanely (in the case of normal mating; embryos from superovulated females should be prepared as morula-stage embryos at 2.5 dpc).
3. Lay the animal on its back and wash it thoroughly with 70% ethanol (*see* **Note 1**).
4. Cut the skin at a lateral position at the midline and hold the skin below and above the incision firmly with your fingers. Pull the skin toward the head and the tail until the peritoneum is exposed.
5. Open the abdominal cavity and push the intestine towards the head and locate the uterus.
6. Grasp the cervix and cut across. Pull the uterus upward by tearing away the ligaments and fat, and cut between the oviduct and ovary.
7. Collect the uterus in a 50 ml tube with approximately 20 ml of M2 medium (*see* **Note 2**). Continue until all uteri are collected.
8. Place a maximum of five uteri in a 6 cm culture dish and insert a 26-gauge needle into the upper part of the uterus near the oviduct. Flush each horn with approximately 0.5 ml of M2 medium towards the cervix. Be careful not to use too much pressure as this could damage the blastocysts.
9. Use a handling pipette to pick up all flushed blastocysts and collect them in a separate drop of M2 or in a centre-well organ dish (recommended).
10. After collecting all blastocysts, wash them through at least three drops of M2 medium and three drops of M16 medium. Transfer the embryos to a micro drop culture dish of M16 overlaid with mineral oil, pre-incubated at 37°C, 5% CO_2 for at least 1 h.

3.3. Cell Culture

In general, embryonic stem cells (ES cells) should be passaged after thawing at least once before injection. Alternatively, at low density they could be trypsinized and re-plated on to the same dish. As most ES cells need to be cultured on a murine embryonic fibroblasts (MEF) layer, make sure that you have enough plates with mitomycin C-treated MEFs before you start. Mitomycin C-treated MEFs can be prepared in advance and frozen.

Following mitomycin C treatment, cells must remain in culture for a further 24 h with fresh medium before they can be frozen. In principle, they can be thawed together with the ES cells and put together on the dish.

Do not forget that ES cells need daily feeding, which means within 24 h at the latest!

1. Warm all media and solutions to 37°C. Thaw the ES cells by taking them out of the liquid nitrogen and place them in a 37°C water bath until the frozen suspension starts to melt. Continue the thawing by rubbing the tube between your hands. Sterilize the vial with 70% ethanol and transfer to a tissue culture hood (*see* **Note 3**).
2. Add 10 ml of pre-warmed ES medium into a 15 ml tube and add the thawed ES cells by placing the tip of the pipette to the wall of the tube and disperse slowly. Do not mix. Centrifuge immediately for 3 min at 4°C, 133 g. This centrifugation step helps to get rid of residual DMSO. Discard the supernatant and loosen up the cell pellet by agitating the tube with your fingers.
3. Re-suspend the cells in an appropriate (e.g. 2 ml) volume of pre-warmed ES cell medium and count the cells if necessary. It is useful to freeze the cells at an ideal number for replating on 6 cm plate, or plate one-third and two-thirds on two different plates and decide after growing which one to propagate further. For service applications, we usually recommend the following strategy when freezing clones for microinjection; per vial freeze one-third of a semiconfluent 10 cm dish and then divide the vial after thawing – 1/3 and 2/3 on to 6 cm dishes and decide later which one to use for the following steps.
4. Add pre-warmed ES cell medium to the MEF dish. Transfer 2×10^4 ES cells per cm^2 of tissue culture dish. In general, cells from a semiconfluent 6 cm dish is enough for injection. Carefully shake the dish to distribute the ES cells evenly (drive the dish in both directions in a figure of 8 shape) and put the cells back to the incubator, 37°C, 5% CO_2, 100% humidity.
5. When the culture reaches approximately 50% to 70% confluence, trypsinize the ES cells using 1 ml trypsin for 5–10 min in the incubator. The time of trypsinization is very much dependent on the cell line in use. For Bruce 4 ES cells: for instance. in our hands, 3 min on average is enough! 129-derived ES cells seem to be much more robust to trypsinization.
6. Add 4 ml pre-warmed ES cell medium and by pressing the tip of the pipette onto the bottom of the dish, pipette up and down three times until a single cell suspension is achieved. Split the cells 1:3–1:10 (depending on the ES cell line in use) on to a new dish with inactivated MEFs. For R1 and E14.1

ES cells, it is recommended to do this split approximately 24–36 h before the injection (e.g. in the late afternoon, 2 days before injection or in the early morning the day before the injection).

7. In the morning of the injection day, check the cells macroscopically. They should be about 50% confluent, nice colonies with sharp edges and only very few flat colonies in the culture (*see* **Fig. 1a**). Wash the cells once with 5 ml PBS and discard the supernatant. It is optimal to feed the ES cells 2–3 h before the injection, but not absolutely necessary.

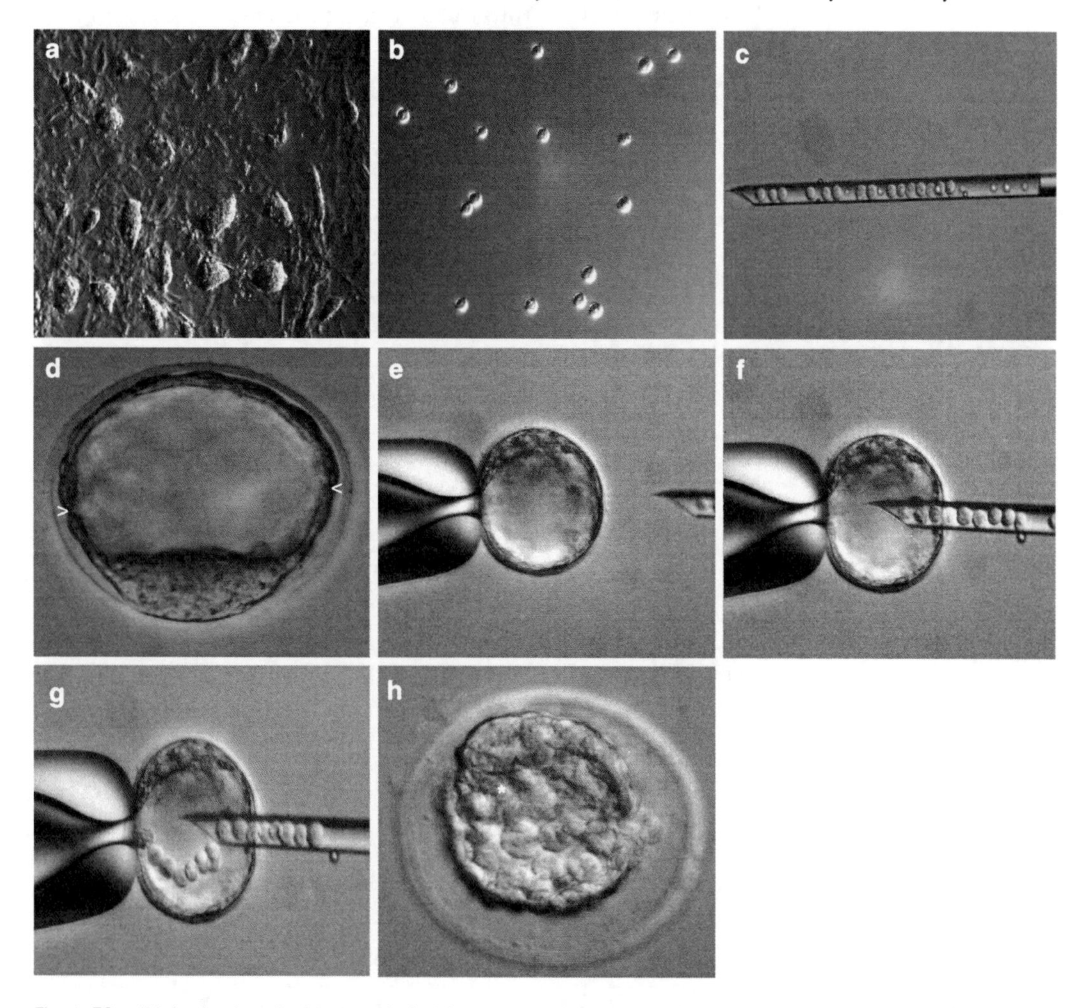

Fig. 1. ES cell injection into the blastocoel of a blastocyst. (**a**) Morphology of embryonic stem cell colonies. The colonies remain composed of a homogenous population of stem cells. Stem cells are comparably small and are tightly packed within the colony. Note the smooth outline of the colony of densely packed cells. (**b**) Single cell suspension of ES cells for injection. (**c**) The injection needle is used to collect ES cells. (**d**) Blastocyst. Arrowheads mark the junctions of trophoblast cells. (**e**) Immobilize the blastocyst on a holding pipette so that the ICM is positioned at either 12 o'clock or 6 o'clock. The tip of the injection needle is brought into the same focal plane as the equator/midpoint of the blastocyst. (**f**) With a swift movement, the needle is introduced into the blastocoel of the blastocyst. (**g**) The cells are released slowly into the cavity. (**h**) After releasing the injected blastocyst from the holding capillary the blastocyst will collapse and the ES cells (star) will come into contact with the ICM.

8. Add 1 ml of pre-warmed trypsin and incubate for 3–10 min in the incubator until the cells have lost contact with each other. Carefully watch this step under the microscope at room temperature when you do it for the first time in order to know the optimal incubation time for your cells. Add 4 ml of pre-warmed ES cell medium and pipette up and down three times, by pressing the tip of the pipette onto the bottom of the dish while releasing the cells, until a single cell suspension is achieved. Check the success of your treatment under the microscope.
9. Put the dish back into the incubator for 15–30 min (*see* **Note 4**). During this time, the MEFs should re-adhere to the plate whereas ES cell re-aggregation will not result in tight connections in this time. After this pre-plating step, take the dish out of the incubator taking care not to disturb the dish too much (*see* **Note 5**); aspirate the supernatant slowly and add it to a 15 ml tube with 5 ml of pre-warmed ES cell medium. Centrifuge for 3 min at 4°C, 133 g in a standard cell culture centrifuge.
10. In the meantime, prepare 10 ml of the HEPES buffered ES cell medium.
11. Discard the supernatant and loosen the pellet by agitating the tube with your fingers. Add 0.5–2 ml of cold HEPES-buffered ES cell medium (depending on the size of the pellet = approximate cell numbers) and resuspend the ES cell pellet. Place the ES cell solution and the buffered ES cell medium on ice in the refrigerator for 30 min or longer (*see* **Note 6**).
12. Discard the supernatant and loosen the pellet again. Resuspend the pellet in 0.5–1 ml cool HEPES-buffered ES medium and pipette up and down 3–5 times. Cells are now ready for injection.
13. This step is optional. Before filling the injection chamber, if necessary you could pipette the cells through drops of cooled HEPES-buffered ES medium by taking the ones from the periphery of the first drop to the next, thereby you could achieve the ideal dilution and we think you could enrich for the successful small cells (*see* **Fig. 1b**).

3.4. Microinjection

3.4.1. ES Cell Injection into Blastocyst Stage Mouse Embryos

1. Set up the injection chamber by placing a micro drop (about 20 μL) of cooled HEPES-buffered ES medium into the centre of the depression of a depression slide and cover it with mineral oil. Alternatively, set up the injection chamber by placing a micro drop of cooled HEPES-buffered ES medium onto a glass slide that is pressed on a metal frame (matching the shape of the glass slide with a rectangular hole cut

by covering the metal frame with a thin layer of grease (*see* **Note 7**). Cover the drop with mineral oil.

2. Place a holding capillary on the left-hand-side micromanipulator. Cross the tubing behind the microscope and connect to the Cell Tram Air, which should be controlled by the right hand. Alternatively, the pressure could be controlled by a mouth pipette.
3. Place an injection capillary in the right-hand-side manipulator. Cross the tubing behind the microscope and connect to the Cell Tram Vario, which should be controlled by the left hand.
4. Transfer several hundred ES cells from the refrigerator into the injection chamber but leave some cell-free areas e.g. at the lower end of the micro drop (*see* **Note 8**) or in the centre.
5. Insert the oil filled (*see* **Note 9**) microinjection capillary into the drop of medium and position it in the centre of the field of view close to the bottom of the injection chamber by using a 2.5 – 5× magnification objective.
6. Change to a 20× magnification objective and pick up only small, round, smooth looking cells with no dark areas (*see* **Fig. 1b**). Pack them as tightly together as possible in the injection capillary (see **Note 10, Fig. 1c**).
7. Transfer ~10 blastocysts to the injection chamber and insert the air-filled holding capillary into the chamber.
8. Prefill the holding pipette with a little medium from the injection chamber (to avoid air bubbles in the chamber when releasing the embryo), then catch a blastocyst with the holding capillary in order to close the pipette and prevent ES cells from being sucked in.
9. Position the embryo at the holding pipette in such a way that the inner cell mass (ICM) is either at 12 o'clock or at 6 o'clock (*see* **Note 11**).
10. Focus on the outer cells of the blastocyst, the so-called throphoblast cells and move the injection needle into the same focal plane.
11. Gently touch the blastocyst. You will see a bulb in the embryo indicating the position of your injection needle relative to the surface of the blastocyst. Move your tip to the midpoint of the embryo. To penetrate the blastocyst it is recommended to aim for a junction between two trophoblast cells (*see* **Fig. 1d**). This will make the injection much easier and it also helps to reduce the damage to the embryo.
12. Inject with one single, continuous movement into the blastocoel and slowly expel 8–15 (*see* **Note 12**) of the collected ES cells into the cavity (*see* **Fig. 1e–g**).

13. Withdraw the injection pipette slowly. When the tip of the injection needle reaches the entry point to the blastocyst, wait a few seconds to allow any overpressure that might have been created inside the embryo to reduce (otherwise, the injected ES cells might be pushed out while withdrawing the needle).
14. After removing the injection needle, the blastocyst will collapse and the ES cells will thereby come into close contact with the ICM (*see* **Fig. 1h**).
15. Repeat with further blastocysts.
16. Move the embryos from one injection group (about 10) back to the incubator as soon as possible after the injection (at least within 1 h).
17. Approximately 1 h after the injection, the blastocysts will re-expand.
18. Transfer the blastocysts into a pseudopregnant recipient female (*see* **Note 13** and chapter 15).

3.4.2. ES Cell Injection into Blastocyst-Stage Embryos by Laser-Supported Injection

1. In the morning of day 3.5 dpc, collect blastocysts as described in **Subheading 3.2** and keep them in M16 medium in the incubator (5% CO2 and 37°C) until injection.
2. Prepare ES cells as described in **Subheading 3.3** and set up the injection chamber as described in **Subheading 3.4.1**.
3. Turn on the laser (e.g. Octax Laser Shot System, XYclone Laser System (*see* **Note 14**).
4. Change to 20× magnification objective and pick up only small, round, smooth looking cells with no dark areas and pack them as tightly together as possible in the injection capillary.
5. Transfer ~10 blastocysts to the injection chamber and catch a blastocyst with the holding capillary.
6. Position the ICM at either 12 o'clock or 6 o'clock and position the crosshairs of the laser on the zona pellucida at the midline of the embryo (*see* **Fig. 2a, b**). Shoot once or twice (this strongly depends on the laser intensity, *see* **Note 14**) with the laser to open the zona pellucida. If you are using blunt capillaries, shoot once to open the zona pellucida and once to drill a hole in the trophoblast cells (try to hit a junction between two trophoblast cells in order to limit any damage by the laser). Insert the needle into the blastocoel and release 8–15 ES cells (*see* **Fig. 2c, d**).
7. Withdraw the needle slowly and release the blastocyst from the holding capillary (*see* **Fig. 2e, f**).

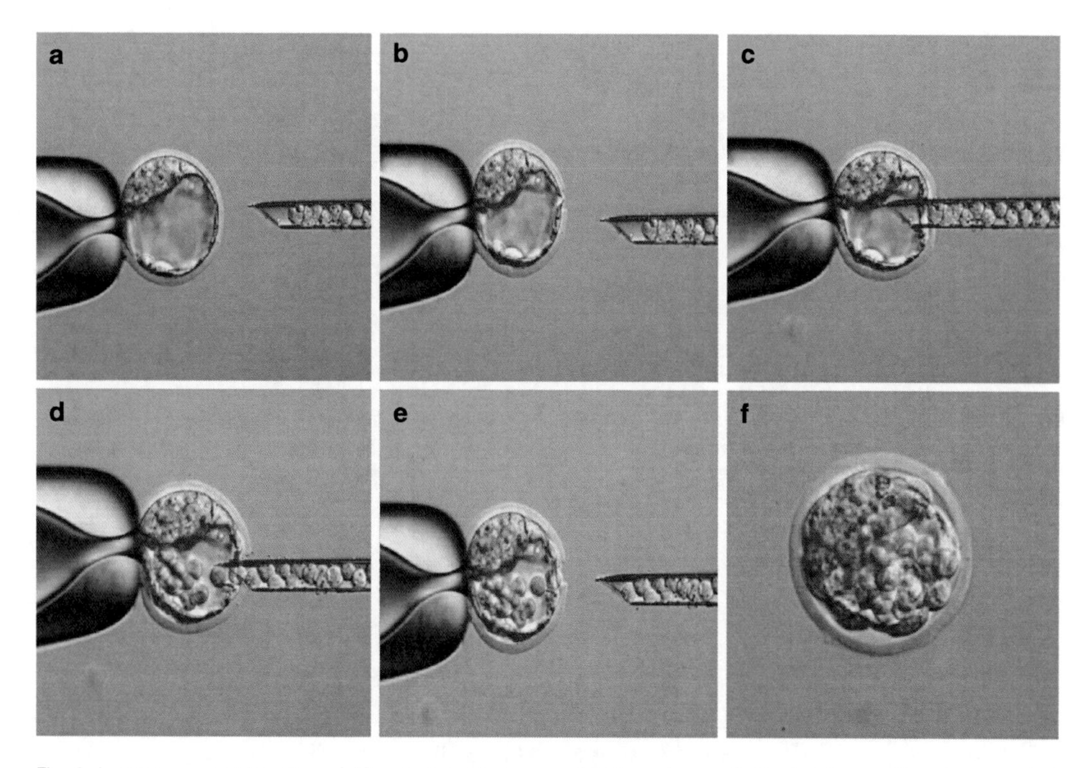

Fig. 2. Laser-supported injection of ES cells into blastocysts. (**a**) Fix embryo with the holding capillary in a way that the ICM is positioned at either 12 o'clock or 6 o'clock. (**b**) Shoot once with the laser to open the zona pellucida and/or trophoblast. (**c**) Insert the injection needle into the blastocoel. (**d**) Release the ES cells. (**e**) Slowly withdraw the injection needle. (**f**) The blastocyst collapses after the injection and will re-expand within 1 h in M16 medium at 5% CO_2, 37°C.

8. Transfer the blastocysts into a pseudopregnant recipient female (*see* **Note 13** and chapter 15).

3.4.3. ES Cell Transfer into Eight-Cell Stage Embryos by Laser-Supported Injection

Recently an alternative procedure to generate mice entirely derived from injected ES cells was described *(23, 24)*.The ES cells are injected into embryos at the eight-cell stage before the formation of the ICM as this is suggested to result in a higher degree of ES cell contribution. The introduction of ES cells before ICM formation might provide a competitive advantage to the injected ES cells. Previously, such studies have been performed but they involved laborious manipulations to penetrate the zona pellucida. Now a laser pulse can be used to perforate the zona pellucida and the ES cells are subsequently injected just under the zona pellucida (perivitteline space) of the eight-cell stage embryo (*see* **Fig. 3a–c**).

The advent of laser-mediated eight-cell injection is expected to solve the problem of comparably lower performance of inbred ES cell lines such as from C57BL/6. Since the purchase of an expensive laser device is the prerequisite, so far not many institutes have fully established this route; however, the use of this technique may increase as more people publish successful results from its use.

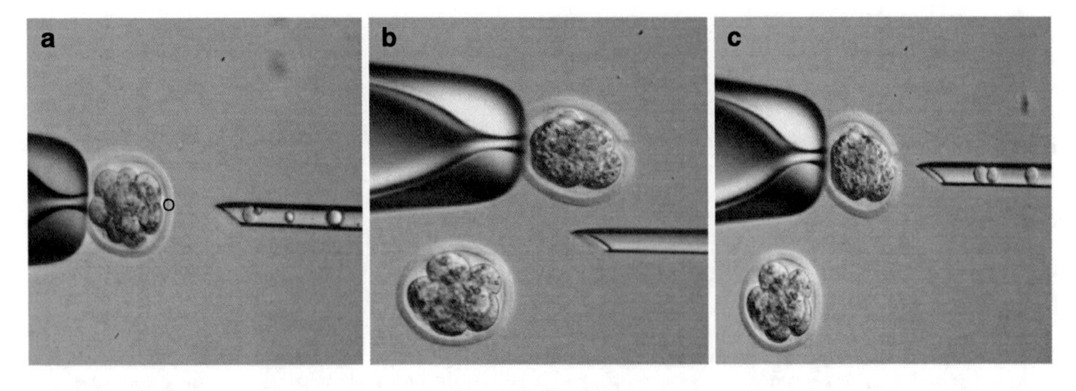

Fig. 3. Laser-supported injection of ES cells into eight-cell stage embryos. (**a**) Immobilize the eight-cell stage embryo with the holding capillary and position the crosshair (*circle*) of the laser right on the zona pellucida as far away from any blastomere as possible. (**b**) Perforate the zona pellucida with one shot of the laser shot system. (**c**) Insert the injection capillary into the perivitelline space of the embryos and release 4–5 ES cells.

1. Isolate eight-cell stage embryos early in the morning of day 2.5 dpc by flushing the oviduct and the upper part of the uterus using a 30-G needle as described below.
2. Sacrifice the mice humanely, lay them on their backs and wash them thoroughly with 70% ethanol.
3. Open the abdominal cavity as described in **Subheading 3.2**. Locate the ovary/oviduct and the uterus. Grasp the uterus approximately 1 cm away from the oviduct with forceps and cut the uterus distal from the forceps.
4. Cut between the ovary and the oviduct and place the oviduct/uterus in a 3 cm culture dish with M2 medium. Continue until all oviducts/uteri are collected.
5. Place 1–5 oviducts/uteri in the lid of a 10 cm dish. Under the stereomicroscope, grab each organ with watchmaker forceps, insert the tip of a 30-G needle attached to a microrsyringe into the infundibulum. Ensure that all sharp edges are removed from the needle using sandpaper. This process may be made easier by bending the needle. Flush each organ with about 200 μL of M2 medium. Successful flushing is indicated by swelling of the oviduct.
6. Collect all eight-cell stage embryos from the individual drops and wash them through three drops of M2 medium and three drops of M16 medium. Transfer them into a micro drop culture dish with M16 overlaid with mineral oil, preincubated at 37°C, 5% CO_2 for at least 1 h.

7. Prepare ES cells as described in **Subheading 3.3** and set up the injection chamber as described in **Subheading 3.4.1**.
8. Turn on the laser (e.g. Octax Laser Shot System, XYclone Laser system, *see* **Note 14**).
9. Change to 20×magnification objective and pick up only small, round, smooth looking cells with no dark areas and pack as tightly together as possible in the injection capillary.
10. Transfer ~20 eight-cell stage embryos to the injection chamber and catch an embryo with the holding capillary.
11. While the eight-cell stage embryo is being restrained by the holding capillary, make a perforation in the zona pellucida using the Octax Laser Shot System with an irradiation time of 6.5 ms at a region as far away as possible from any blastomeres (*see* **Fig. 3a**).
12. Use two laser shots for making a larger hole if you want to use blunt-ended needles. Only one shot is necessary when using injection capillaries with a tip (as for injections without using a laser).
13. Insert an injection needle through the perforation in the zona pellucida and introduce approximately five ES cells into the perivitelline space. Withdraw your needle carefully and release the embryos from the holding capillary (*see* **Fig. 3**).
14. Transfer the injected embryos into 0.5 dpc pseudopregnant females the same day or culture them overnight to blastocyst stage and transfer them to 2.5 dpc pseudopregnant female (*see* Chapter 15).

3.4.4. Piezo-Supported Microinjection of Soft Tetraploid Blastocyst-Stage Embryos

During the past 10 years, the generation of fully ES cell derived mice (ES mice) through either tetraploid embryo injection or aggregation *(25)* using F1 embryonic stem cells (i.e. V6.5) has shown that the number of animals with slight abnormalities is quite high. In addition, the technique requires comparably more space, labour and co-ordination in the animal facility because of caesarean sections and fostering; this means that it is not routinely applied. It is advantageous for experiments where a single mutation can be introduced in a homozygous fashion in ES cells and where direct analysis of the mutant is the final goal of the experiment. However, in experiments with conditional mutants where the targeted mutation is not immediately active and the strain will be crossed to other strains carrying different mutations, the use of this technique is not recommended. For the generation of tetraploid embryos, see the following chapter.

4. Notes

1. *Decontamination of the animal.* This important step reduces the risk of contaminating the dissection with mouse hair and will reduce the risk of contamination with ectodermal parasites (especially, if the donor mice does not have SPF (specific pathogen free) health status).
2. *Organ collection.* Alternatively, you could always collect organs in PBS and use the medium for flushing the embryos, you could also collect organs in dishes if you prepare the organs right next to where you flush the embryos, for transport its better to use the tubes for collection.
3. *Media warm up.* Alternatively, you could place the media at room temperature in advance since the small volume of medium in the 6 cm dish finally warms up in the incubator very quickly. This will omit the potential contamination risk through using water baths and keep the compounds in your media longer. In this case, warm the vial quickly in your hands.
4. *Preplating.* This step is called preplating. The MEF cells will loosely attach to the bottom of the dish quicker than the ES cells. This is used to get rid of most of the feeders prior to injection without having to culture the ES cells in feeder-free conditions for one passage. The exact timing for a sufficient preplating strongly depends on the ES cell line in use and secondly on their concentration (number of ES cells in the solution). The more ES cells you have, the longer you can preplate without fearing to be left with too few ES cells to inject.
5. *Preplating.* As the MEF cells are only loosely attached to the bottom of the culture dish, too much shaking would lift them off again.
6. *ES cell purification.* During this incubation, only the viable cells will sediment leaving the dead cells in the supernatant. This step, therefore, further purifies the ES cells for injection.
7. *Metal frame.* It consists of a hollow frame fitted to the stage of the microscope and was developed at EMBL by Kristina Vintersten. It can be purchased from Biomedical Instruments (www.pipettes.de), or from the EMBL, or can be homemade by a good mechanical workshop.
8. *Cell-free area.* This is used as the injection area in the micro drop. Residual feeder cells (their size can be quite big) or groups of sticky ES cells can stick to the tip of the holding pipette during the process of catching the blastocyst and thus can be very annoying if they are obstructing the area.

9. *Installation of injection capillary.* If you are using the Eppendorf TransferTip, place the empty capillary in the right-hand side manipulator, which is connected to the CellTram Vario. Fill the capillary with mineral oil by turning the wheel of the CellTram Vario. Stop the filling process close to the tip of the needle and move the capillary into the injection drop onto the bottom of the injection chamber. Focus on the needle and continue filling the capillary with oil until the very end of the injection needle. Take care that no air bubbles can be found in the whole system (capillary, tubing, CellTram Vario) as this can result in uncontrollable movements.
10. *ES cells within the injection capillary.* Less medium will be co-injected into the blastocyst if the cells are packed chain-like inside the capillary. Thus, less overpressure is created and it is assumed that essential factors in the fluid in the blastocoel are not diluted. If there are larger cell-free gaps inside the injection needle, exit the blastocyst leaving the very tip of the needle inside, expel the cell-free medium and move the capillary inside the blastocyst again as soon as cells are close to the tip. If the shape of the needle does not allow the collection of enough cells for at least 5–10 embryos at a time (shorter stretch of small diameter) it is recommended to insert the holder capillary and embryos first to the injection chamber (**steps** 7 and **8**) and close the holder by adopting one embryo first. Otherwise, ES cells are soaked into the holder needle and thereby block it from the inside.
11. *Blastocyst position at the holding needle.* Take great care not to harm the ICM as it gives rise to the embryo proper.
12. *Number of injected ES cells.* The optimal number to generate high percentage chimeric mice depends on the ES cell line and the passage number you are using, and needs to be determined experimentally for each cell line. Even with the best culture conditions and with the greatest care, the ES cells will accumulate mutations during the culture and as a result the potency to generate high chimeric animals will be lowered. In general, the higher the passage number the more ES cells need to be injected. In general, 8–15 ES cells is a good number to start off with.
13. In case you do not get enough 2.5 dpc pseudopregnant recipient females on the day of injection, the blastocysts can be either cultured overnight and then transferred into 2.5 dpc recipients (usually service facilities are injecting every day during the week and so it is likely that 2.5 dpc recipients will be available the next day). Otherwise, you can transfer the blastocysts into the infundibulum of a 0.5 dpc pseudopregnant recipient female the same day (if available) or the next day (*see* Chapter 15).

14. The laser systems were invented for the field of assisted reproductive technology (ART) and are based on an infrared laser system. The OCTAX Laser Shot System uses an indium–gallium–arsenic–phosphorus laser diode emitting at a wavelength of 48 μm. A live video of the cells/embryos is displayed on the computer monitor via a digital camera. The video is overlaid with a crosshair and the laser beam is triggered either by mouse or by foot switch. The irradiation time for the laser is adjusted within the software. The laser beam causes a tangential thinning or opening of the zona pellucida of the embryos by a localized photothermal process, which lyses the glycoprotein matrix. Trench-like openings with even walls are created, which have a circular appearance in the inverted microscope. The diameter of the drilling can be adapted by varying the irradiation time of the laser.

Additional Notes and Trouble Shooting

The wall of the blastocyst is difficult to penetrate

- The spike of the injection capillary is not sharp enough or broken. Change the injection needle.
- The injection capillary has been inserted through a cell in the blastocyst wall. Try to find a connection between two trophoblast cells and insert the injection needle there.
- The blastocyst appears very soft. Blastocysts from superovulated embryo donor mice may have been used. Use blastocysts from natural mated mice. The females should be 8–12 weeks of age and the males 2–6 months old.

The blastocysts have developed too far

- Recover and inject the embryos earlier during the day
- Check the light/dark cycle within your animal facility or animal room. It is well known and documented that the light:dark cycle has a profound impact in regulating circadian rhythms and stimulating and synchronizing breeding cycles. The interruption of the dark part of the cycle can be very disruptive. Even short light pulses can confuse the circadian clock. Most animal facilities have an automated 12 h dark: 12 h light cycle. If there are problems with plug rates or embryo yield, then check if something went wrong with the automated system or someone entered the animal facility during the dark phase.

References

1. Thomas, K. R. and Capecchi, M. R. (1987). Site directed mutagenesis by gene targeting in mouse embryo-derived stem cells. *Cell*, **51**, 503–512
2. Bradley, A., Evans, M., Kaufman, M. H. and Roberston, E. J. (1984). Formation of germ line chimaeras from embryo-derived teratocarcinoma cell lines. *Nature*, **309**, 255–256

3. Robertson, E. J. (1987). Teratocarcinomas and embryonic stem cells: a practical approach. *In*:D. Rickwood and B. D. Hames (series eds.) *Practical Approach*. IRL Press, Oxford, 254 pp
4. Jaenisch, R. (1988). Transgenic animals. *Science*, **240**, 1468–1475
5. Joyner, A. L., ed. (1993). Gene targeting: a practical approach. *In*:D. Rickwood and B. D. Hames (series eds.) *Practical Approach*. IRL Press, Oxford, New York
6. Pinkert, C. A. (2002). *Transgenic Animal Technology: A Laboratory Handbook*, 2nd edition. Academic Press, San Deigo
7. Nagy, A., Gertsenstein, M., Vintersten, K. and R. Behringer (2003). *Manipulating the Mouse Embryo - A Laboratory Manual*, 3rd edition. Cold Spring Harbor Laboratory Press, Cold Spring Harbor, NY
8. Gu, H., Zou, Y. R. and Rajewsky, K. (1993). Independent control of immunoglobulin switch recombination at individual switch regions evidenced through Cre-loxP-mediated gene targeting. *Cell*, **73**, 1155–1164
9. Gu, H., Marth, J. D., Orban, P. C., Mossmann, H. and Rajewsky K. (1994). Deletion of a DNA polymerase β gene segment in T cells using cell-type-specific targeting. *Science*, **265**, 103–106
10. Gardner, R. L. (1968). Mouse chimeras obtained by the injection of cells into the blastocyst. *Nature*, **220**, 596–597
11. Martin, G. R. (1981) Isolation of pluripotent cell line from early mouse embryos cultured in medium conditioned by teratocarcinoma stem cells. *Proc Natl Acad Sci U S A*, **78**, 7634–7636
12. Evans, M. J. and Kaufmann, M. H. (1981). Establishment in culture of pluripotential cells from mouse embryos. *Nature*, **292**,154–156
13. Smithies, O., Gregg, R. G., Boggs, S. S., Koralewski, M. A. and Kucherlapati, R. S. (1985). Insertion of DNA sequences into the chromosomal β-globin locus by homologous recombination. *Nature*, **317**(6034), 230–234
14. Thomas, K. R. and Capecchi, M. R. (1990). Targeted disruption of the murine int-1 proto-oncogene resulting in severe abnormalities in midbrain and cerebellar development. *Nature*, **346**(6287), 847–850
15. Kühn, R., Schwenk, F., Aguet, M. and Rajewsky, K. (1995). Inducible gene targeting in mice. *Science*, **269**(5229), 1427–1429
16. Logie, C. and Stewart, F. (1995). Ligand-regulated site-specific recombination. *Proc Natl Acad Sci U S A*, **92**, 5940–5944
17. Gossen, M., Freundlieb, S., Bender, G., Muller, G. Hillen, W. and Bujard, H. (1995). Transcriptional activation by tetracyclines in mammalian cells. *Science*, **268**(5218), 1766–1769
18. Feil, R., Brochard, J., Mascrez, B., LeMeur, M., Metzger, D. and Chambon, P. (1996). Ligand-activated site-specific recombination in mice. *Proc Natl Acad Sci U S A*, **93**, 10887–10890
19. Kellendonk, C., Tronche, F., Monaghan, A. P., Angrand, P. O., Stewart, F. and Schutz, G. (1996). Regulation of Cre recombinase activity by the synthetic steroid Ru486. *Nucleic Acids Res*, **24**(8), 1404–1411
20. Koentgen, F. et al. (1993). Targeted disruption of the MHC class II Aa gene in C57BL/6 mice. *Int Immunol*, **5**, 957–964
21. Eggan, K., Akutsu, H., Loring, J., Jackson-Grusby, L., Klemm, M., Rideout III, W. M., Yanagimachi, R. and Jaenisch, R. (2001). Hybrid vigor, fetal overgrowth, and viability of mice derived by nuclear cloning and tetraploid embryo complementation. *Proc Natl Acad Sci U S A*, **98**, 6209–6214
22. Eggan, K., Rode, A., Jentsch, I., Samuel, C., Hennek, T., Tintrup, H., Zevnik, B., Erwein, J., Loring, L., Jackson-Grusby, L., Speicher, M. R., Kuehn, R. and Jaenisch, R. (2002). Male and female mice derived from the same embryonic stem cell clone by tetraploid embryo complementation. *Nat Biotechnol*, **20**, 455–459
23. Gridley, T. and Woychik, R. (2007). Laser surgery for mouse geneticists. *Nat Biotechnol*, **25** (1), 59–60
24. Poueymirou, W. T. et al. (2007). F0 generation mice fully derived from gene-targeted embryonic stem cells allowing immediate phenotypic analyses. *Nat Biotechnol*, **25**(1), 91–99
25. Schwenk, F., Zevnik, B., Brüning, J., Röhl, M., Willuweit, A., Rode, A., Hennek, T., Kauselmann, G., Jaenisch, R. and Kühn, R. (2003). Hybrid embryonic stem cell-derived tetraploid mice show apparently normal morphological, physiological and neurological characteristics. *Mol Cell Biol*, **23**(11), 3982–3989

Chapter 14

Generation of Chimeras by Morula Aggregation

Anne Plück and Christian Klasen

Summary

This chapter describes the tools and the experimental route of targeted manipulation by aggregation in the mouse using targeted embryonic stem cells (ES cells). Instead of injecting ES cells into the blastocoel of a diploid blastocyst-stage embryo (3.5 dpc) ES cells can be brought together with diploid morula-stage embryos (2.5 dpc). The zona pellucida of the embryo needs to be removed and one or two embryos (sandwich aggregation) are put together with ES cells into an indentation well of a cell culture grade dish overnight for aggregation. This can be performed manually using a stereomicroscope and does not require any special training or expensive instrumentation.

The next day, the embryo would have developed into a blastocyst in vitro and can be transferred to a pseudopregnant female mouse (*see* Chapter 15).

The use of tetraploid embryos generated by electrofusion will lead to entirely ES cell-derived fetuses.

Key words: Gene targeting, ES cell transfer techniques, Morula aggregation, Sandwich aggregation, Preimplantation stage embryos, Stud males, Embryo donor mice, Germline transmission, Electrofusion, Tetraploid embryos

1. Introduction

The first aggregation chimeras between two diploid morula-stage embryos were generated to study the developmental potential of the two different embryos *(1)*. Germline transmitting chimeras can be produced by both microinjection *(2)* and aggregation techniques *(3)*.

Elizabeth J. Cartwright (ed.), *Transgenesis Techniques*, Methods in Molecular Biology, vol. 561
DOI 10.1007/978-1-60327-019-9_14, © Humana Press, a part of Springer Science+Business Media, LLC 2009

However, the success of these techniques is very much dependent on ES cell quality. For excellent quality clones, you can produce germline transmitting chimeras with both techniques equally well. However, if the quality of the ES cells is somehow suboptimal, you will find advantages with the microinjection technique. By observing the single cell suspension under high magnification, it is possible to select individual ES cells for injection by morphological criteria, whereas at much lower magnification, aggregates of ES cells are picked essentially randomly, and therefore cannot select individual ES cells. However, it is possible that there is still some kind of selection, just by the potential of the ES cells to reaggregate and take part in an aggregate at all, since not all ES cells in suspension will re-aggregate. Moreover, it is known that at most three ES cells contribute to the formation of the embryo proper *(4)*.

It is possible that during all procedures in cell culture such as transfection, selection, and subcloning, some ES cells change their genetic program upon exposure to suboptimal culture conditions thereby losing their capacity to contribute at a high level to multiple cell lineages and may not colonize the germline of the resulting animal. This event could also alter parameters like the morphology of the cell (size, shape) and thus could be counter selected under high magnification. However, neither technique can completely compensate for low-quality ES cells; it would simply not work at all with ES cells that had been mistreated.

By applying both techniques to the same ES cell clones in parallel, we found that it was necessary to aggregate more embryos than were injected in order to generate the same number of good chimeras. However, since you can aggregate many more embryos the same time compared to injection you can produce the same number of good chimeras per experimental day. The aggregation experiment simply needs more space in the animal unit since you need to accommodate more transfer cages.

This applies if outbred stock animals are used as embryo donors. Working with inbred strains here is less efficient and not recommended. We have experienced good results using morula aggregation to produce chimeras when using ES cells generated from 129 substrains (E14.1, R1).

2. Materials

2.1. Animals

In general, the same considerations as described for the use of animals for blastocyst injection apply to the use of animals for the aggregation (*see* Chapter 13).

Embryo donors are used from outbred stock mice like CD1.

2.2. Cell Culture

1. Humidified incubator with 5% CO_2 and 37°C dedicated for ES cells only.
2. ES cells.
3. MEF cells (inactivated).
4. ES and MEF cell media – *see* **Subheading 2.3** of Chapter 13.
5. 0.05% Trypsin–EDTA in sterile PBS (without Ca^{2+} and Mg^{2+}).
6. 1 M HEPES buffer.
7. 2-Mercaptoethanol.
8. Mitomycin C (Sigma).
9. 70% ethanol.
10. 0.1% Gelatin solution.
11. Center-well organ culture dish (Falcon).

2.3. Embryo Culture

1. Humidified incubator with 5% CO_2 and 37°C dedicated for embryos only.
2. Dissecting microscope.
3. M2 and M16 media (*see* Chapter 13).
4. Embryo–handling pipette made from glass capillaries (Blaubrand Micropipettes).
5. Mouth pipette (Aspirator tube assemblies, Sigma).
6. Embryo-tested mineral oil; stored in the dark (Sigma).
7. Micro drop culture dishes (Falcon).

2.4. Aggregation

1. Eight-cell stage to compact morula-stage embryos (2.5 dpc).
2. Two-cell stage embryos (1.5 dpc).
3. Semiconfluent ES cells grown on a 6- or 10-cm dish with MEF or STO feeder layer.
4. ES cell medium.
5. Darning needle (BLS).
6. Pulse generator (CF-150/B Cellfusion/BSL or Multiporator/Eppendorf).
7. Dissecting microscope.
8. Electrode chamber (GSS-1000 Fusion Electrode/BLS 3FE-0057).
9. 0.2-mm Microfusion chamber (Eppendorf).
10. Incubator, humidified, 5% CO_2 and 37°C.
11. Mouth pipette assembly.
12. 0.05% Trypsin–EDTA.
13. 0.1% Gelatin solution (Merck) in cell culture grade water.

14. Phosphate-buffered saline, Ca^{2+} and Mg^{2+} free.
15. 0.3 M Mannitol.
16. BSA.
17. Ultrapure water.
18. 0.22-Micrometer Semiconfluent Millipore filter.
19. M2 medium.
20. M16 medium.
21. Tyrode's acid solution.

3. Methods

3.1. Aggregation of Two Diploid Embryos (Sandwich Aggregation)

The sandwich aggregation technique can be used to aggregate two genetically identical embryos with ES cells, or two embryos of different genotypes.

1. Prepare the aggregation dish. Set up a microdrop culture with M16 medium in a 6-cm culture dish containing three to four rows of drops (depending on the size of your drops (*see* **Note 1**)); place only one drop in the first row and four drops in each of the remaining rows.
2. Completely overlay the micro drops with mineral oil.
3. Clean the needle to be used for indentations (*see* **Note 2**) with 70% ethanol and create indentations in the drops by pressing and turning the needle into the plastic. Create six indentations in each micro drop, except for the one in the first row. In the beginning, make sure that the indentations are deep enough, smooth (*see* **Note 3**) and at the periphery of the drop. Prepare sufficient aggregation plates for your experiment. Place the plates into the incubator (5% CO_2 and 37°C) at least several hours before you start. It is recommended that the plates are prepared in the late afternoon the day before aggregation.
4. Collect eight-cell stage to morula-stage embryos of the two genotypes you want to aggregate by flushing oviducts at 2.5 dpc. *See* Chapter 13 – **step 3** of **Subheading 3.4** for full details. Briefly, cut out the oviduct and approximately 1 cm of the uterus; insert a ground, blunt-tipped 30-gauge needle into the infundibulum of the oviduct and flush each oviduct with approximately 0.2 ml of M2 medium. Collect all eight-cell stage to compact morula-stage embryos (*see* **Note 4**) and wash them through three drops of fresh M2 medium to get rid of the debris.

5. Place the flushed embryos (keep the different genotypes separately) in a center-well organ culture dish with M16 in the incubator until the next step.
6. Prepare the dish for removal of the zona pellucida. Set up drops of M2 and Tyrode's acid solution in a lid (surface tension of the drops is kept) of a 10-cm culture or Petri dish as follows: set up five lanes that consist (from top to bottom) of one drop of M2 medium followed by three drops of Tyrode's acid solution and two drops of M2 medium at the end.
7. Take the embryos of one genotype out of the incubator, wash them in a drop of M2 medium and transfer them to the top row of the 10-cm dish containing the M2 medium. (The number of embryos in each drop strongly depends on the speed of manipulations; if the embryos stay too long in acid Tyrode's they could be damaged!). It is recommended to start off with 10–20 embryos in each drop.
8. To remove the zona pellucida pick up the embryos in the first lane with as little medium as possible (*see* **Note 5**) and wash them through two drops of Tyrode's acid solution as fast as possible. Then transfer the embryos to the third drop and observe them for the dissolution of the zona pellucida (*see* **Fig. 1b**). If the zona pellucida does not dissolve within ~ 10 s move the embryos carefully up and down with your pipette.
9. Immediately remove the embryos after the zona pellucida dissolves and transfer them to the next drop containing M2 medium.
10. Transfer the embryos to the last drop of M2 medium and repeat the procedure with the remaining embryos (*see* **Note 6**).

Now, either you can transfer the zona-free embryos back into M16 in a center-well organ culture dish, place in the incubator and continue to remove the zona pellucida from the embryos

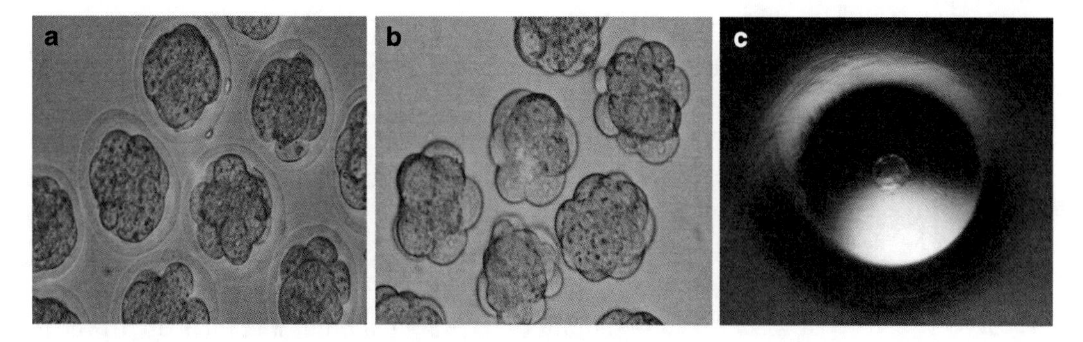

Fig. 1. Removal of the zona pellucida and placing the zona-free embryo into the depression of the aggregation plate. (**a**) Embryos with intact zona pellucida. (**b**) Embryos without zona pellucida. (**c**) An aggregate after overnight culture in a depression well.

of the second genotype as described above or you can place the zona-free embryos of the first genotype individually into the prepared indentations in the aggregation plate. In both cases, it is important to transfer the embryos back into M16 medium in the incubator. You should not leave the embryos in a group together for more than 1 h, otherwise they will aggregate together into a big clump.

11. After the removal of the zona pellucida, place the embryos of the first genotype individually into the indentations of the aggregation plate.
12. Repeat this step with the embryos of the second genotype and make sure that the embryos in each indentation are physically attached to each other.
13. Move the aggregation plate back to the incubator and culture the aggregates overnight. The next day, the aggregates should have formed a single embryo at the late morula or blastocyst stage (*see* **Fig. 1c**).
14. For transfer to a pseudopregnant recipient female mouse, the embryos from the indentations (*see* **Note 7**) need to be harvested carefully; since they are zona-free embryos, they are very fragile. It is recommended to use handling pipettes with a slightly bigger diameter. Do not harvest too many at a time since they are sticky and will potentially form a clump. Also, when preparing the transfer needle it is recommended to use a little bit more medium between the embryos, compared to the very close chain that is recommended when loading a transfer pipette with embryos with intact zona pellucida.
15. Transfer the embryos to a pseudopregnant recipient female mouse (*see* Chapter 15).

3.2. Aggregation of Diploid Embryos and ES Cells

1. Set up an aggregation plate as described in **Subheading 3.1** in the late afternoon the day before aggregation and place it in the incubator overnight.
2. Collect eight-cell stage to morula-stage embryos by flushing the oviduct of 2.5 dpc mice and remove the zona pellucida (*see* **Subheading 3.1**).
3. Transfer the zona-free embryos individually into the indentations of the aggregation plate.
4. Wash the semiconfluent 6-cm ES cell culture (*see* **Note 8**) with PBS and discard the supernatant (ES cells need to be cultured once without a feeder layer of MEF).
5. Add a minimum of trypsin (~ 0.5 ml) and incubate the cells for approximately 2 min at 5% CO_2 and 37°C.
6. Observe the cells under a microscope. When the first cells start to lift up from the culture dish, gently tap against the

dish several times to loosen the majority of the cells from the bottom of the culture dish.

7. Add 3 ml of ES cell culture medium to stop the trypsin reaction. Check under the microscope if the cells have formed clumps (*see* **Note 9**).
8. Alternatively, create a single cell suspension and then put the cells back into the incubator for about 0.5–1 h and wait for reaggregation of ES cells. As soon as large enough aggregates have formed, you can put the dish on ice or into the refrigerator. Reaggregation ensures that only live cells are aggregated within the clumps and you can make sure that no MEF cells are in the aggregates, since even the last ones will stick first to the bottom of the dish even before aggregates are formed.
9. Pick ~200 clumps of cells and transfer them to the drop without indentations in the aggregation plate. Pick again from this drop and try to enrich for clumps with the right number of cells (10–15) and transfer the cells into the middle of the first drop in the second row.
10. Pick one clump of the right size and place it into the first indentation together with the zona-free embryo. Make sure that the clump of cells and the embryo are touching each other. Ensure that all excess ES cells are removed from the drop.
11. Repeat this with all the remaining indentations in that drop and then in all the remaining drops.
12. Move the aggregation plate back into the incubator and culture the aggregates for 24 h.
13. After overnight culture, the aggregates should have formed a single embryo at the late morula or blastocyst stage (*see* **Fig. 1c**).
14. For harvesting, *see* **step 14** of **Subheading 3.1**.
15. Transfer the embryos into a pseudopregnant recipient female (*see* Chapter 15). Do not transfer more than six embryos each side in a double-sided transfer, or ten in a single-sided transfer.

3.3. Generation of Tetraploid Embryos

1. Collect two-cell stage embryos (the parameters below apply to embryos from the CD1 outbred strain) by flushing the oviduct in the morning of day 1.5 dpc as described in Chapter 13 – **step 3** of **Subheading 3.4** and store them in a center-well organ culture dish with M2 until all preparations for the fusion are carried out (*see* **Note 10**, **Fig. 2a**).

2a. For the BLS cell fusion, set the pulse generator to nonelectrolyte fusion (*see* **Note 11**) and to the following parameters: GSS-1000 fusion electrode; voltage, 147 V; duration, 26 μs; repeat number, 2; AC voltage, 2.

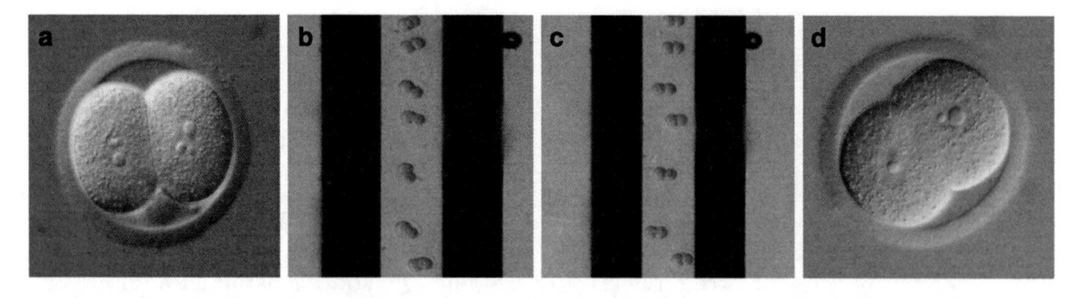

Fig. 2. Fusion of two-cell stage embryos. (**a**) Two-cell stage embryo before fusion. (**b**) Non-orientated embryos in the fusion chamber. (**c**) Orientated embryos in the weak A/C field (after pre-alignment) lined up between the electrodes. (**d**) Embryo undergoing fusion of the two blastomeres.

2b. For the Eppendorf Multiporator, set mode to tetraploid fusion. For pre-alignment: voltage, 3 V; duration: 10 s. For pulse: voltage, 70 V; duration, 35μs, no. of pulses, 1.

Please be aware that the electrofusion parameters for the BLS and the Multiporator apply to the GSS-1000 fusion electrode and the 0.2-mm fusion chamber respectively. The parameters cannot be transferred directly to other electrodes with different electrode gaps.

3. Place the electrode chamber in a 10-cm cell culture dish under the dissecting microscope and connect it to the pulse generator (*see* **Note 12**).
4. Place a big drop of mannitol in the electrode chamber and a small one in the dish. Place two drops of M2 in the dish.
5. Rinse up to 30 embryos through the small drop of mannitol and place them between the electrodes.
6. After all the embryos have been properly oriented, apply the fusion pulse (for BLS). For the Eppendorf Multiporator, the orientation need not be so exact because of the pre-alignment pulse (*see* **Note 13**, **Fig. 2b, c**).
7. Transfer the embryos to the M2 drop.
8. Transfer the embryos to M16 under oil and check approximately 30 min later to see which embryos have undergone fusion. Separate the fused embryos and recheck every 5–10 min (*see* **Fig. 2d**). At least 80% of the embryos should be fused if optimal parameters were used.
9. Culture the fused embryos for 24 h. The next day, you should expect to see four-cell stage tetraploid embryos (at least 80% of the fused embryos should develop). At this stage, the embryos are used for aggregation. For injection, culture them for an additional 24 h.

3.4. Aggregation of Tetraploid Embryos and ES Cells

1. Set up an aggregation dish as described in **Subheading 3.1** in the late afternoon the day before aggregation and place it in the incubator (5% CO_2 and 37°C).
2. Remove the tetraploid embryos generated in **Subheading 3.3** from the incubator where they had been for the last 24 h. The majority of the tetraploid embryos should be at the four-cell stage (*see* **Note 14**). Remove the zona pellucida from the embryos (*see* **Subheading 3.1**) and place the zona-free embryos individually into the indentations of the aggregation plate.
3. Prepare the ES cells (*see* **steps 4–9 of Subheading 3.2**) and place clumps of cells of the right size (10–15) into the indentations containing the embryos as described in **Subheading 3.2**.
4. Culture the aggregates for 24 h in 5% CO_2 and 37°C.
5. After overnight culture, the aggregates should have formed a single embryo at the late morula or blastocyst stage (*see* **Fig. 1c**).
6. For harvesting, *see* **step 14** of **Subheading 3.1**.
7. Transfer the embryos to pseudopregnant recipient females (*see* Chapter 15).

4. Notes

1. *Aggregation plate.* Take care not to put two drops too close to each other. During the movement of the dish in and out of the incubator, the drops might come into contact with each other and run together. Larger drops will be more affected by the movement and vibration and this can lead to fluid movement in the individual indentations.
2. *Darning needle.* The needle to create indentations is a medium-sized darning needle. In order to be able to press it vertically into the plastic and turn it, it is necessary that your workshop modify the end of it; a plastic handle (such as that used for opening valves) should be added to the end.
3. *Shape of indentations.* The wall of the indentation wells should be smooth, as fractions could damage the embryos during harvesting and the embryos may stick. If the indentations are too steep, it makes harvesting of the aggregates more difficult. Try to blow the aggregates out of the well using the medium in the preloaded capillary. If the indentations are not deep enough, the aggregates might not stay in place.

4. *Selection of embryos.* In order to achieve a good efficiency, it is essential to use only morphologically perfect-looking eight-cell stage to morula-stage embryos for aggregation.
5. *Removal of zona pellucida.* It is essential to transfer a minimal volume of M2 medium to the drops of Tryode's acid solution as the pH of the Tyrode's acid solution will change immediately and will inhibit fast and efficient dissolving the zona pellucida.
6. *Sticky embryos.* After the removal of the zona pellucida, the embryo will be sticky and thus it is important to work very quickly. For removal of the zona, it is recommended to use the lids of sterile cell culture grade plastic dishes, since the drops stay nicely round, because of the surface tension, even after entry of the pipette.
7. *Harvesting zona-free embryos.* This can be done by carefully blowing some medium into the indentation. The aggregates should move out of the indentation as one complete embryo, and the blastocoel should not collapse.
8. *MEF-free ES cell aggregates.* As you cannot influence the process of reaggregation and select ES cells in any way, it is very important to grow the ES cells in their last passage free of all MEF cells. It is recommended to passage the cells at least once before aggregation. The last split should be done in a "feeder-free" manner on cell culture dishes coated with 0.1% gelatin with an extensive period of preplating to get rid of all MEF cells.
9. *Size of ES cell aggregates.* The clumps should contain about 10–15 cells (depending on the ES cell line used). If the majority of the clumps are still too big, pipette up and down several times using a 1-ml pipette. If the clumps are too small, put the culture dish back in the incubator and leave the cells there to reaggregate. Check every 15 min.
10. *Developmental differences between mouse strains.* Different mouse strains with different genetic background develop at different rates. The exact time to harvest the embryos has to be determined empirically. But, in general, two-cell embryos are found in the early morning of 1.5 dpc.
11. *Embryo orientation.* You can switch to electrolyte fusion, by which you can fuse in the growth medium, e.g., M16 or M2. But, then, all embryos have to be orientated manually as a weak A/C orientation field is lost.
12. *Fusion electrode.* Always keep the electrode clean of salts as this can affect the electrical conductance across the electrode and this will cause embryo lysis.

13. *Mannitol.* Leave the embryos for as short a time as possible in mannitol, keep the mannitol cold and replace the mannitol after every fusion.

14. *Fused embryos.* You will also find one-cell, two-cell and eight-cell stage embryos. Discard the one-cell stage embryos. The two-cell stage embryos are delayed and can be used later in the day when they have developed to the four-cell stage. The eight-cell stage embryos can be used right away for aggregation.

Acknowledgments

Dr. Alfonso Gutierrez-Adan, Departamento de Reproduccion Animal y Conservacion de Recursos Zoogeneticos, INIA, Ctra de la Coruna Km 5,9, Spain, agutierr@inia.es)

Ronald Naumann, Transgenic Core Facility, MPI of Molecular Cell Biology and Genetics, Pfotenhauerstrasse 108, D-01307 Dresden, Germany, naumann@mpi-cbg.de

Kristina Vintersten, Mount Sinai Hospital, Samuel Lunenfeld Research Institute, 600 University Ave, Toronto, Canada, vintersten@mshri.on.ca

References

1. Tarkowsky, A. K. (1961) Mouse chimaeras from fused eggs. *Nature* **190**, 857–860.
2. Thomas, K. R., and Capecchi M. R. (1990) Targeted disruption of the murine int-1 proto-oncogene resulting in severe abnormalities in midbrain and cerebellar development. *Nature* **346**, 847–850.
3. Wood, S. A., Allen, N. D., Rossant, J., Auerbach, A., and Nagy, A. (1993) Noninjection methods for the production of chimeric stem cell–embryo chimeras. *Nature* **365**, 87–89.
4. Wang, Z., and Jaenisch, R. (2004) At most three ES cells contribute to the somatic lineages of chimeric mice and of mice produced by ES-tetraploid complementation. *Dev. Biol* **275**, 192–201.

Chapter 15

Surgical Techniques for the Generation of Mutant Mice

Anne Plück and Christian Klasen

Summary

The previous two chapters have described the generation of chimeric embryos by blastocyst microinjection and morula aggregation. This chapter describes the reimplantation of these embryos into pseudopregnant female recipient mice (foster mice) in order for the embryos to develop to term. Four different surgical techniques will be described in detail.

The transfer of blastocyst-stage embryos back into the uterus is most commonly used for chimeric embryos generated by microinjection or aggregation, or for rederivation purposes when strains are transferred to a clean environment. Fertilized oocytes manipulated by pronuclear microinjection or two-cell-stage embryos are reintroduced to the infundibulum of the oviduct of pseudopregnant female mice.

To generate pseudopregnant females, they need to be mated to sterile male mice. These males can be either generated by vasectomy of stud males or naturally sterile males can be used. Two different techniques for the vasectomy are shown: the vas deferens is accessed through an incision in the scrotal sac or through an abdominal incision.

Key words: Uterus transfer, Oviduct transfer, Vasectomy, Abdominal incision, Incision in the scrotal sac, Rederivation, Aseptic conditions

1. Introduction

While all preimplantation-stage embryos were handled in vitro in order to introduce the mutation (in vitro development), in vivo development from the blastocyst stage on is required for embryos to develop to term.

Elizabeth J. Cartwright (ed.), *Transgenesis Techniques*, Methods in Molecular Biology, vol. 561
DOI 10.1007/978-1-60327-019-9_15, © Humana Press, a part of Springer Science+Business Media, LLC 2009

To achieve the best possible conditions, two different techniques are applied: transfer of fertilized oocytes or all later stages up to the blastocyst stage into the infundibulum of the oviduct of a 0.5 dpc (days postcoitum) recipient female and transfer of blastocyst-stage embryos into the uterus of a 2.5 dpc recipient ("oviduct transfer" and "uterus transfer").

Pseudopregnancy of the recipient mouse ("foster mouse") is achieved by mating female mice with vasectomized males or by mating to naturally sterile male mice.

In principle, the embryos should be transferred to the place where they originally came: the early preimplantation stages (days 0.5–2.5) are put back to the oviduct to a 0.5-day pseudopregnant mouse, whereas the blastocyst-stage embryos are reintroduced to the uterus of a 2.5-day pseudopregnant female mouse (this is an easier procedure to carry out, and can even be performed without a microscope). However, there are also facilities that transfer all preimplantation-stage embryos into the oviduct of 0.5-day pseudopregnant females, which works at comparable efficiencies.

Reimplantation of embryos into the reproductive tract of 1.5-day pseudopregnant fosters is not recommended because of very low efficiency and unpredictable results.

Astonishingly, for the survival of the mouse after surgery, aseptic conditions are not as essential as for the rat. However, it is good practice to routinely perform any surgery on a clean workbench using sterile tools; this will also maintain the hygiene levels required in any SPF or SOPF status animal facility.

Since embryos that have been transferred have at least received a penetration of their zona pellucida or the zona pellucida was removed during the in vitro manipulation, many facilities perform an intensive health screen on the foster animals and negative (nonmutant) littermates before releasing those animals to users' rooms.

The embryo transfer techniques described below can also be used in order to get rid of pathogens carried by existing mouse strains when moved to a different facility, or if embryos were used for transport preferentially to island countries such as the UK and Australia. The exchange of strains between labs is a major task for facilities in our days, because of the strong force of globalism specifically in science.

However, it is not clear whether infections of all known pathogens can be transmitted vertically back to the mother and lead to seroconversion. In immunological terms, pregnancy is a state of tolerance; the immune response may be suppressed by certain mechanisms, or contact between the recipient mother and the pathogen may be inhibited by the placental barrier and therefore would not necessarily lead to an immune response (which is essential for health monitoring by ELISA).

Of course, this testing could become very expensive in a highly productive facility, and experience with the conventional injection techniques over the last 20 years in many places has shown that, to our knowledge, there was no relevant cases of pathogen transmission (with the exception of *Pasteurella pneumotropica* in two cases).

In our opinion, the most dangerous germ in today's business might be the transmission of Parvo virus, since it is extremely stable in the environment and appears at very low incidence (specifically in B6 mice which remain seronegative). It is also possible that it is transmitted even during in vivo fertilization (authors' own experience) from infected male mice to clean wild-type female mice *(1)*. Any decontamination of the embryos would not help in this case, since the virus might enter during fertilization. However, it is still very unlikely to retrieve it in single experiments in particular using C57BL/6 mice because of the low prevalence within a given positive population. (We found one positive within 12 samples, and think we were simply lucky.)

2. Materials

2.1. Surgical Tools

All tools used should be sterile. All stainless steel tools can be autoclaved in a stainless steel toolbox with a silicon pad inside and reused for several surgeries inside the clean bench by washing in sterile PBS and 70% ethanol or using a glass bead sterilizer.

Glass capillaries for embryo transfer must be sterilized by dry heat: 180°C for 2 h. Autoclaving is not recommended because of precipitation of contents of the steam within the capillaries.

Syringes and needles are used sterile as they come from the supplier (sterilized by irradiation) (*see* **Notes 1** and **2**).

1. Thick, straight blunt-ended forceps.
2. Large scissors – for opening the skin during embryo preparation procedure.
3. Fine, straight blunt-ended forceps.
4. Fine, bent blunt-ended forceps.
5. Two pairs of straight watchmaker forceps No. 5 (*see* **Note 3**).
6. Bent watchmaker forceps No. 7.
7. Small scissors.
8. Fine scissors (Iris scissors).
9. Specialized forceps for sewing.
10. Small box for sewing needles.
11. Three sewing needles.

12. Sewing silk.
13. Seraphine clamp.
14. Two pairs of extremely fine watchmaker forceps (Inox No. 5) exclusively used for oviduct transfer.
15. Wound clips.
16. Wound clip applier.
17. Wound clip remover.
18. Heating platform for postoperative warming of the animal.
19. Glass bead sterilizer.
20. Cauterizer (alternatively: simple forceps and camping burner).
21. Stereomicroscope with light box and additional cold light lamp.
22. 1-ml syringes.
23. 0.45-mm needles.
24. Self-made, size-sorted glass capillaries at 100–120 μm diameter.
25. 50-ml tube with sterile PBS.
26. 50-ml tube with 70% EtOH.
27. Spray bottle with 96% EtOH.

2.2. Animals

1. Recipient females.
   Female recipients should be at least 8 weeks of age and weigh between 25 and 35 g depending on the strain (hybrids: 25–32 g, CD1: 28-35 g).

   Outbred stock females (e.g. CD1) or F1 hybrids (e.g. C57BL/6 × CBA, C57BL/6 × Balb/c, or C57BL/6 × DBA/2) are good mothers as they carry the embryos to term and nurse the newborns to weaning age. They also allow inspection of the newborn pups and accept pups from another mother. Pseudopregnant females are mated to vasectomized males and the day of the vaginal plug is considered as day 0.5 of pseudopregnancy (0.5 dpc). Use 0.5 dpc pseudopregnant females for oviduct transfer and 2.5 dpc pseudopregnant females for uterine transfer.
2. Vasectomized males.
   Vasectomized males receive the surgery best at 5-6 weeks of age, since they do not have so much body fat at that age and need to rest after the surgery for 3 weeks. Usually the genetic background of the vasectomized males is not critical as these males should not produce any offspring, but they should be good breeding performers. To achieve a reasonably good plug rate for the females, each day the plug rate of an individual male is recorded on its cage card and the male is exchanged if it does not perform for more than 3 weeks, when mated for three consecutive nights.

In the case of F1 females, sometimes plug rates in normal mating are very high (50–60%) on the first night. If this is the case, just mate them 1:1. Sawdust priming has no relevant effect in these females; it may even lower the plugging rates.

In the case of CD1 females, plug rates are usually in the range of 20% on the first night. Sawdust priming works very well in order to synchronize females and receive reproducible plug rates with CD1 females (Whitten effect) *(2, 3)*.

2.3. Anaesthesia

The administration of a combination of ketamine/xylazine is a very reliable anaesthesia for mouse surgery *(4)*. The recommended dose varies between 100 and 200 mg/kg body weight for ketamine and between 5 and 16 mg/kg body weight for xylazine.

We recommend the use of a ketamine/xylazine mix at a concentration of 100-mg ketamine, 20-mg xylazine, in a final volume of 10 ml (in sterile PBS). The final concentration is: 10 μl/g body weight of this solution. Store for a maximum of 2 weeks at 4°C. It can also be stored as a 10× frozen stock.

The combined use of ketamine/xylazine has been well studied and has been shown to be very reliable because ketamine has no effect on the peripheral nervous system, and due to this its effect on the respiration will be indifferent (e.g. swallowing reflex will be untouched).

Please note that higher doses of ketamine will not necessary lead to a deeper anaesthesia, rather a prolonged anaesthesia will be achieved. Anaesthetics containing ketamine must be combined with other substances (e.g. xylazine) to obtain a suitable muscle relaxation for surgery as ketamine blocks pain perception and consciousness, but catalepsy (muscular rigidity) will be activated.

The ketamine/xylazine mix is administered intraperitoneally. The above recipe provides an onset of approximately 5 min with ~45 min duration of surgical anaesthesia.

After completion of the surgical procedure, we recommend that animals are kept at 36°C on a heated plate and monitored carefully until they regain reflex responsiveness.

3. Methods

3.1. Single-Sided Uterus Transfer: Embryo Transfer into 2.5-Day Pseudopregnant Foster Female

1. Cover the eyes with a drop of eye ointment or wet the eyes with a drop of PBS.
2. Place the mouse on its side, facing to the left. Remove the hair from the body wall between the last rib and the hind leg, and wet with 70% EtOH. Alternatively, you can wet the fur with 96% EtOH and do a parting in the hair (ask your veterinarian for approved protocols).

3. Cut approximately 3-5 mm in the skin right above where the ovary is expected to be found (*see* **Fig. 1a**).
4. In some cases, there is a fat layer underneath the skin that will need to be opened as well in order to locate the reproductive organ within the peritoneum (orange spot of the ovary).
5. Cut approximately 3-4 mm through the peritoneal wall, where the fat pad attached to the ovary can be located.
6. Pull out the fat pad using anatomic forceps. Some of the ligaments attached between the ovary and the kidney should be cut to prevent excessive pulling on the kidney. Never touch the ovary or oviduct itself with your forceps, just take the fat pad.
7. Pull approximately 1 cm of the ovary/uterus outside of the body in such a way that the first part of the uterus can be easily accessed at a right angle.
8. Secure the fat pad with a seraphine clamp (*see* **Fig. 1b**).
9. Make a hole through the uterus wall with a 25G needle, and insert the glass capillary which has been preloaded with embryos for transfer (*see* **Fig. 1c, d, Note 4**).
10. Place the ovary/uterus back behind the body wall by pushing back the fatty pad with anatomic forceps.
11. Make sure that all tissues associated with the ovary/uterus are back inside the peritoneum (this is easily achieved by holding the peritoneal wall up on both sides of the incision using forceps like a "shopping bag"). Seal the skin with a wound clip (*see* **Fig. 1e**). Some people like to sew the peritoneal wall with one stitch. This is recommended when the incision is larger than 3–5 mm. This stitch is not necessary for small incisions. Pregnancy rates are not negatively influenced by either procedure.
12. Seventeen days later potentially chimeric animals will be born. Coat colour chimerism can be detected about 4-7 days postnatal. In some combinations of ES cells and carrier embryos (129 derived ES cells in embryos from C57BL/6 mice), the colour of the eye, although not opened, can be the first indicator of chimerism even one day after birth (B6 derived phenotypic eyes would look like dark spots, 129 derived eye colour would look much lighter).

If ES cells are derived from a 129 inbred mouse strain, which is albino, pale yellow coloured or agouti (C57BL/6 J mice are black), the resulting chimeric pups will therefore display a mixed coat colour. The exact coat colour distribution in the chimeric mouse is further complicated by the fact that the colour of each individual hair is defined by the genotype of the follicle cell and the genotype of the neighbouring melanocyte. So different

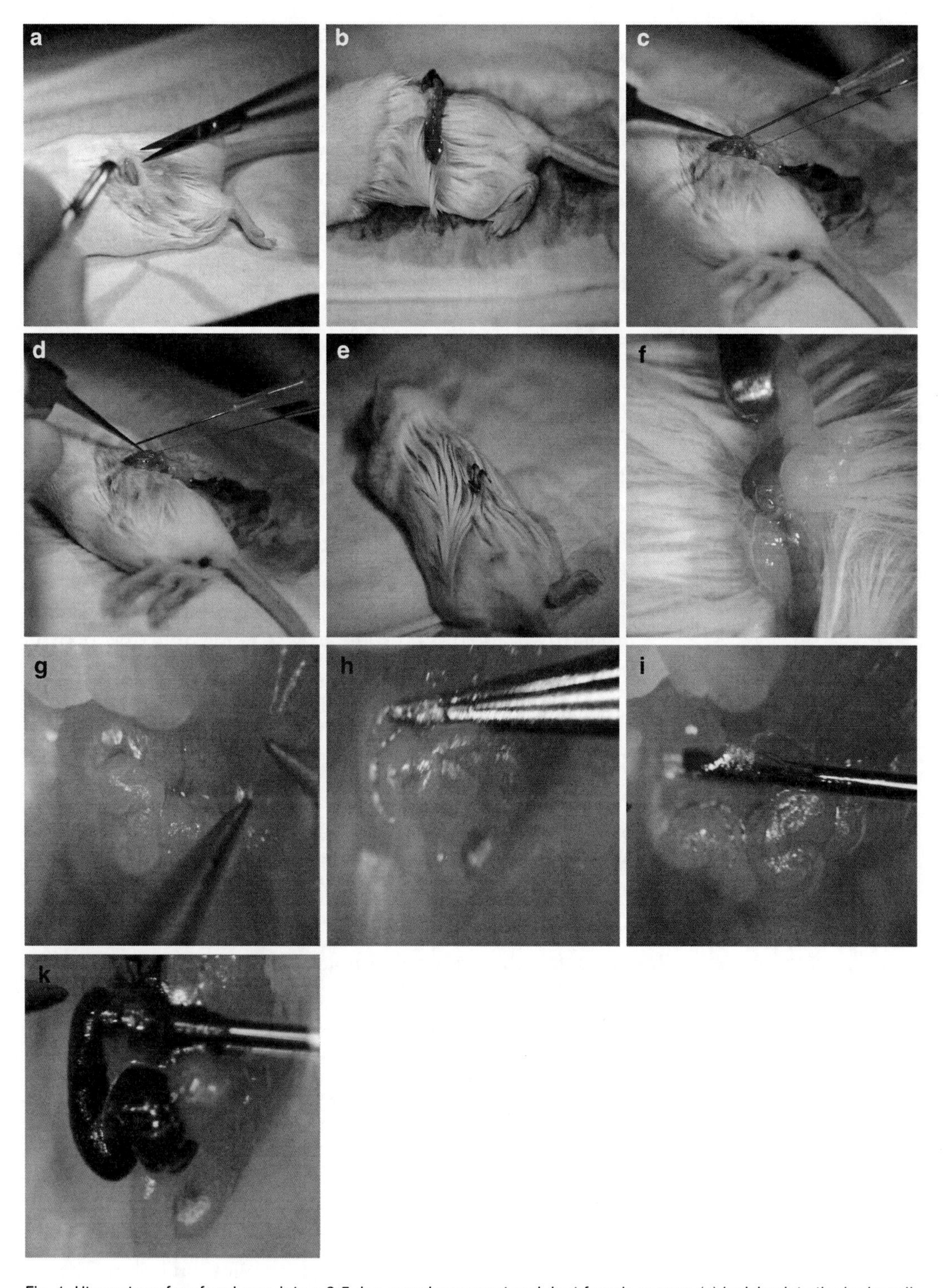

Fig. 1. Uterus transfer of embryos into a 2.5 dpc pseudopregnant recipient female mouse. (**a**) Incision into the body wall, where the ovary is expected to be found. (**b)** Ovary/uterus placed outside and kept in position with a seraphine clamp. (**c**) A 25G needle is used to make a hole in the uterus. (**d**) The embryos, preloaded in a glass capillary, are transferred to the uterus by inserting the glass capillary and expelling the embryos. (**e**) The skin is sealed with a wound clip. Oviduct transfer of embryos into a 0.5 pseudopregnant recipient female mouse. (**f**) Ovary/oviduct kept in position by a tissue holder at the fat pad. (**g**) The bursa surrounding the ovary/oviduct is torn apart. (**h**) The infundibulum is located. (**i**) The glass capillary is inserted into the infundibulum. (**k**) With a gentle blow, the embryos are released into the oviduct.

combinations of genotypes are possible here. The higher amount of ES-cell-derived coat colour the chimera has, the higher the chance that the germ cells are also derived from ES cells. Hence, the chimera will be more likely to pass on the recombinant genotype to the next generation. Chimeras are in this case mated to C57BL/6 J female mice. The following offspring will be black if wild type and agouti if germ line transmission (the so called coat colour transmission) has occurred. Among the agouti offspring, half will carry one recombinant allele in case the ES cell clone was heterozygous for the mutation.

3.2. Oviduct Transfer: Embryo Transfer into 0.5-Day Pseudopregnant Fosters

1. Cover the eyes with a drop of eye ointment or wet the eyes with a drop of PBS.
2. Place the mouse on its side, facing to the left. Remove the hair from the body wall between the last rib and the hind leg, and wet with 70% EtOH. Alternatively, you can wet the fur with 96% EtOH and do a parting of the hair.
3. Cut approximately 3–5 mm in the skin right above where the ovary is expected to be found (*see* **Fig. 1a**).
4. Where necessary, open the fat layer underneath the skin in order to locate the reproductive organ within the peritoneum (orange spot of the ovary).
5. Cut approximately 3–4 mm through the peritoneal wall, where the fat pad attached to the ovary can be located.
6. Pull out the fat pad with anatomic forceps. Cut some of the ligaments attached between the ovary and the kidney to prevent excessive pulling on the kidney. Never touch the ovary or oviduct itself with your forceps, just take the fat pad.
7. Place approximately 1 cm of the ovary/uterus outside of the body (*See* **Fig. 1b**).
8. Use a seraphine clamp to hold the fat pad in position (*see* **Fig. 1f**).
9. Using two watchmaker forceps, tear the bursa surrounding the ovary/oviduct between the two main vessels located vertically left and right to the region where the infundibulum is visible underneath (*see* **Fig. 1g**). Be careful not to destroy one of the vessels, since the area will flood with blood and the operation will become very difficult or even impossible.
10. The exact position of the infundibulum is located by moving in with closed watchmaker forceps (*see* **Fig. 1h**). If you can clearly target the inside of the infundibulum, enter the same opening with your preloaded glass capillary (*see* **Fig. 1i**) and expel the air bubbles in front of the embryos, the embryos themselves, and the following air bubble.

Carefully hold the fat away from the inserted capillary and infundibulum with one watchmaker forceps in your left hand while transferring the preloaded embryos (**Note 4**) into the ampulla. Use the position of the air bubbles to indicate where your embryos are, and move them using two closed watchmaker forceps forward inside the oviduct so that they definitely behind one or two curves of the duct. For teaching reasons here we injected blue dye (Evans blue) to make the structure of the oviduct visible (*see* **Fig. 1k**).

11. The ovary/uterus is carefully released from the seraphine clamp and placed back inside the peritoneum by pushing back the fat pad with anatomic forceps.
12. Make sure that all tissues associated with the ovary/uterus are back inside the peritoneum (this is easily achieved by holding the peritoneal wall up on both sides of the incision using forceps like a "shopping bag"). Seal the skin with a wound clip (*see* **Fig. 1e**). Some people like to sew the peritoneal wall with one stitch if the incision was large. This is not necessary for small incisions. Pregnancy rates are not negatively affected by either procedure.

3.3. Vasectomy: Abdominal Operation

The male mice are ideally operated at 5–6 weeks of age, from a strain with a good breeding performance (e.g. CD1). They will then need to recover from surgery for about 3 weeks.

1. Anaesthetize the mouse as mentioned above.
2. Cover the eyes with a drop of ointment or wet the eyes with a drop of PBS.
3. Place the mouse on its back, facing away from the operator. Shave the abdomen and swab the shaved area with 70% EtOH. Alternatively, you can wet the fur with 96% EtOH and do a parting of the hair at the place of the incision.
4. Make a small (approximately 5 mm) vertical incision in the skin (*see* **Fig. 2a**) at the midline.
5. Make a similar-sized vertical incision in the peritoneal wall on the silver line. Both testes can be reached through the same incision.
6. Localize the testicular fat pad on the right side and pull it through the incision using blunt forceps (*see* **Fig. 2b**). Just for teaching reasons we took out the testes as well. In a real surgery situation, we try to avoid pulling out the testes themselves.
7. With sharp fine-pointed forceps clean an approximately 1-cm strand of the vas deferens from the attached membrane (*see* **Fig. 2c**).

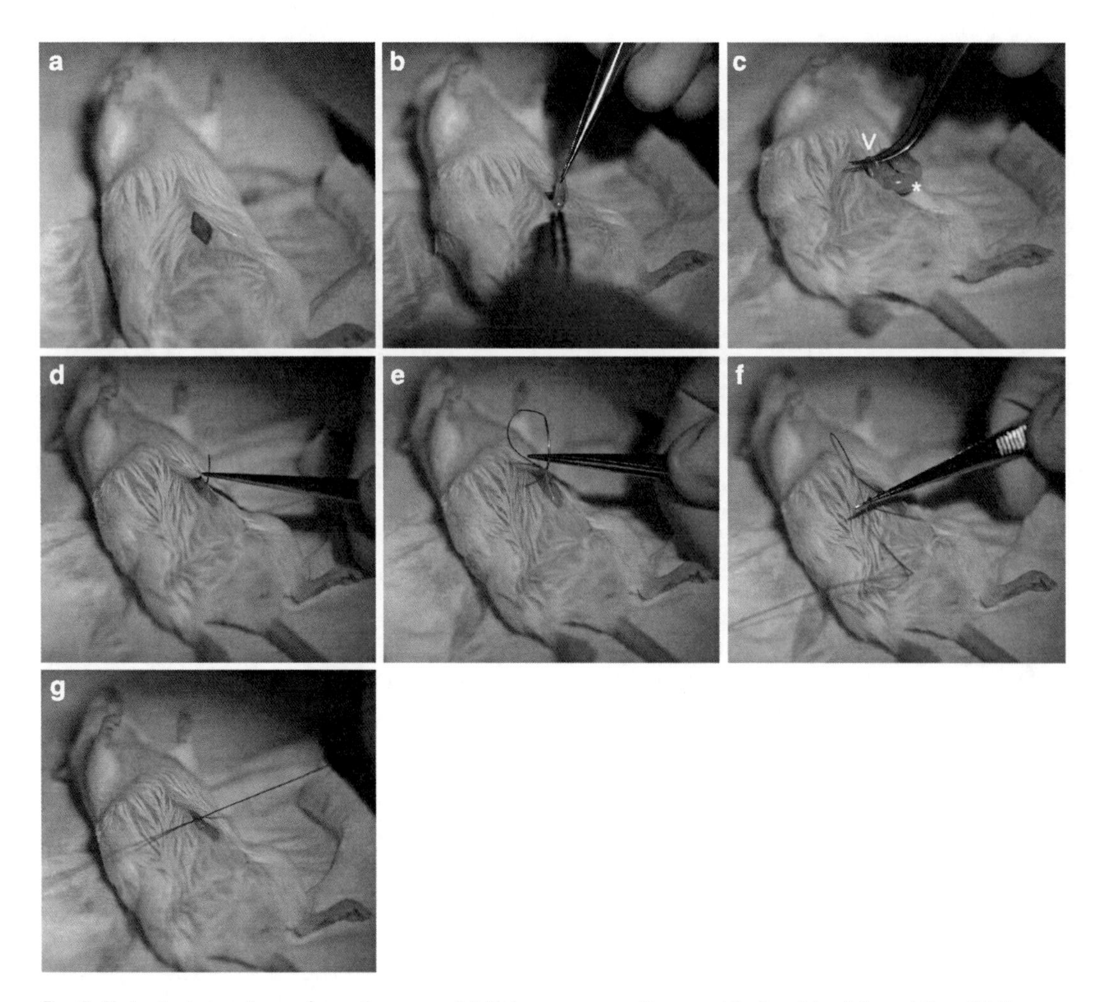

Fig. 2. Abdominal vasectomy of a male mouse. (**a**) Make an approx. 8-mm cut in the skin at the midline. (**b**) The vas deferens is reached through a transverse incision of the body wall. (**c**) The testis (*star*) and the vas deferens (*arrowhead*). Cauterize an approx. 0.5-cm portion of the vas deferens and remove it. Repeat with the other vas deferens through the same incision. (**d**) The two sides of the body wall are stitched together with a curved surgical needle. (**e–g**) Make two stitches in the body wall and then sew up the skin or clip the skin with a wound clip.

8. Using a cauterizer cut out approximately 1 cm of the vas deferens and heat seal both ends of the vas deferens, carefully put the fat pad back into the peritoneum (*see* **Note 5**).
9. Repeat the procedure with the vas deferens on the left side.
10. Close the muscle layer using two resorbable 5-0 sutures (*see* **2d–g**).
11. Seal the skin with a wound clip (*see* **Fig. 1e**).

3.4. Vasectomy: Scrotal Operation

The male mice are ideally operated at 5–6 weeks of age, from a strain with a good breeding performance (e.g. CD1) and require about 3 weeks to recover from the operation.

1. Anaesthetize the mouse as mentioned above.
2. Cover the eyes with a drop of ointment or wet the eyes with a drop of PBS.
3. Place the mouse on its back, facing away from the operator.
4. Swap the scrotum with 70% EtOH.
5. Gently press down the testicles into the scrotum.
6. Make a 3-5-mm cut in the skin in the midline just under the penis, between the scrotal sacks (*see* **Fig. 3a**).
7. Carefully loosen the skin slightly from the underlying tissues.
8. Make a 3-mm cut through the cremaster muscle tissue covering the right side testicle. This cut should be placed exactly between the midline "wall" and the testicle.
9. With fine-pointed forceps (Dumont Number 5 straight) locate the vas deferens which lies underneath the testis and can be recognized by a blood vessel running along (*see* **Fig. 3b**).
10. Cauterize the vas deferens using a small vessel cauterizer or with heated (red glowing) forceps close to the cauda epididymidis (*see* **Fig. 3d**, **e**).

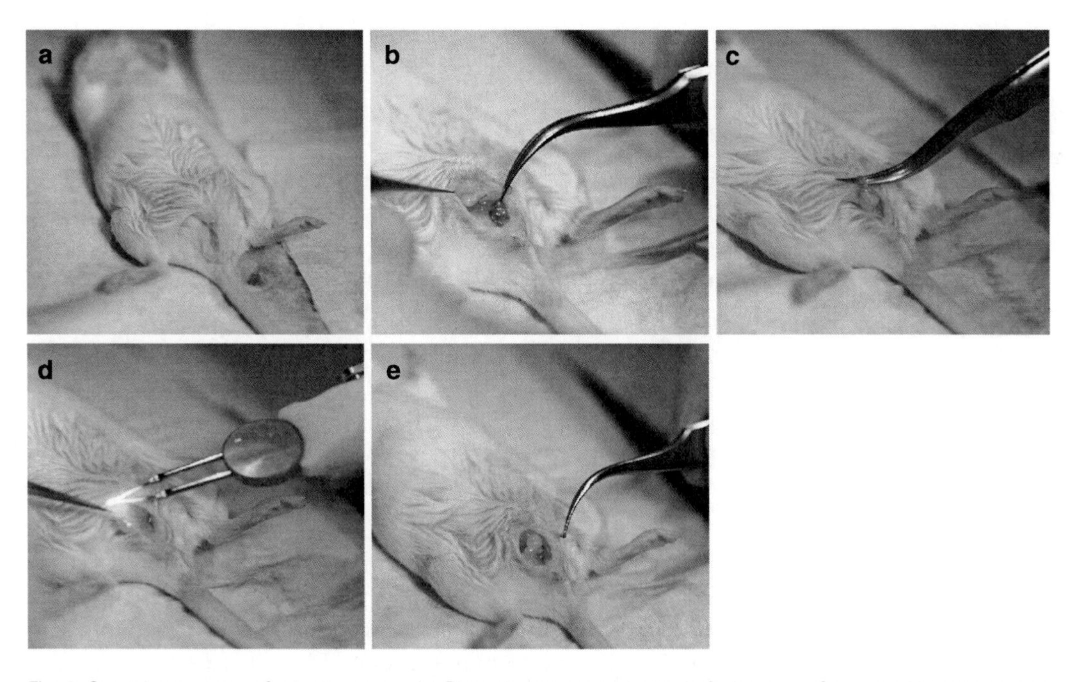

Fig. 3. Scrotal vasectomy of a male mouse. (**a**) Place the mouse on its back, facing away from you. Make an approx. 3-5mm cut in the skin between the scrotal sacks just under the penis. (**b**) After making an approx. 5-mm cut of the muscle covering the right testicle, the vas deferens is pulled outside. (**c**) The vas deferens is cauterized. (**d**, **e**) Approx. 3-mm piece of the vas deferens is removed. Same procedure is used for the left side.

11. Pull out an approximately 5-mm stretch of the vas deferens, and cut using the same procedure (*see* **Fig. 3d**, **e**).
12. Place all tissues associated with the testicle back inside the covering tissue.
13. Repeat the procedure on the left testicle (**steps 6–12**).
14. Close the muscle layer using 1 or 2 resorbable 5–0 sutures.
15. Seal the skin with a wound clip.
16. Check the male the day after surgery to make sure no bleeding has occurred from the broken blood vessels.
17. Leave the wound clip in for 4–6 days, then remove it.
18. Test breeding can be done if you are not sure about your procedure.

Mate the males about 21 days after surgery, check for plugs and for pregnancies in the group of plug-positive females. Continue until each male has produced three plug-positive females (*see* **Note 6**).

4. Notes

1. *Check instruments.* Always check all instruments under the microscope before use. It is essential to determine that they are still functional before beginning surgery. This is particularly important for watchmaker forceps which are very easy to damage.
2. *Cleaning of tools.* Washing used tools with sterile PBS or sterile water before contact with ethanol is essential to avoid protein precipitation on the tools. If this is not done very soon after the procedure, expensive surgery tools become useless, since it becomes difficult to remove blood residues or tissue particles.
3. *Broken watchmaker forceps.* Watchmaker forceps that have bent tips can no longer be used during embryo transfer operations. However, these can be repaired using a grinding device; they will be suitable for use during embryo preparation.
4. *Preloaded.* Preloaded means that all capillaries for transfers are already loaded with the embryos and stored in a wet chamber. The capillary should be loaded as follows: approximately 3-5-mm medium, 1 air bubble, 3-5-mm medium, 1 air bubble, embryos in medium as tightly packed as possible, 1 air bubble, and 1–2-mm medium.

5. *Cauterizer versus heated forceps*. The use of the cauterizer instead of cheap simple forceps heated in a Bunsen flame has, in our experience, led to the regeneration of fertility at the age of about 6 months of the vasectomized male twice within 2 years. We switched to the cauterizer for safety and good handling reasons. A naked flame inside the sterile workbench is not allowed in many countries and is more complicated and dangerous for the operating staff to use compared to the cauterizer. However, the regeneration of fertility in males that had been shown by test breeding to be sterile was detected when pups with the wrong coat colour were born. Postmortem dissection confirmed that the two ends of the vas deferens had rejoined, and was visible by a thickened part within the vas deferens. To prevent this situation from occurring, it is essential to remove a substantial portion of the vas deferens.
6. *Vasectomized males*. We use vasectomized males according to their performance (plug rate is monitored on the cage card) or up to 12 months of age.

References

1. Agca, Y., Bauer, B.A., Johnson, D.K., Critser, J.K., and Riley, L.K. (2007). Detection of mouse Parvovirus in *Mus musculus* gametes, embryos, and ovarian tissues by polymerase chain reaction assay. *Comp. Med.* **57**, 51–56.
2. Whitten, W.K. (1958). Modification of the oestrous cycle of the mouse by external stimuli associated with the male. Changes in the oestrous cycle determined by vaginal smears. *J. Endocrinol.* **17**, 307–313.
3. Whitten, W.K., Bronson, F.H., and Greenstein, J.A. (1968). Estrus-inducing pheromone of male mice: transport by movement of air. *Science* **161**, 584–585.
4. Erhardt, W., Hebestedt, A., Aschenbrenner, G., Pichotka, B., and Blümel, G. (1984). A comparative study with various anesthetics in mice (pentobarbitone, ketamine-xylazine, carfentanyl-etomidate). *Res. Exp. Med.* **184**, 159–169.

Chapter 16

Site-Specific Recombinases for Manipulation of the Mouse Genome

Marie-Christine Birling, Françoise Gofflot, and Xavier Warot

Summary

Site-specific recombination systems are widespread and popular tools for all scientists interested in manipulating the mouse genome. In this chapter, we focus on the use of site-specific recombinases (SSR) to unravel the function of genes of the mouse. In the first part, we review the most commonly used SSR, Cre and Flp, as well as the newly developed systems such as Dre and PhiC31, and we present the inducible SSR systems. As experience has shown that these systems are not as straightforward as expected, particular attention is paid to facts and artefacts associated with their production and applications to study the mouse genome. In the next part of this chapter, we illustrate new applications of SSRs that allow engineering of the mouse genome with more and more precision, including the FLEX and the RMCE strategies. We conclude and suggest a workflow procedure that can be followed when using SSR to create your mouse model of interest. Together, these strategies and procedures provide the basis for a wide variety of studies that will ultimately lead to the analysis of the function of a gene at the cellular level in the mouse.

Key words: Site-specific recombinases, Conditional knockout, Cre, Flp, PhiC31, Inducible systems, FLEX, RMCE, Mouse model engineering

1. Introduction

Site-specific recombinases (SSR) perform rearrangements of DNA segments by binding to specific short DNA sequences, at which they cleave the DNA backbone, exchange the two DNA strands involved and rejoin the DNA strands with conservation of the phosphodiester bond energy *(1)*. In nature, site-specific recombination systems are highly specific, fast and efficient. As such, they are employed in a number of processes such as: bacterial

Elizabeth J. Cartwright (ed.), *Transgenesis Techniques*, Methods in Molecular Biology, vol. 561
DOI 10.1007/978-1-60327-019-9_16, © Humana Press, a part of Springer Science+Business Media, LLC 2009

genome replication, differentiation and pathogenesis, movement of genetic elements such as transposons, plasmids, phages and integrons.

SSR recombinases have been adapted to be used as a tool in mice to modify genes and chromosomes *(2)*. In this approach, the recombinase is used to delete, insert or invert a segment of DNA flanked by recombination sites in the experimental animal. SSR are thus used as "switches" that allow genes to be activated or silenced whereever and whenever the investigator decides. It has been used to generate animals with mutations limited to certain cell types (tissue-specific knockout) or animals with mutations that can be activated by drug administration (inducible knockout) *(3, 4)*. In addition, it can be used to activate lineage tracers in specific progenitor populations to map cell fate within wild-type or mutant mice *(5)*.

In this chapter, we present the different SSR systems that have been developed, focusing on their use to direct site-specific DNA rearrangements in transgenic mice to create spatial and temporal control of gene knockout (generally referred to as conditional knockout). These technologies allow "side effects" in knockouts such as embryonic lethality, cell-non autonomous or systemic effects of the mutation that were precluding accurate analysis of gene function to be circumvented.

2. SSRs Systems and Their Use in Mouse

2.1. Cre/LoxP

2.1.1. Origin and Target Sites

The 38 kDa Cre (*cy*clization *re*combination) recombinase is a tyrosine recombinase, member of the λ integrase family of SSR, from bacteriophage P1 that catalyses a conservative DNA recombination event between two 34 bp recognition sites, the *loxP* sites (*lo*cus of crossing over [*X*] of *P*1). The main interest of the Cre that attracted the 'mouse community' is the autonomy of the reaction: the Cre recombinase does not require accessory proteins or high-energy cofactors, thus making it suitable for use in mammalian cells such as mouse cells. The *loxP* sequence is composed of two 13-bp inverted repeats separated by an 8-bp core region (*see* **Table 1**). Upon binding to the inverted repeats, Cre synapses with a second *loxP* site and then cleaves the DNA in the spacer region to initiate strand exchange with the synapsed *lox* partner *(1)*. Recombination products are dependent on the location and relative orientation of the *loxP* sites. Two DNA species containing single *loxP* sites will be fused, whilst a DNA fragment between *loxP* sites in the same orientation will be excised in circular form, and DNA between opposing *loxP* sites will be inverted with respect to the rest of the DNA (*see*

Table 1
Cre, FLP and ΦC31 recognition sites: classical sites and examples of alternative sites

Target site	Inverted repeat 1	Spacer	Inverted repeat 2
*lox*P	ATAACTTCGTATA	ATGTATGC	TATACGAAGTTAT
*lox*511	ATAACTTCGTATA	ATGTATaC	TATACGAAGTTAT
*lox*5171	ATAACTTCGTATA	ATGTgTaC	TATACGAAGTTAT
FRT	GAAGTTCCTATTC	TCTAGAAA	GTATAGGAACTTC
F3	GAAGTTCCTATTC	TtcAaAtA	GTATAGGAACTTC
F5	GAAGTTCCTATTC	TtcAaAAg	GTATAGGAACTTC
Rox	TAACTTTAAATAAT	GCCA	ATTATTTAAAGTTA
attB	TCGAGTGAGGTGGAGTACGCGCCCGGGGAGCCCAA GGGCACGCCCTGGCACCCGCA		
attP	CTAGACCCTACGCCCCCAACTGAGAGAACT-CAAAGGT TACCCCAGTTGGGGCACG		

In alternative sites, nucleotides that are different from the wild-type sites are *lowercase*

The core nucleotide triplets, where the ΦC31-mediated recombination occurs, are *underlined*

Fig. 1a). To implement the panel of DNA rearrangements, variant target sites have been engineered that modulate recombination efficiency, stability and issue (*see* **Table 1**; reviewed in *(5)*). Due to its size, the 34-bp *loxP* site is unlikely to occur at random in the mouse genome, and yet, it is small enough to be effectively neutral towards gene expression when inserted in chromosomal DNA by genetic manipulation. All these characteristics made this Cre recombinase really attractive to mouse geneticists as a tool for "tailoring" the mouse genome *(4)*.

2.1.2. Development and Applications to the Mouse

Despite the development of other SSR systems such as the Flp-FRT (see below), the Cre-LoxP system remains the most frequently used for genetic manipulation in mice as it was the first well-studied recombination system. Professor Brian Sauer and his collaborators are credited with the development of this system, showing that the Cre recombinase can efficiently catalyze recombination in mammalian cells *(2, 6)*. Soon after these publications, Cre-LoxP recombination was successfully used for targeted activation of genes in transgenic mice and for conditional gene knockout *(7–9)*. For these applications, transgenic strains expressing a Cre transgene are produced and mated with a second transgenic

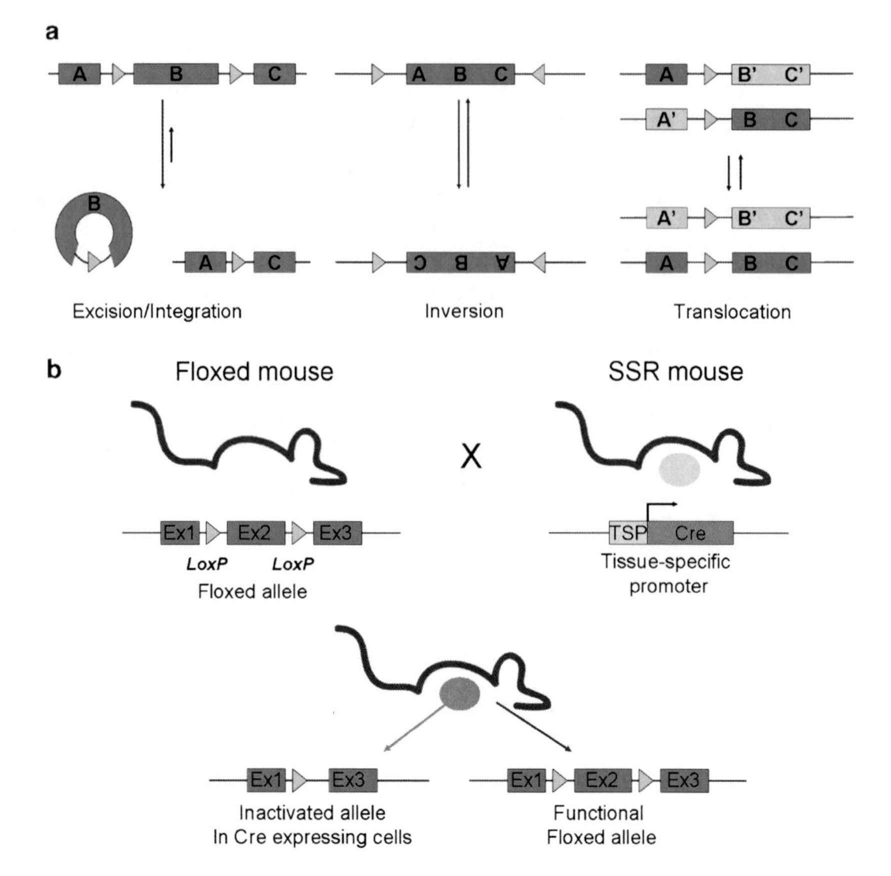

Fig. 1. (a) The Site-specific recombination system. Depending on the relative orientation and location of the recombination site (triangle), the recombination event will result in DNA excision or integration; DNA inversion or DNA translocation. (b) Classical Cre-loxP strategy for conditional, tissue-specific, gene knockout. The heterozygous floxed mouse is obtained using classical homologous recombination in ES cells, where loxP sites are introduced around an essential exon of the gene of interest. The SSR mouse, in which the Cre recombinase is under the control of a tissue-specific promoter (TSP), is usually a transgenic line produced by pronuclear microinjection. Breeding of the two mice will lead to excision of the floxed exon only in Cre-expressing cells or tissue, while the gene of interest remains functional in other cells.

line carrying the target gene *fl*anked by two *loxP* sites (so called "floxed" allele) in the desired orientation. The Cre transgene can be expressed under the control of an ubiquitous promoter (e.g. CMV enhancer *(10)*, to create a constitutional knockout (cKO), but is generally driven by a cell-specific promoter for spatial control of gene inactivation or activation (e.g. *(11)*) **(Fig. 1b)**. As many Cre transgenic lines have been produced with diverse promoters, valuable resources are developing at the international level, gathering information on Cre-expressing mice lines, aiming to help investigators in selecting the adequate lines for conditional mutagenesis (e.g. the website developed by Andras Nagy, http://www.mshri.on.ca/nagy/; MouseCre website under development at the Mouse Clinical Institute, Strasbourg, France).

2.2. Flp/FRT

2.2.1. Origin and Target Sites

Flp recombinase is another tyrosine recombinase, member of the λ integrase superfamily of SSR, originally isolated from the yeast *Saccharomyces cerevisiae* and named for its ability to invert or "flip" a DNA segment *(12)*. Flp has been used in various systems, e.g. mice, to create gene deletions, insertions, inversions and exchanges.

Flp can recombine specific sequences of DNA without the need for cofactors at target sites termed FRT (*F*lp recombinase *R*ecognition *T*arget). Cre and Flp share a common mechanism of DNA recombination, and their respective recognition sites share common features: as with *loxP* sites (see above), the structure of FRT sites includes two 13-bp palindromic sequences or inverted repeats that are separated by an 8-bp asymmetric core or spacer sequence **(Table 1)**. The relative orientation of target sites with respect to one other will determine the outcome of the recombination reaction i.e. excision or inversion of the DNA segment flanked by the two FRT sites, as exemplified for Cre/*loxP* system **(Fig. 1a)**. The finding that Flp can tolerate certain variation in its target recognition sequences has broadened the spectrum of genetic manipulations that can be engineered **(Table 1)**.

2.2.2. Development and Applications to the Mouse

Initial use of Flp in mammalian cells revealed inefficient recombinase activity because of thermal instability of the protein. Indeed, Flp recombinase, derived from the 2 μm plasmid of *S. cerevisiae* which grows at 30 °C, has no biological pressure to be active at higher temperature. Subsequent screening for thermostable mutants resulted in the identification of Flpe which exhibits a fourfold increase in recombination efficiency at 37 °C *(13)*. Flpe has been shown to be highly effective in mice and proved to be a good complementary tool to the Cre/*loxP* system. More recently, a mouse codon-optimized Flp gene has been engineered (Flpo) which shows DNA recombination efficiency similar to that of Cre *(14)*.

Transgenic mice harbouring Flpe under the control of various promoters have been established. Three Flpe deleter lines (lines in which the Flpe is under the control of a ubiquitous promoter to create a line with early and generalized expression of Flpe) have been published so far and work very well *(15–17)*. Such lines can be used in the conditional targeting strategy design to efficiently remove the selection marker used for homologous recombination.

2.3. Other Recombinases

Apart from Cre and Flp, other recombinases have been identified that will provide additional tools for genetic manipulation of the mouse genome.

2.3.1. Dre Recombinase

In the search of Cre-like recombinases from P1 related phages, the Dre recombinase from the bacteriophage D6 has been isolated.

Like Cre, Dre is a tyrosine recombinase with a distinct DNA specificity for a 32 bp DNA site, termed *rox* (**Table 1**). Dre has been shown to be able to catalyze site-specific recombination in mammalian cells. Of particular interest is the fact that Cre and Dre are heterospecific: Cre does not catalyze recombination at *rox* sites and Dre does not catalyze recombination at *lox* sites *(18)*.

2.3.2. PhiC31 Integrase

Origin and Target Sites

The integrase of the phage ΦC31 was isolated from *Streptomyces lividans* and was the first large serine recombinase described *(19)*. It uses a fundamentally different recombination mechanism than Cre, Dre and Flp. Indeed, ΦC31 mediates recombination between the heterotypic sites *attP* (34 bp in length) and *attB* (39 bp in length) to produce *attL* and *attR*, the latter being effectively inert to further ΦC31-mediated recombination (**Table 1**; *(20)*). The minimal *attP* and *attB* sites include imperfect inverted repeats and a three-nucleotide recombination site that imposes directionality. Like Cre and Flpe, ΦC31 excises and inverts DNA intervals flanked by *attP* and *attL* in direct and indirect orientation, respectively. ΦC31 action is irreversible as *attL* and *attR* cannot recombine. ΦC31 has been successfully applied for insertional mutagenesis *(21)* and Recombinase-*M*ediated Cassette *E*xchange (RMCE; *(20)*; see details below). ΦC31 is more efficient at catalyzing intermolecular reactions than intramolecular reactions and is therefore ideal to facilitate site-specific insertions into the mammalian genome (transgene insertion or cassette exchange).

Development and Application to the Mouse

Andreas et al. *(22)* have showed that the efficiency of ΦC31 is significantly enhanced by the C-terminal but not the N-terminal addition of a nuclear localization signal, and becomes comparable with that of the Cre/loxP system. More recently, Raymond et al. *(14)* developed a de novo synthesis of mouse codon-optimized ΦC31 (ΦC31o) which shows DNA recombination efficiencies similar to that of Cre in embryonic stem (ES) cells. Interestingly, both ΦC31 heterozygous and homozygous mice are viable and neither show obvious deleterious effects from the widespread expression of the SSR.

2.4. Spatio-Temporal Knockout in the Mouse: Inducible Systems

An additional level of accuracy in the analysis of gene function can be introduced with the alteration of gene activity not only in a particular cell lineage (spatial control) but also at a particular time (temporal control), using inducible SSR systems. The main interest of these systems is to circumvent severe or lethal phenotype that would impair analysis of late gene function. However, they can also avoid the risk of impaired fertility, generalized systemic disorders, or compensatory responses developed to adapt to the genetic alteration that would mask the phenotype.

The first approach developed to add a temporal switch was the use of an inducible promoter to express the recombinase *(23)*. Today, temporal regulation is achieved mainly by two ways: either by combining the Cre-LoxP system with the tetracyclin-dependent regulatory system (TetR) or by using Cre/Flp variants that function only in the presence of an exogenous inducer. These two systems are outlined below.

2.4.1. TetR-Based Systems

This system is based on the transactivation system developed by Gossen and Bujard *(24)*, in which the first transgene expresses a fusion protein containing the VP16 transactivation domain and the *E. coli* tetracycline repressor (TetR), which specifically binds to and represses the tet operon (*tetO*) associated with the promoter of the target gene (second transgene). In the presence of tetracycline, the repressor cannot bind anymore to the *tetO* sequence, which results in transcription of the target gene. Of interest, to switch on Cre or Flp expression at a specific time is the reverse system, the (rTetR), in which the tetracyline-controlled transactivator binds DNA only when the inducer is present. To achieve this control, a transgenic line is produced in which the promoter driving SSR expression is engineered to contain the *tetO* DNA sequences (e.g. Cre^{tetO} allele). This transgenic line is then crossed with two other lines: one that encodes the reverse transactivator (rtTA) and one that contains the floxed allele of the gene of interest. In the absence of the tetracycline, no recombinase is transcribed, while addition of the inducer can rapidly induce SSR expression, thus allowing target gene modification (**Fig. 2a**). Initially designed to be used with tetracycline, the favoured inducer is now doxycyline (Dox) because of its low cost and efficiency in activating the rtTA at doses that are well below cytotoxic levels. Although it has been successfully used *(25–28)*, the main disadvantage of this system is the complexity of the tertiary system, that necessitates the production, careful characterization, and mating of three different transgenic lines, and heavily complicates mouse husbandry.

2.4.2. Ligand-Inducible Recombinases

In this system, the recombinase coding region is fused with the ligand-binding domain (LBD) of either the oestrogen, glucocorticoid or progesterone receptor (ER, GR and PR, respectively; *see* **Table 2**) *(29–33)*. According to current understanding, the fusion protein produced is trapped in the cytoplasm because of Hsp90 sequestration and becomes localized to the nucleus only in the presence of the adequate steroid that releases the fusion protein, freeing it to enter the nucleus and to mediate recombination (**Fig. 2b**). To preclude the binding and interference of endogenous steroids, mutations have been introduced in the LBD, and the experimental inducers are synthetic steroids, such as tamoxifen or 4-hydroxytamoxifen (OHT) for the ER, dexamethasone for

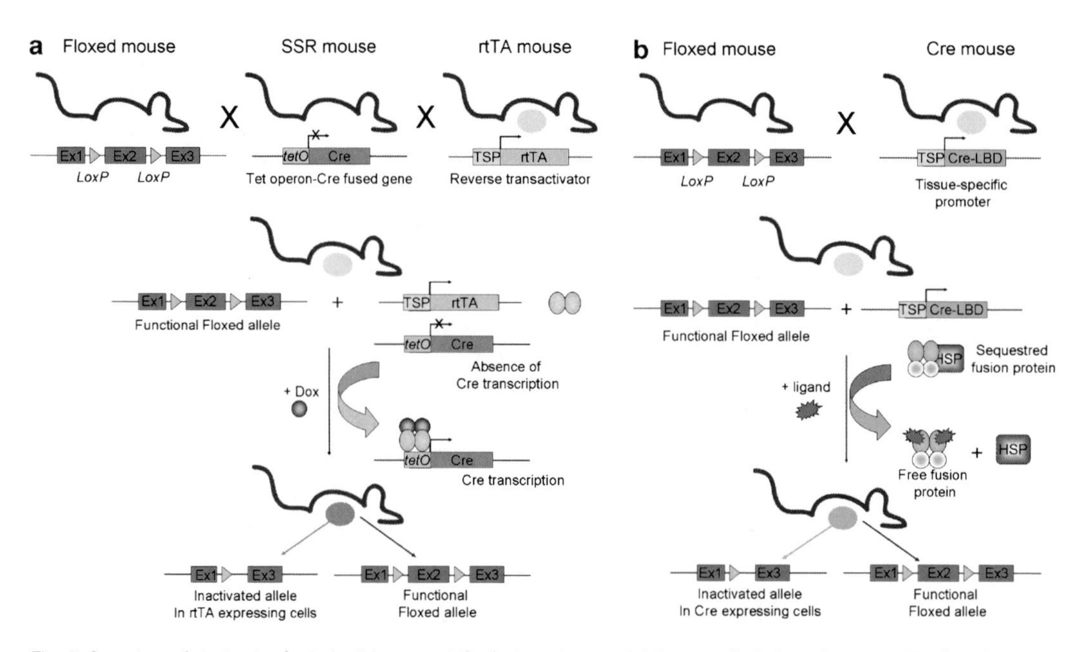

Fig. 2. Overview of strategies for inducible gene cKO of a target gene. **(a)** Reverse TetR-based system. The floxed mouse is obtained using classical homologous recombination in ES cells, and loxP sites are introduced around an essential exon of the gene of interest. In the illustrated strategy, the SSR mouse expresses a tetO-Cre transgene, in which the promoter driving Cre expression is engineered to contain the Tet operon (*tetO)* DNA sequences. The rtTA mouse is a transgenic line in which the "reverse" tetracyline-controlled transactivator (rtTA) is under the control of a tissue-specific promoter (TSP). Breeding of the three mice will result in excision of the floxed exon only in the presence of the inducer, doxycycline (Dox), and only in rtTA-expressing cells or tissue, while the gene of interest remains functional in other cells. **(b)** Ligand-inducible SSR system. In this strategy, the Cre mouse is a transgenic which encodes the Cre (*circles*) and a mutated ligand-binding domain (LBD-ovals) from the estrogen or progesterone receptor (see text for details). The tissue-specific expression of the transgene will result in a fusion protein that is thought to be sequestered into the cytoplasm through interaction with heat-shock protein (HSP). Breeding of the two mice will result in excision of the floxed exon only in the presence of the ligand (Tam, OHT or RU486), and only in Cre-expressing cells or tissue, while the gene of interest remains functional in other cells.

GR or RU486 for PR (**Table 2**). Amongst these various systems, the CreER recombinases have proved to be particularly useful. A first generation was produced by fusion of the Cre recombinase with a mutant LBD of the human ER (G521R), resulting in a chimeric protein $CreER^{T}$; named because of its ability to bind the synthetic ER agonist Tamoxifen and OHT *(30)*. Recombination has been observed with these fusion proteins using both ubiquitous and tissue-specific promoters *(3, 30)*, although with variable levels of efficiency depending on tissue type. Shortly after, a second generation of CreER was produced by the same group, the $CreER^{T2}$, which contains a C400 V/M453A/L544A triple mutation in the human ER LBD. This variant was shown to be tenfold more sensitive to OHT as compared to $CreER^{T}$ *(31, 34)* and was thus favoured for the production of inducible tissue-specific Cre

Table 2
Ligand-inducible recombinases

SSR	LBD	Inducer	Procedure	References
Cre				
CreER	Human ER	Estradiol, OHT	In vitro	*(60)*
$CreER^{T}$	Mutated human ER (G521R)	OHT	In vivo	*(30)*
$CreER^{TBD}$	Mutated mouse ER (G525R)	Tam, OHT	In vitro	*(61)*
$CreER^{T1}$	Mutated human ER (G400V/L539A/L540A)	ICI (OHT)	In vitro	*(31)*
$CreER^{T2}$	Mutated human ER (G400V/M543A/L540A)	OHT	In vitro/In vivo	*(29, 31)*
$CreER^{EDB}$	Mutated human ER (G521R)	OHT, raloxifen	In vivo	*(38)*
$CreGR^{Dex}$	Mutated human GR (I747 T)	Dex, TA, RU486	In vitro	*(29)*
CrePR1	Mutated human PR (hPR891)	RU486	In vitro/In vivo	*(32, 33)*
$CreER^{TM}$	Mutated mouse ER (G525R) (from **Ref.** *63*)	Tam	In vivo (gestation)	*(45)*
$CreER^{Tam}$	Mutated mouse ER (G525R)	Tam	In vivo	*(46)*
Flp				
FlpER	Human mutated ER (Gly400Val)	Estradiol	In vitro	*(62)*
$FlpeER^{T2}$	Mutated human ER (G400V/M543A/L540A)	OHT, Tam	In vivo (gestation)	*(36)*

OHT 4-hydroxytamoxifen; *Tam* tamoxifen; *Dex* Dexamethasone, *TA* Triamcinolone acetonide; *ICI* ICI 164,384

transgenic mouse lines. Many transgenic mouse lines that express $CreER^{T2}$ in specific somatic tissues and/or cell lineages have thus been produced and successfully used to address biological questions (e.g. *see (11, 35)*). Recently, a $FlpeER^{T2}$ recombinase has been developed, adding a new tool for temporal control of the Flp/FRT system *(36)*. The main advantage of this system compared to the TetR-based technology is the requirement of only one transgenic line to be crossed with the floxed allele mouse line to obtain conditional alteration of the target gene.

Despite the fact that other alternatives have been and are still currently developed, the $CreER^{T2}$ is at the moment the main tool available for the mouse community for the generation of spatially and temporally controlled mutagenesis. With the three major mice knockout programs underway worldwide *(37)*, there is an urgent need for the production of validated transgenic lines that express $CreER^{T2}$ in specific somatic tissues and/or cell lineages. However, the main bottleneck in the production of these inducible tissue-specific Cre/Flp transgenic lines is the accurate determination of the recombinase expression and activity profile required for proper phenotypic analysis of these conditional mutant mice.

2.5. Facts and Artefacts

Although the SSR and inducible SSR have demonstrated efficiency as tools for genome engineering in mice, there are still a number of caveats in their applications, and users should pay attention to specific facts and artefacts when programming their experiments.

2.5.1. SSR Expression/Activity Patterns

All these systems rely on appropriate promoters and/or enhancers to control the expression of SSR in a specific cell lineage. However, recombination could occur in other cells than the desired target, because of inappropriate or unexpected expression of the transgene. In most cases, validation of SSR-mediated recombination at anatomical levels is based on published information on the gene whose promoter is used to drive recombinase expression. These data are often incomplete, and hence this analysis is often biased as it does not allow the evaluation of additional expression, misexpression due to position effect, or "leakage effect" of the inducible construct in other tissues. Published information on specific Cre/Flp line may thus be incomplete depending on how accurate the characterization of SSR expression and activity has been performed. In addition, classic approaches such as traditional transgenesis and BAC engineering have caveats, such as insertion site side effects or passenger genes within the BAC. When using a specific Cre/Flp line constructed by BAC transgenesis, users should always verify the BAC map to check for the presence of potentially interfering genes. Finally, tissue-specific SSR transgenic lines generated by pronuclei injection often result in mosaic expression of the recombinase within the selected cell type *(38)*. This artefact represents a limitation to certain experiments, such as those involving gene knockouts of non-cell autonomous factors.

2.5.2. Variable Recombination Efficiency

Although for most published Cre/Flp lines, recombination properties have been validated by reporter gene studies, users should be aware that the efficiency of SSR recombination can be locus-dependent, and therefore the recombination pattern obtained with a particular reporter gene does not necessarily predict that of other floxed genes (*see* *(39)*; J. Becker and B. Kieffer, personal communication). Indeed, the chromatin structure at the locus of interest, the state of DNA methylation and the transcriptional activity seem to affect the efficiency of recombination.

It has also been reported that the ability of a floxed target gene to be recombined could also vary from one cell type to another *(32)*. This was potentially explained by differential accessibility of Cre to *loxP* sites because of cell type and development-specific chromatin conformations.

Before starting a long-term phenotypic analysis, it is thus mandatory to monitor recombination at the target locus by PCR/qPCR or Southern blot.

2.5.3. Cre/Flp Toxicity

Originally, the expression of Cre/Flp in mice was expected to be innocuous because of the absence of endogenous *loxP* and FRT sites in the mouse genome. However, it has been shown that these two SSR can mediate recombination between degenerated *loxP* and FRT sequences, although with reduced efficiency, and some pseudo-*loxP* sites have been identified in the mouse genome *(40)*. Cre-mediated recombination at these sites could thus explain the reports of Cre toxicity in vitro and in vivo *(41–43)*. When using Cre/Flp lines controlled by strongly expressed promoter, especially in the brain and testis, one should be aware of potential toxicity of the recombinase in those cell lineages. However, we should mention that these reports are rare, and 10 years of experience using Cre transgenic lines suggest that Cre/Flp toxicity is more likely the exception than the rule.

2.5.4. Inducible SSR and Inducer Toxicity

In parallel to their benefits for genome engineering, the development of strategies allowing inducible gene recombination has also brought new complexity and new potential side effects.

- *Inducer toxicity:* In parallel to the development of inducible ligand-activated Cre/Flp, several protocols of induction have been tested. As the $CreER^{T2}$ is the inducible Cre used most often, particular interest has been paid to the effects of Tam and OHT. An early report showed that short-term Tam treatments to adult mice have low acute toxicity and cause no severe abnormalities in mice *(44)*. Since then, toxicity of oral administration of Tam to adult mice and injection to pregnant females has been reported *(45, 46)*. Current procedures involving the injection of 1 mg Tam or OHT for five consecutive days in adult mice aged 8–10 weeks do not seem to be associated with obvious toxicity although no in-depth analysis has been performed.
- *Recombination efficiency after induction:* Despite the testing of different ways of administration and various doses of inducers, recombination efficiency never achieves 100%, and appears to vary from low to >80% *(38, 46, 47)*. This could be due to an insufficient level of Cre expression in those cells. Alternatively, the intracellular level of Tam may be the limiting factor. Indeed, inducer access to the target tissue may vary according to tissue specific properties such as vascularization. For example, access of Tam to the brain cells has often been questioned because of the presence of the blood-brain barrier. Efficient induction of recombination has been observed in the brain when analysis was performed 1 month after the last injection

of Tam. This variation in recombination efficiency, associated with mosaic expression of the recombinase in the target cells, is an important feature to keep in mind when analyzing a phenotype.

- *Leakiness of recombinase activity:* A necessary requirement for ligand-inducible recombinase is the absence of background recombination in the absence of the inducer. This background level may vary from one Cre/Flp transgenic line to another, but also for the same Cre/Flp transgenic line on different floxed alleles (for example, see *(48)* and *(49)*, in which the same Cre line was used and shown to work with or without leakiness depending of the floxed allele). The so-called "leakiness" is presumably due to proteolysis of the fusion protein, yielding a low level of free active recombinase. According to the accessibility of the target floxed gene, this low level of recombinase will be sufficient or not to mediate excision. It is notable that the stronger the promoter used, the higher the basal level of activity observed *(39)*, also suggesting that high level of recombinase production may overwhelm the capacity of Hsp90 sequestration system within the cytoplasm.

3. Further Applications of SSRs to Engineer the Mouse Genome: Examples

3.1. FLEX System

The FLEX strategy, a SSR-mediated genetic switch, uses two pairs of inversely oriented heterotypic recombinase target sites (RTs) that flank a cassette of interest (**Fig. 3**). Recombination between either pair of homotypic RTs inverts the cassette and places the other homotypic RTs pair in a direct orientation. Recombination between this pair of directly repeating RTs excises one of the other heterotypic RTs, hence, 'locking' the recombination product against reinversion to the original orientation. The FLEX design was originally developed to monitor Cre recombination at the cellular level in transgenic mice *(50, 51)*.

FLEX arrays have been used recently in gene trap vectors developed for high-throughput conditional mutagenesis in ES cells *(52)*.

3.2. RMCE

Recombinase-mediated cassette exchange (RMCE) is based on the replacement of gene cassettes flanked by two non-interacting recombinase target sites (RTs) (**Fig. 4**) *(53)*.

Two steps are needed to perform RMCE. First, the genomic locus needs to be 'tagged' by the introduction of heterospecific

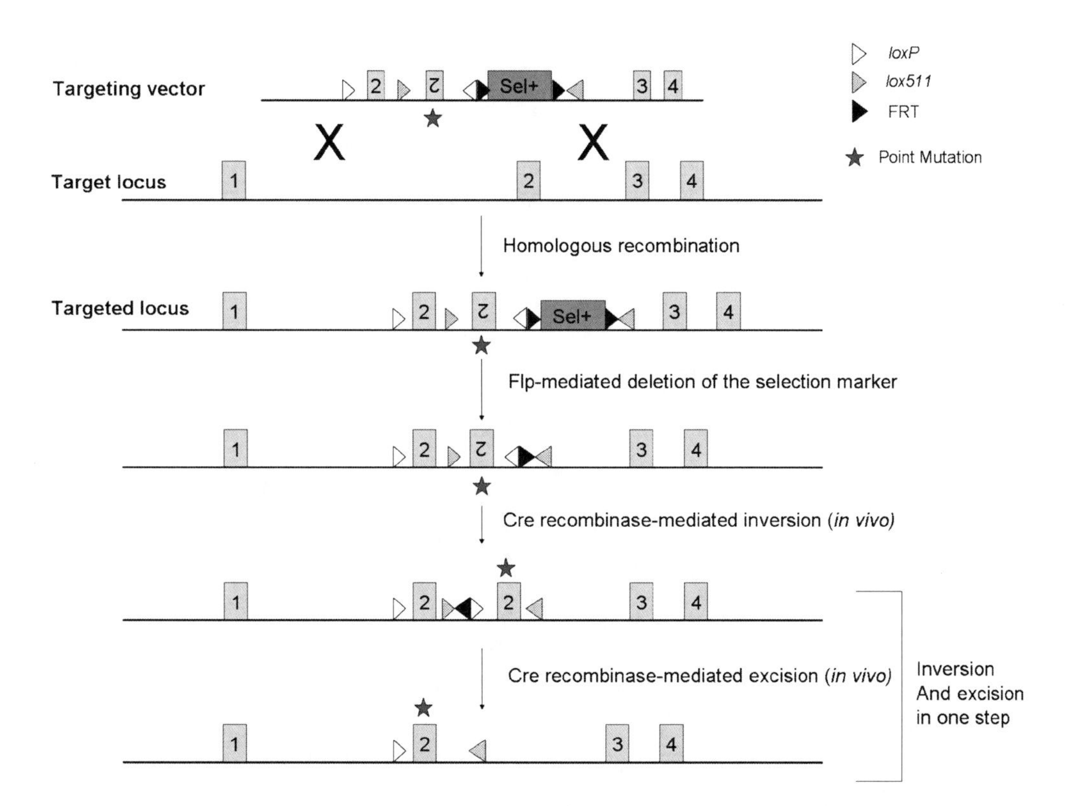

Fig. 3. Application of the SSR system to engineer the mouse genome: the FLEX switch using Cre, *loxP* and *lox511* sites. The locus of interest is engineered by homologous recombination to introduce the two pairs of heterotypic *loxP* and *lox511* sites, each in reverse orientation, surrounding the non selectable mutation (indicated by a star) present in the exon in the antisense orientation. The selection marker (Sel+) is flanked by FRT sites in direct orientation. After targeting, the Flp recombinase is first used to remove the selection marker. Cre-mediated recombination between the *loxP* sites and subsequently between the two *lox511* sites (or vice versa) will result in the introduction of the mutation in the target gene. *Sel+* positive selection marker.

RTs into the host genome (by homologous recombination in most of the cases). Second, the 'targeting' of the tagged site is mediated by recombinase exchange of the tagging cassette with the desired transgene cassette. Integration sites have been identified by intensive screening and have been shown to provide stable expression levels with at least comparable efficiency to classical multiple integration protocols *(54–56)*. The Rosa26 and the β-actin locus have been shown to be accessible to the RMCE procedure in ES cells *(57, 58)*.

The International Genetrap Consortium (IGTC, www.genetrap.org) and the European Conditional Mouse Mutagenesis Program (www.eucomm.org) provide information about all trapped genes where vectors with RMCE-compatible heterospecific RTs have been employed.

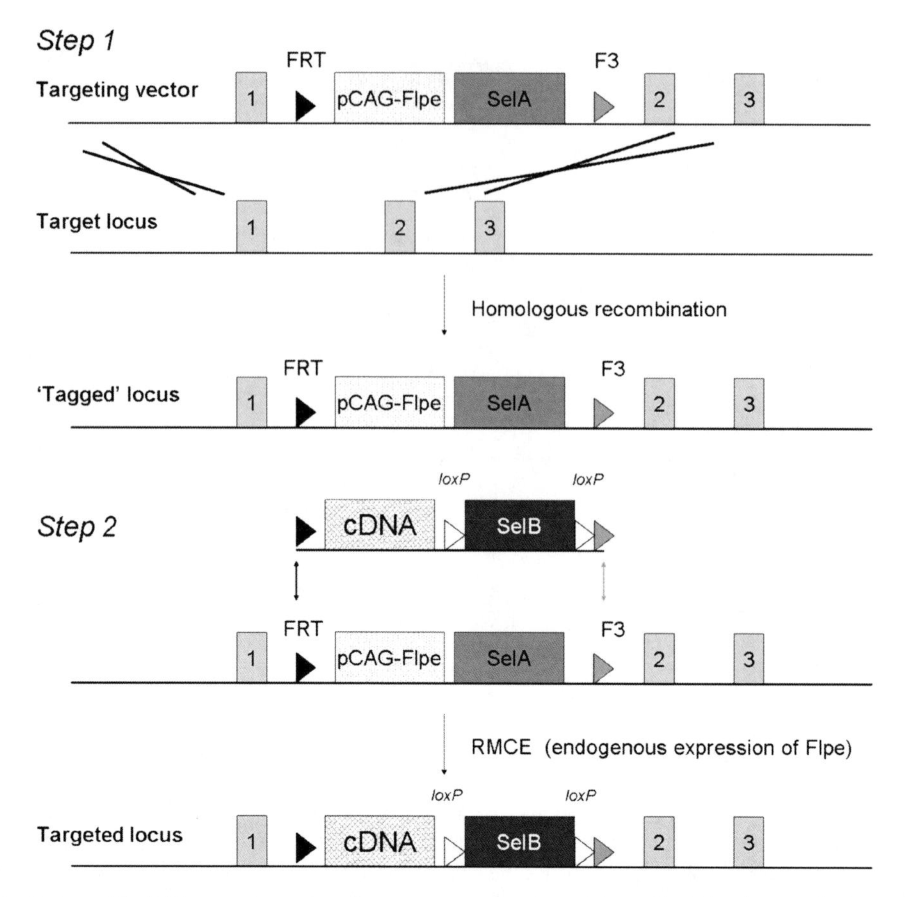

Fig. 4. Application of the SSR system to engineer the mouse genome: Recombinase-Mediated Cassette Exchange (RMCE) using Flp, FRT and F3 sites. *Step 1* The locus of interest is modified by homologous recombination to introduce one FRT site, one F3 site, an expression cassette for the Flp (pCAG-Flpe) and a selection marker A to "tag" the locus. *Step 2* The cassette with the cDNA to insert in the locus flanked by FRT and F3 sites, together with a second selection marker B, is introduced and exchange mediated by the endogenous expression of Flpe will take place resulting in insertion of the cDNA of interest. The use of two different selection markers allows the selection of the event of interest in step 1 and step 2. The selection B is flanked by *loxP* sites to enable further removal. SelA and SelB: selection marker A and B, respectively; pCAG: Chicken beta-actin with CMV enhancer elements promoter.

3.3. Engineering Conditional Mutations: Towards More and More Subtleties

More and more sophisticated mice models are now being generated: introduction of conditional mutations, introduction of conditional stop codons and conditional KI (knock-ins) have become standard (**Fig. 5**).

In contrast to classical gene inactivation, the conditional locus before excision/reversion by SSR needs to be functional with respect to gene regulation and function. Several points have to be taken into account, including, minimizing the potential interference of the inserted floxed or flipped fragment with the regulation of the gene (position in intron, avoiding enhancer elements or splice sites, size of the intron chosen for insertion).

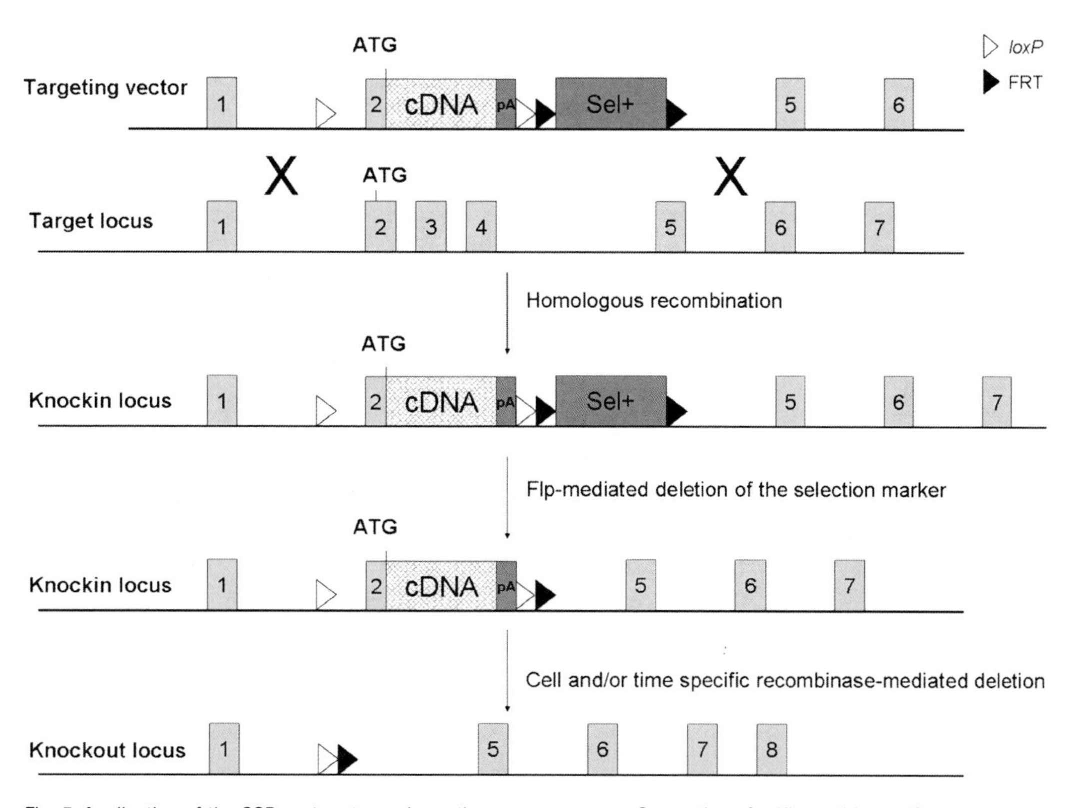

Fig. 5. Application of the SSR system to engineer the mouse genome: Generation of a KI-conditional KO model. In this approach the cDNA of interest is flanked by two *loxP* sites and introduced into the locus of interest by homologous recombination resulting in a knockin of the cDNA in the locus of interest. Subsequent action of Cre will result in excision of the cDNA resulting in a knockout of the gene of interest. *Sel+* positive selection marker. *pA* polyadenylation signal.

Creating a mouse model is largely dependant on the target gene and its localization in the genome. One should keep in mind that the two recombinase target sites (RTs, e.g. *loxP*) should be separated by 3 kb or less, in order to avoid loss of one site during homologous recombination. Indeed, in the majority of designs of conditional targeting constructs one RT site is closely associated with the selection marker used for the selection of the homologous recombination event and the second RT is located up to 3 kb apart from the first RT. As there is only selection pressure on one RT site for integration, the other RT can be lost after homologous recombination. The probability of integrating the selection marker with the two RTs declines as the distance between these sites increases. Torres and Kühn *(59)* reported a much better score (up to 10 kb between both RTs). In our hands, the loss of one RT at a distance as short as 1 kb between the two RTs was observed in the targeting of some loci (unpublished data).

4. How to Use Recombinases to Create Your Mouse Model of Interest

As stated in **Subheading 2**, the use of a SSR system to create mouse models for conditional knockout of genes requires production and mating of two transgenic animals. The overall strategy involves several steps with a series of controls to be carried out.

- Establishment of the "floxed" mouse by the introduction of *loxP* sites around an essential exon of the gene of interest. This step is achieved by homologous recombination in ES cells and further microinjection of the targeted clones to generate chimeras.
- Removal of the selection marker introduced in the gene of interest for selection of the homologous recombination events. The presence of the selection marker can disturb the transcription of the target gene and lead to inappropriate conclusions on the function of the gene. The selection marker can be flanked by FRT sites and removal can be achieved by crossing the chimera with Flp deleter animals.
- Validation of the conditional model.
  - The introduction of *loxP* sites should not disturb the gene transcription. The mRNA of the targeted gene in the conditional mutants is detected by RT-PCR and quantified by qRT-PCR and compared to the level of mRNA found in wild-type animals.
  - When the mouse line is established, it is worth establishing a complete knockout of the gene of interest by crossing with a Cre-deleter strain (e.g. CMV-Cre) and analyzing the outcome of the excision of the floxed exon (presence/absence of mRNA and/or protein by RT-PCR and/or Western-Blot).
- Breeding between the "floxed" mouse and a given "Cre" mouse of interest. Generation and characterization of "Cre" mouse lines are described in the next chapter of this book.
- Prior to embarking on specific phenotyping experiments, it is crucial to characterize the Cre-excision pattern of the target gene. Indeed, it is well recognized that Cre activity and efficiency in excision is dependent on the target of the Cre and that the pattern of excision of a Cre line usually characterized by using an "indicator" or "reporter" mouse line cannot be fully transposed to any target gene. Excision patterns of the target gene should be analyzed by PCR/qPCR on DNA extracts from various tissues and can be followed by qRT-PCR to assess the level of mRNA still present in the target tissue (*see* **Subheading 2.5**).

5. Conclusion

The use of multiple, highly efficient SSRs in ES cells and mice now opens possibilities for a broader range of molecular manipulations. These not only include the potential for additional genetic alterations during ES cell expansion (such as removal of the selection markers) for performing multiple general or tissue specific gene knockouts, but also for studying gene function in overlapping domains of expression. Additionally, the use of multiple SSRs could potentially be used for controlling gene expression in a temporal 'off-on' manner in single or multiple cell types or tissues.

References

1. Grindley, N.D., Whiteson, K.L. and Rice, P.A. (2006) Mechanisms of site-specific recombination. *Annu Rev Biochem*, **75**, 567–605
2. Sauer, B. and Henderson, N. (1988) Site-specific DNA recombination in mammalian cells by the Cre recombinase of bacteriophage P1. *Proc Natl Acad Sci U S A*, **85**, 5166–5170
3. Metzger, D. and Chambon, P. (2001) Site- and time-specific gene targeting in the mouse. *Methods*, **24**, 71–80
4. Nagy, A. (2000) Cre recombinase: the universal reagent for genome tailoring. *Genesis*, **26**, 99–109
5. Branda, C.S. and Dymecki, S.M. (2004) Talking about a revolution: the impact of site-specific recombinases on genetic analyses in mice. *Dev Cell*, **6**, 7–28
6. Sauer, B. and Henderson, N. (1989) Cre-stimulated recombination at loxP-containing DNA sequences placed into the mammalian genome. *Nucleic Acids Res*, **17**, 147–161
7. Gu, H., Marth, J.D., Orban, P.C., Mossmann, H. and Rajewsky, K. (1994) Deletion of a DNA polymerase beta gene segment in T cells using cell type-specific gene targeting. *Science*, **265**, 103–106
8. Lakso, M., Sauer, B., Mosinger, B., Jr., Lee, E.J., Manning, R.W., Yu, S.H., Mulder, K.L. and Westphal, H. (1992) Targeted oncogene activation by site-specific recombination in transgenic mice. *Proc Natl Acad Sci U S A*, **89**, 6232–6236
9. Orban, P.C., Chui, D. and Marth, J.D. (1992) Tissue- and site-specific DNA recombination in transgenic mice. *Proc Natl Acad Sci U S A*, **89**, 6861–6865
10. Dupe, V., Davenne, M., Brocard, J., Dolle, P., Mark, M., Dierich, A., Chambon, P. and Rijli, F.M. (1997) In vivo functional analysis of the Hoxa-1 3′ retinoic acid response element (3′RARE). *Development*, **124**, 399–410
11. Gaveriaux-Ruff, C. and Kieffer, B.L. (2007) Conditional gene targeting in the mouse nervous system: insights into brain function and diseases. *Pharmacol Ther*, **113**, 619–634
12. O'Gorman, S., Fox, D.T. and Wahl, G.M. (1991) Recombinase-mediated gene activation and site-specific integration in mammalian cells. *Science*, **251**, 1351–1355
13. Buchholz, F., Angrand, P.O. and Stewart, A.F. (1998) Improved properties of FLP recombinase evolved by cycling mutagenesis. *Nat Biotechnol*, **16**, 657–662
14. Raymond, C.S. and Soriano, P. (2007) High-efficiency FLP and PhiC31 site-specific recombination in mammalian cells. *PLoS ONE*, **2**, e162
15. Farley, F.W., Soriano, P., Steffen, L.S. and Dymecki, S.M. (2000) Widespread recombinase expression using FLPeR (flipper) mice. *Genesis*, **28**, 106–110
16. Kanki, H., Suzuki, H. and Itohara, S. (2006) High-efficiency CAG-FLPe deleter mice in C57 BL/6J background. *Exp Anim*, **55**, 137–141
17. Rodriguez, C.I., Buchholz, F., Galloway, J., Sequerra, R., Kasper, J., Ayala, R., Stewart, A.F. and Dymecki, S.M. (2000) High-efficiency deleter mice show that FLPe is an alternative to Cre-loxP. *Nat Genet*, **25**, 139–140
18. Sauer, B. and McDermott, J. (2004) DNA recombination with a heterospecific Cre homolog identified from comparison of the

pac-cl regions of P1-related phages. *Nucleic Acids Res*, **32**, 6086–6095

19. Thorpe, H.M. and Smith, M.C. (1998) In vitro site-specific integration of bacteriophage DNA catalyzed by a recombinase of the resolvase/invertase family. *Proc Natl Acad Sci U S A*, **95**, 5505–5510
20. Belteki, G., Gertsenstein, M., Ow, D.W. and Nagy, A. (2003) Site-specific cassette exchange and germline transmission with mouse ES cells expressing phiC31 integrase. *Nat Biotechnol*, **21**, 321–324
21. Sclimenti, C.R., Thyagarajan, B. and Calos, M.P. (2001) Directed evolution of a recombinase for improved genomic integration at a native human sequence. *Nucleic Acids Res*, **29**, 5044–5051
22. Andreas, S., Schwenk, F., Kuter-Luks, B., Faust, N. and Kuhn, R. (2002) Enhanced efficiency through nuclear localization signal fusion on phage PhiC31-integrase: activity comparison with Cre and FLPe recombinase in mammalian cells. *Nucleic Acids Res*, **30**, 2299–2306
23. Kuhn, R., Schwenk, F., Aguet, M. and Rajewsky, K. (1995) Inducible gene targeting in mice. *Science*, **269**, 1427–1429
24. Gossen, M. and Bujard, H. (1992) Tight control of gene expression in mammalian cells by tetracycline-responsive promoters. *Proc Natl Acad Sci U S A*, **89**, 5547–5551
25. Lewandoski, M. (2001) Conditional control of gene expression in the mouse. *Nat Rev Genet*, **2**, 743–755
26. Saam, J.R. and Gordon, J.I. (1999) Inducible gene knockouts in the small intestinal and colonic epithelium. *J Biol Chem*, **274**, 38071–38082
27. St-Onge, L., Furth, P.A. and Gruss, P. (1996) Temporal control of the Cre recombinase in transgenic mice by a tetracycline responsive promoter. *Nucleic Acids Res*, **24**, 3875–3877
28. Utomo, A.R., Nikitin, A.Y. and Lee, W.H. (1999) Temporal, spatial, and cell type-specific control of Cre-mediated DNA recombination in transgenic mice. *Nat Biotechnol*, **17**, 1091–1096
29. Brocard, J., Feil, R., Chambon, P. and Metzger, D. (1998) A chimeric Cre recombinase inducible by synthetic, but not by natural ligands of the glucocorticoid receptor. *Nucleic Acids Res*, **26**, 4086–4090
30. Feil, R., Brocard, J., Mascrez, B., LeMeur, M., Metzger, D. and Chambon, P. (1996) Ligand-activated site-specific recombination in mice. *Proc Natl Acad Sci U S A*, **93**, 10887–10890
31. Feil, R., Wagner, J., Metzger, D. and Chambon, P. (1997) Regulation of Cre recombinase activity by mutated estrogen receptor ligand-binding domains. *Biochem Biophys Res Commun*, **237**, 752–757
32. Kellendonk, C., Tronche, F., Casanova, E., Anlag, K., Opherk, C. and Schutz, G. (1999) Inducible site-specific recombination in the brain. *J Mol Biol*, **285**, 175–182
33. Kellendonk, C., Tronche, F., Monaghan, A.P., Angrand, P.O., Stewart, F. and Schutz, G. (1996) Regulation of Cre recombinase activity by the synthetic steroid RU 486. *Nucleic Acids Res*, **24**, 1404–1411
34. Indra, A.K., Warot, X., Brocard, J., Bornert, J.M., Xiao, J.H., Chambon, P. and Metzger, D. (1999) Temporally-controlled site-specific mutagenesis in the basal layer of the epidermis: comparison of the recombinase activity of the tamoxifen-inducible Cre-ER(T) and Cre-ER(T2) recombinases. *Nucleic Acids Res*, **27**, 4324–4327
35. Feil, R. (2007) Conditional somatic mutagenesis in the mouse using site-specific recombinases. *Handb Exp Pharmacol*, **178**, 3–28
36. Hunter, N.L., Awatramani, R.B., Farley, F.W. and Dymecki, S.M. (2005) Ligand-activated Flpe for temporally regulated gene modifications. *Genesis*, **41**, 99–109
37. Collins, F.S., Rossant, J. and Wurst, W. (2007) A mouse for all reasons. *Cell*, **128**, 9–13
38. Schwenk, F., Kuhn, R., Angrand, P.O., Rajewsky, K. and Stewart, A.F. (1998) Temporally and spatially regulated somatic mutagenesis in mice. *Nucleic Acids Res*, **26**, 1427–1432
39. Vooijs, M., Jonkers, J. and Berns, A. (2001) A highly efficient ligand-regulated Cre recombinase mouse line shows that LoxP recombination is position dependent. *EMBO Rep*, **2**, 292–297
40. Thyagarajan, B., Guimaraes, M.J., Groth, A.C. and Calos, M.P. (2000) Mammalian genomes contain active recombinase recognition sites. *Gene*, **244**, 47–54
41. Forni, P.E., Scuoppo, C., Imayoshi, I., Taulli, R., Dastru, W., Sala, V., Betz, U.A., Muzzi, P., Martinuzzi, D., Vercelli, A.E. et al. (2006) High levels of Cre expression in neuronal progenitors cause defects in brain development leading to microencephaly and hydrocephaly. *J Neurosci*, **26**, 9593–9602
42. Loonstra, A., Vooijs, M., Beverloo, H.B., Allak, B.A., van Drunen, E., Kanaar, R., Berns, A. and Jonkers, J. (2001) Growth inhibition and DNA damage induced by Cre recombinase in mammalian cells. *Proc Natl Acad Sci U S A*, **98**, 9209–9214

43. Schmidt, E.E., Taylor, D.S., Prigge, J.R., Barnett, S. and Capecchi, M.R. (2000) Illegitimate Cre-dependent chromosome rearrangements in transgenic mouse spermatids. *Proc Natl Acad Sci U S A*, **97**, 13702–13707
44. Furr, B.J. and Jordan, V.C. (1984) The pharmacology and clinical uses of tamoxifen. *Pharmacol Ther*, **25**, 127–205
45. Danielian, P.S., Muccino, D., Rowitch, D.H., Michael, S.K. and McMahon, A.P. (1998) Modification of gene activity in mouse embryos in utero by a tamoxifen-inducible form of Cre recombinase. *Curr Biol*, **8**, 1323–1326
46. Vasioukhin, V., Degenstein, L., Wise, B. and Fuchs, E. (1999) The magical touch: genome targeting in epidermal stem cells induced by tamoxifen application to mouse skin. *Proc Natl Acad Sci U S A*, **96**, 8551–8556
47. Brocard, J., Warot, X., Wendling, O., Messaddeq, N., Vonesch, J.L., Chambon, P. and Metzger, D. (1997) Spatio-temporally controlled site-specific somatic mutagenesis in the mouse. *Proc Natl Acad Sci U S A*, **94**, 14559–14563
48. Schuler, M., Dierich, A., Chambon, P. and Metzger, D. (2004) Efficient temporally controlled targeted somatic mutagenesis in hepatocytes of the mouse. *Genesis*, **39**, 167–172
49. Mataki, C., Magnier, B.C., Houten, S.M., Annicotte, J.S., Argmann, C., Thomas, C., Overmars, H., Kulik, W., Metzger, D., Auwerx, J. et al. (2007) Compromised intestinal lipid absorption in mice with a liver-specific deficiency of liver receptor homolog 1. *Mol Cell Biol*, **27**, 8330–8339
50. Schnutgen, F., Doerflinger, N., Calleja, C., Wendling, O., Chambon, P. and Ghyselinck, N.B. (2003) A directional strategy for monitoring Cre-mediated recombination at the cellular level in the mouse. *Nat Biotechnol*, **21**, 562–565
51. Schnutgen, F. and Ghyselinck, N.B. (2007) Adopting the good reFLEXes when generating conditional alterations in the mouse genome. *Transgenic Res*, **16**, 405–413
52. Schnutgen, F., De-Zolt, S., Van Sloun, P., Hollatz, M., Floss, T., Hansen, J., Altschmied, J., Seisenberger, C., Ghyselinck, N.B., Ruiz, P. et al. (2005) Genomewide production of multipurpose alleles for the functional analysis of the mouse genome. *Proc Natl Acad Sci U S A*, **102**, 7221–7226
53. Wirth, D., Gama-Norton, L., Riemer, P., Sandhu, U., Schucht, R. and Hauser, H. (2007) Road to precision: recombinase-based targeting technologies for genome engineering. *Curr Opin Biotechnol*, **18**, 411–419
54. Coroadinha, A.S., Schucht, R., Gama-Norton, L., Wirth, D., Hauser, H. and Carrondo, M.J. (2006) The use of recombinase mediated cassette exchange in retroviral vector producer cell lines: predictability and efficiency by transgene exchange. *J Biotechnol*, **124**, 457–468
55. Schucht, R., Coroadinha, A.S., Zanta-Boussif, M.A., Verhoeyen, E., Carrondo, M.J., Hauser, H. and Wirth, D. (2006) A new generation of retroviral producer cells: predictable and stable virus production by Flp-mediated site-specific integration of retroviral vectors. *Mol Ther*, **14**, 285–292
56. Toledo, F., Liu, C.W., Lee, C.J. and Wahl, G.M. (2006) RMCE-ASAP: a gene targeting method for ES and somatic cells to accelerate phenotype analyses. *Nucleic Acids Res*, **34**, e92
57. Seibler, J., Kleinridders, A., Kuter-Luks, B., Niehaves, S., Bruning, J.C. and Schwenk, F. (2007) Reversible gene knockdown in mice using a tight, inducible shRNA expression system. *Nucleic Acids Res*, **35**, e54
58. Seibler, J., Kuter-Luks, B., Kern, H., Streu, S., Plum, L., Mauer, J., Kuhn, R., Bruning, J.C. and Schwenk, F. (2005) Single copy shRNA configuration for ubiquitous gene knockdown in mice. *Nucleic Acids Res*, **33**, e67
59. Torres, R.M. and Kühn, R. (1997) In: Torres, R.M. and Kühn, R. (eds.), *Laboratory protocols for conditional gene targeting*. Oxford University Press, Oxford, pp. 167
60. Metzger, D., Clifford, J., Chiba, H. and Chambon, P. (1995) Conditional site-specific recombination in mammalian cells using a ligand-dependent chimeric Cre recombinase. *Proc Natl Acad Sci U S A*, **92**, 6991–6995
61. Zhang, Y., Riesterer, C., Ayrall, A.M., Sablitzky, F., Littlewood, T.D. and Reth, M. (1996) Inducible site-directed recombination in mouse embryonic stem cells. *Nucleic Acids Res*, **24**, 543–548
62. Logie, C. and Stewart, A.F. (1995) Ligand-regulated site-specific recombination. *Proc Natl Acad Sci U S A*, **92**, 5940–5944
63. Littlewood, T.D., Hancock, D.C., Danielian, P.S., Parker, M.G. and Evan, G.I. (1995) A modified oestrogen receptor ligand-binding domain as an improved switch for the regulation of heterologous proteins. *Nucleic Acids Res*, 23, 1686–1690
64. Weber, P., Metzger, D. and Chambon, P. Temporally controlled targeted somatic mutagenesis in the mouse brain. European Journal of Neuroscience, 14, 1777–1783

Chapter 17

Cre Transgenic Mouse Lines

Xin Wang

Summary

With the development of the Cre-LoxP system, conditional gene targeting has rapidly become a powerful technology that facilitates the study of gene function. This advanced technique circumvents three major concerns sometimes levelled against conventional transgenic and gene-targeting approaches. First of all, gene ablation may exert its effect in multiple cell and tissue types, creating a complex phenotype in which it is difficult to distinguish direct function in a particular tissue from secondary effects resulting from altered gene function in other tissues. Secondly, a gene deletion expressed in the germ line may cause embryonic lethality, thereby precluding analysis of gene function in the adult tissues. Thirdly, the transgenic approach represents a somewhat surreal over-expression of a given protein often causing spurious phenotypes. The generation of conditional knockout mice is a multiple-step process, which involves mating the flox mutant mouse line (essential exon/s of the gene of interest are flanked by two LoxP sites) and the Cre-expressing mouse line. Over the past few years many inducible and/or tissue-specific Cre mouse lines have been developed. This chapter will give a brief review of the generation of Cre-expressing mouse lines and will discuss the strategy of using these Cre lines. In addition, information regarding established Cre-expressing mouse lines will be provided.

Key words: Mouse, Cre recombinase, Transgenesis

1. Introduction

Cre transgenic mice can be produced by two methods (1) standard transgenesis with DNA pronuclear injection or, (2) knock-in approach with homologous recombination in ES cells. The former approach is a relatively rapid procedure with a straight-forward concept in terms of the amount of time and the effort involved in the construction of the Cre-transgene vector and inserting exogenous DNA into the genome. In many cases this method has been proven as the most popular strategy to generate very efficient Cre lines.

Elizabeth J. Cartwright (ed.), *Transgenesis Techniques*, Methods in Molecular Biology, vol. 561
DOI 10.1007/978-1-60327-019-9_17, © Humana Press, a part of Springer Science+Business Media, LLC 2009

The core part of this method is to choose a tissue-specific and/or inducible promoter which drives Cre recombinase expression specifically, as required, and efficiently. With the availability of the complete sequence of the mouse genome, defining and utilizing a specific promoter and its regulatory elements for transgenesis has become relatively easy. However, this standard transgenesis approach involves random integration of multiple copies of the Cre-transgene into the genome, which sometimes results in unwanted Cre expression patterns or low Cre expression levels that prohibit any further study of the gene function. Therefore, a number of founder lines must be generated, in order to identify a suitable strain. In 1995, Kühn and colleagues first introduced inducible gene targeting using a temporal Cre-expressing mouse line *(1)*. Many genes have different functions from development stages to adulthood; therefore, the study of gene function at discrete time points may be required. For example, the Cre-ER line is widely used to achieve such temporal control of genetic recombination in which Cre has been fused to a mutated form of the ligand binding domain of the estrogen receptor *(2)*. The mutant ER does not bind its natural ligand at physiological concentrations, but binds to a synthetic ligand, tamoxifen, and several of its metabolites (e.g. 4-hydroxytamoxifen). Thus, Cre-ER activity can be induced following tamoxifen injection, thereby allowing the temporal deletion of the LoxP flanked gene.

As an alternative to conventional transgenesis, Cre recombinase can be introduced downstream of the targeted promoter by homologous recombination in ES cells. The advantage of this approach is that the Cre expression pattern reflects 'truthful' endogenous gene expression, because all regulatory elements surrounding the transcriptional start site of the targeted gene are present at their native chromosomal position. However, this approach is a long and expensive procedure; moreover, replacement of an endogenous gene with Cre gene (Knock-in/Knock-out) causes a null-allele of the gene. Whether such a non-functional-allele results in any phenotypes in the Cre-expressing mice needs to be carefully screened, and care must be taken to ensure that the knock-in Cre line is included in the gene function study as a control group in order to correctly interpret any phenotypes caused by the conditional knock-outs.

The principal and technical protocols for designing the Cre-transgene vector, DNA pronuclear injection, constructing Cre-targeting vector and gene targeting, are the same as those for generating conventional transgenic and gene-targeting mice, which have been described in previous chapters of this book. The use of Cre-expressing mice and information regarding readily available, established Cre-expressing lines will be given in detail in the methods section below.

2. Materials

2.1. Monitoring Cre Expression Using the Rosa 26 Reporter Line

1. Rosa 26 mouse reporter line (available from Jackson Lab. Stock # 003504. Strain name: B6; 129-Gt(ROSA)26Sortml/Sho/J (*see* **Note 1**).
2. *LacZ* fixing solution: 0.2% glutaraldehyde, 50 mM EGTA, 100 mM $MgCl_2$ in 100 mM sodium phosphate, pH 7.3.
3. *LacZ* washing buffer: 2 mM $MgCl_2$, 0.01% sodium deoxycholate, 0.02% Nonidet-P40 (NP40) in 100 mM sodium phosphate, pH 7.3.
4. *LacZ* staining solution: 0.5mg/mL X-gal, 5 mM potassium ferrocyanide, 5 mM potassium ferricyanide in *LacZ* washing buffer.

2.2. Generation of Conditional Gene-Knockout Mice

PCR primer sequences to amplify Cre transgene
Forward primer: 5′-GACGGAAATCCATCGCTCGACCAG-3′
Reverse primer: 5′-GACATGTTCAGGGATCGCCAGGCG-3′

3. Methods

3.1. Monitoring Cre Expression Using the Rosa 26 Reporter Line

Although many Cre-expressing mouse lines have been characterized by the producers, it is recommended for users to monitor Cre expression prior to crossing the Cre line with the flox mice, because the recombination efficiency and specificity heavily depends on the pattern and level of Cre expression. The Rosa 26 reporter strain is widely used for this purpose; it has a 'stop signal' containing polyadenylation sequences flanked by two LoxP sites placed upstream of *lacZ* in order to prevent its expression *(3)*. Upon commencement of activity of Cre, the 'stop signal' is removed, and *lacZ* is expressed according to the pattern of Cre expression (*see* **Fig. 1**).

1. To detect *lacZ* expression, fix the tissue or embryo samples in the fixing solution for 1 h at 4°C with shaking (*see* **Note 1**).
2. Wash the samples three times briefly with the wash buffer.
3. Cover the sample tubes with foil to exclude light and stain samples in 0.5 mg/mL X-gal in the staining solution overnight at room temperature.
4. The following day, wash the samples in PBS three times for 10 min each, and then store in PBS at 4°C until ready for photography.

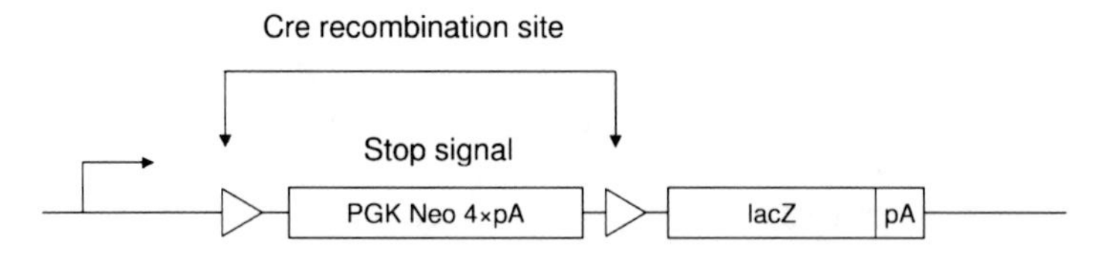

Fig. 1. Strategy for creating the Rosa 26 reporter mouse line. The 'stop signal' comprises the PGK promoter, neomycin-resistant gene, and four copies of polyadenylation sequences flanked by two LoxP sites (small triangles). Upon the Cre-mediated recombination, the 'stop signal' will be removed; therefore *Lac Z* is expressed representing the Cre expression pattern.

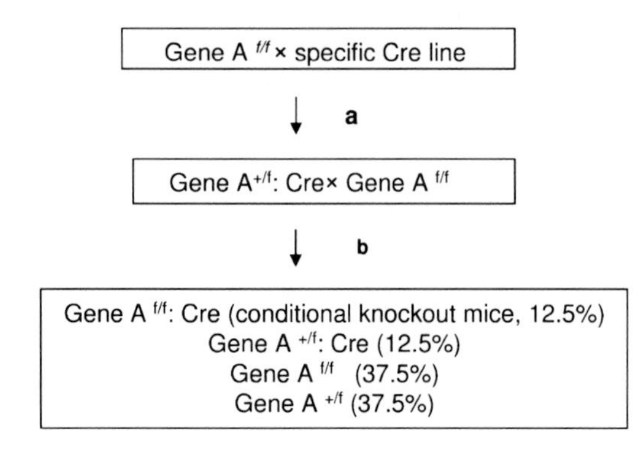

Fig. 2. Breeding scheme for the production of conditional knockout mice.

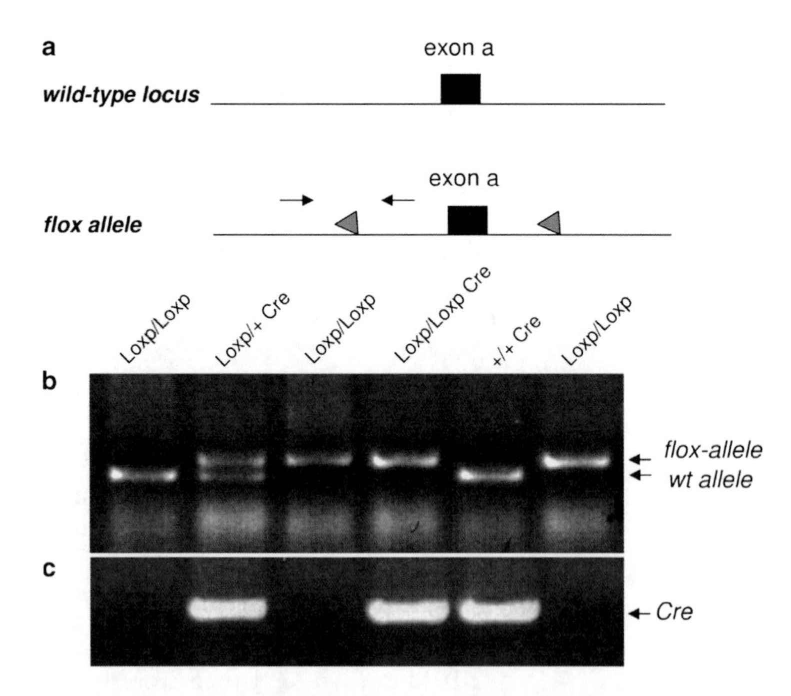

Fig. 3. Genotype determination of conditional knockouts (Loxp/Loxp:Cre). (**a**) Strategy for designing PCR primers to detect the flox allele, the primers are shown as *arrows*. (**b**) PCR amplification of the genomic DNA using primers for the flox gene; the lower band indicates the wild-type allele and the higher band indicates the flox allele. (**c**) PCR amplification of the genomic DNA using primers to detect the Cre transgene.

5. To examine Cre activity at the cellular level, section the tissue or embryo samples using a cryostat and following the same procedure for *LacZ* staining as described in **steps 1–4** above.

3.2. Generation of Conditional Gene-Knockout Mice

Once the specificity of Cre expression has been confirmed, the Cre transgenic line can be mated with the floxed (flanked with loxP) mice to produce conditional knockout mice. Two rounds of breeding are required (*see* **Fig. 2** and **Note 3**). The first cross (a) of a LoxP homozygous mutant mouse mated with a Cre transgenic mouse will yield the following pups: LoxP heterozygous mice carrying the Cre gene (Gene $A^{+/f}$: Cre). In the second breeding step (b), Gene $A^{+/f}$: Cre mice are mated with Gene $A^{f/f}$. Pups born at the expected Mendelian ratio will be of a number of different genotypes, including the desired conditional knockouts (Gene $A^{fl/fl}$: Cre).

Genotype determination of conditional knockouts is achieved by two separate PCR reactions on tail or ear snip genomic DNA (*see* **Fig. 3** and **Note 4**). One PCR will identify the flox allele, using gene specific primers. The second PCR is designed to detect the Cre transgene (*see* primer sequences in **Subheading 2.2**).

3.3. Detecting Gene Deletion

Tissue-specific gene deletion can be assessed by PCR or Southern blot on genomic DNA extracted from the targeted tissues. PCR primers should be designed to flank the two LoxP sites; this will amplify a large DNA fragment for the flox allele, but only a short fragment for the deleted allele (*see* **Fig. 4a**). Using the same

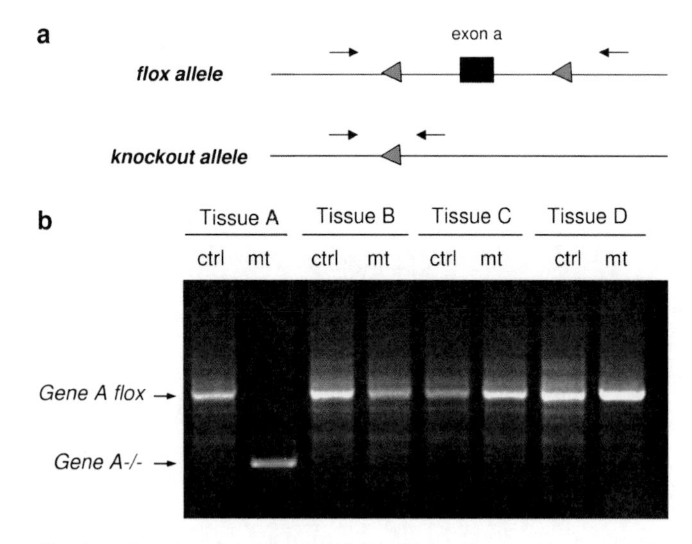

Fig. 4. The detection of gene deletion by PCR. (**a**) The primers are designed to flank the two LoxP sites to distinguish the floxed allele from the knockout allele. (**b**) PCR amplification of the genomic DNA extracted from various tissue samples and amplified with the primers to detect the gene deletion in specific tissues.

Table 1
Examples of some established tissue/or cell-specific Cre-expressing mice

Promoter	Tissue/cell of Cre expression	References
Nes	Neuronal and glial cell precursors	*(4)*
Nes-ER	Neural stem cells, inducible	*(5)*
CamKIIa	Forebrain	*(6)*
CamKIIa-ER	Forebrain, inducible	*(7)*
GFAP	Astrocytes	*(8)*
L7/pcp-2	Cerebellar Purkinje cells	*(9)*
MHC	Cardiomyocytes	*(10)*
MerCreMer-MHC	Cardiomyocytes, inducible	*(11)*
MLC2v	Cardiomyocytes	*(12)*
MLKC1f	Skeletal muscle	*(13)*
smMHC	Smooth muscle	*(14)*
Alb	Hepatocytes	*(15)*
Ins2	Pancreatic β-cells	*(16)*
Probasin	Prostate	*(17)*
ApoE	Kidney	*(18)*
Pax6	Retina	*(19)*
Tie 1	Endothelial cells	*(20)*
Tie 2	Endothelial cells	*(21)*
Lck	T cells	*(22)*
LysM	Macrophage, neutrophil	*(23)*
K5	Basal cells of epidermis	*(24)*
K14-ER	Epidermis, inducible	*(25)*
WAP	Mammary gland	*(26)*
Col2a-1	Chondrocyte	*(27)*
OG2	Osteoblast	*(28)*
aP2	Adipose tissue	*(29)*

principle, Southern blot analysis should be performed to confirm the gene deletion by using a specific probe to detect the flox allele and the deletion allele on the basis of the size difference of the DNA fragments. Western blot or immunohistochemistry should then be used to assess the level of protein expression (*see* **Note 5** and **Note 6**).

It is important to prepare and analyse RNA/DNA/protein from various tissues in order to verify the specificity of the gene deletion (*see* **Fig. 4b**). Particular care must be taken, when analysing Cre induced gene deletion in a specific cell lineage, only as opposed to the whole tissue.

3.4. Established Cre-Expressing Mouse Lines

Over the past ten years many tissue-specific and/or inducible Cre expressing mouse lines have been developed which have greatly facilitated the study of gene function. Established and published Cre-expressing mouse lines can be searched in the database (http://www.mshri.on.ca/nagy/Cre-pub.html). **Table 1** summarises some of the well established Cre lines, covering a wide range of tissues. This list is in no-way comprehensive, but should be used as a point of reference.

4. Notes

1. Also see www.fhcrc.org/science/labs/soriano/rosa26.html for useful information about this strain.
2. Whole-mount *LacZ* staining. Large tissues or embryos need to be fixed in 2% paraformaldehyde together with 0.2% glutaraldehyde for 30 min at 4°C with shaking and then dissected into half for further 30 min for 4 h fixing to allow the fixative to penetrate into the samples. The fixing solution always needs to be prepared freshly and the samples need to be wrapped in foil to avoid light during the staining. The staining solution with 0.5mg/mL X-gal can be re-used after filtering.
3. Generation of conditional gene-knockout mice. It normally takes 14-15 weeks to obtain the conditional gene-knockout mice (Gene $A^{f/f}$: Cre) through two rounds of mating. We normally set up two mating cages for each round of breeding.
4. For genotyping, 100 μg genomic DNA is recommended to use for PCRs as high concentration of DNA would inhibit the PCR reaction, thereby leading to the wrong interpretation of results.
5. If a good antibody is available, western blot or immunochemistry will give a clear answer as to whether the gene (protein) deletion occurred in the specific tissue. However, if such an antibody is not available, real-time quantitative RT-PCR can be used to detect the gene deletion at the transcript level.

6. It is expected that a small residue of protein expression may be detected in some cases, as the efficiency of Cre-mediated recombination varies depending on the chromosomal position of the targeted gene and therefore, Cre activity can not be achieved by 100% in such circumstances.

References

1. Kühn, R., Schwenk, F., Aguet, M., and Rajewsky, K. (1995). Inducible gene targeting in mice. *Science*, **269**, 1427–1428
2. Feil, R., Brocard, J., Mascrez, B., LeMeur, M., Metzger, D., and Chambon, P. (1996). Ligand-activated site-specific recombination in mice. *Proc. Natl. Acad. Sci. U.S.A.*, **93**, 10887–10890
3. Soriano, P. (1999). Generalized lacZ expression with the ROSA26 Cre reporter strain. *Nat Genet.*, **21**, 70–71
4. Guy, J., Hendrich, B., Holmes, M., Martin, J.E., and Bird, A. (2001). A mouse Mecp2-null mutation causes neurological symptoms that mimic Rett syndrome. *Nat Genet.*, **27**(3), 322–326
5. Imayoshi, I., Ohtsuka, T., Metzger, D., Chambon, P., and Kegeyama, R. (2006). Temporal regulation of Cre recombinase activity in neural stem cells. *Genesis*, **44**(5), 233–238
6. Casanova, E., Fehsenfeld, S., Mantamadiotis, T., Lemberger, T., Greiner, E., Stewart, A.F., and Schütz, G. (2001). A CamKIIalpha iCre BAC allows brain-specific gene inactivation. *Genesis*, **31**(1), 37–42
7. Casanova, E., Fehsenfeld, S., Lemberger, T., Shimshek, D.R., Sprengel, R., and Mantamadiotis, T. (2002). ER-based double iCre fusion protein allows partial recombination in forebrain. *Genesis*, **34**(3), 208–214
8. Fraser, M.M., Zhu, X., Kwon, C.H., Uhlmann, E.J., Gutmann, D.H., and Baker, S.J. (2004). Pten loss causes hypertrophy and increased proliferation of astrocytes in vivo. *Cancer Res.*, **64**(21), 7773–7779
9. Barski, J.J., Dethleffsen, K., and Meyer, M. (2000). Cre recombinase expression in cerebellar Purkinje cells. *Genesis*, **28**(3-4), 93–98
10. Agah, R., Frenkel, P.A., French, B.A., Michael, L.H., Overbeek, P.A., and Schneider, M.D. (1997). Gene recombination in postmitotic cells. *Targeted expression of Cre recombinase provokes cardiac-restricted, site-specific rearrangement in adult ventricular muscle in vivo. J. Clin. Invest.*, **100**(1), 169–179
11. Sohal, D.S., Nghiem, M., Crackower, M.A., Witt, S.A., Kimball, T.R., Tymitz, K.M., Penninger, J.M., and Molkentin, J.D. (2001). Temporally regulated and tissue-specific gene manipulations in the adult and embryonic heart using a tamoxifen-inducible Cre protein. *Circ Res*, **89**, 20–25
12. Chen, J., Kubalak, S.W., Minamisawa, S., Price, R.L., Becker, K.D., Hickey, R., Ross, J. Jr., and Chien, K.R. (1998). Selective requirement of myosin light chain 2v in embryonic heart function. *J. Biol. Chem.*, **273**(2), 1252–1256
13. Gerald, W.M.B., Haspel, J.A., Smith, X.C.L., Wiener, H.H., and Burden, S.J. (2000). Selective expression of Cre recombinase in skeletal muscle fibers. *Genesis*, **26**, 165–166
14. Xin, H.B., Deng, K.Y., Rishniw, M., Ji, G., and Kotlikoff, M.I. (2002). Smooth muscle expression of Cre recombinase and eGFP in transgenic mice. *Physiol. Genomics*, **10**(3), 211–215
15. Kellendonk, C., Opherk, C., Anlag, K., Schütz, G., and Tronche, F. (2000). Hepatocyte-specific expression of Cre recombinase. *Genesis*, **26**(2), 151–153
16. Gannon, M., Shiota, C., Postic, C., Wright, C.V.E., and Magnuson, M. (2000). Analysis of the Cre-mediated recombination driven by rat insulin promoter in embryonic and adult mouse pancreas. *Genesis*, **26**, 139–141
17. Maddison, L.A., Nahm, H., DeMayo, F., and Greenberg, N.M. (2000). Prostate specific expression of Cre recombinase in transgenic mice. *Genesis*, **26**, 154–156
18. Leheste, J.R., Melsen, F., Wellner, M., Jansen, P., Schlichting, U., Renner-Müller, I., Andreassen, T.T., Wolf, E., Bachmann, S., nykjaer, A., and Willnow, T.E. (2003). Hypocalcemia and osteopathy in mice with kidney-specific megalin gene defect. *FASEB J.*, **17**(2), 247–249
19. Marquardt, T., Ashery-Padan, R., Andrejewski, N., Scardigli, R., Guillemot, F., and Gruss, P. (2001). Pax6 is required for the multipotent state of retinal progenitor cells. *Cell*, **105**, 43–55

20. Gustafsson , E., Brakebusch, C., Hietanen, K., and Fassler, R. (2001). Tie-1-directed expression of Cre recombinase in endothelial cells of embryoid bodies and transgenic mice. *J. Cell Sci.*, **114**, 671–676
21. Kisanuki, Y.Y., Hammer, R.E., Miyazaki, J., Williams, S.C., Richardson, J.A., and Yanagisawa, M. (2001). Tie2-Cre transgenic mice: a new model for endothelial cell-lineage analysis in vivo. *Dev. Biol.*, **230**(2), 230–242
22. Gu, H., Marth, J.D., Orban, P.C., Mossmann, H., and Rajewsky, K. (1994). Deletion of a DNA polymerase beta gene segment in T cells using cell type-specific gene targeting. *Science*, **265**(5168), 103–106
23. Takeda, K., Clausen, B.E., Kaisho, T., Tsujimura, T., Terada, N., Förster, I., and Akira, S. (1999). Enhanced Th1 activity and development of chronic enterocolitis in mice devoid of Stat3 in macrophages and neutrophils. *Immunity*, **10**(1), 39–49
24. Sano, S., Itami, S., Takeda, K., Tarutani, M., Yamaguchi, Y., Miura, H., Yoshikawa, K., Akira, S., and Takeda, J. (1999). Keratinocyte-specific ablation of Stat3 exhibits impaired skin remodeling, but does not affect skin morphogenesis. *EMBO J.*, **18**(17), 4657–4668
25. Vasioukhin, V., Degenstein, L., Wise, B., and Fuchs, E. (1999). The magical touch: genome targeting in epidermal stem cells induced by tamoxifen application to mouse skin. *Proc. Natl. Acad. Sci. U.S.A.*, **96**(15), 8551–8556
26. Wagner, K.U., Wall, R.J., St-Onge, L., Gruss, P., Wynshaw-Boris, A., Garrett, L., Li, M., Furth, P.A., and Hennighausen, L. (1997). Cre-mediated gene deletion in the mammary gland. *Nucleic Acids Res.*, **25**(21), 4323–4330
27. Ovchinnikov, D.A., Deng, J.M., Ogunrinu, G., and Behringer, R.R. (2000). Col2a1-directed expression of Cre recombinase in differentiating chondrocytes in transgenic mice. *Genesis*, **26**, 145–146
28. Castro, C.H., Stains, J.P., Sheikh, S., Szejnfeld, V.L., Willecke, K., Theis, M., and Civitelli, R. (2003). Development of mice with osteoblast-specific connexin43 gene deletion. *Cell Commun. Adhes.*, **10**(4-6), 445–450
29. Barlow, C., Schroeder, M., Lekstrom-Himes, J., Kylefjord, H., Deng, C.X., Wynshaw-Boris, A., Spiegelman, B.M., and Xanthopoulos, K.G. (1997). Targeted expression of Cre recombinase to adipose tissue of transgenic mice directs adipose-specific excision of loxP-flanked gene segments. *Nucleic Acids Res.*, 25(12), 2543–2545

Chapter 18

Large-Scale Mouse Mutagenesis

Elizabeth J. Cartwright

Summary

Determining gene function by using transgenesis in the mouse has been immensely popular amongst researchers since the introduction of the techniques in the 1980s. However, we are still a long way from knowing the function of the majority of the genes in the genome. It is becoming increasingly accepted that there needs to be a coherent programme of systematic mutation of every protein-coding gene in the mouse genome. The target is therefore to generate approximately 20,000-25,000 gene knockout mouse models. This is an ambitious aim but one which is vital to enhance our understanding of physiology, development and disease. Recently a number of programmes have been established to meet this aim. A particular feature of these large co-ordinated methods for generating genetically modified mice is that they may obviate the need for individual researchers to perform the onerous process of generating the mutant mouse of their gene of interest - in the future we may be able to order our mouse model of interest with the same ease with which we order laboratory reagents and supplies today.

Key words: Mouse knockout, ENU mutagenesis, Phenotype, Mutagenesis, Embryonic stem cells, Gene targeting

1. Introduction

Since the 1980s, researchers world-wide have been genetically modifying the mouse genome, using the many transgenic techniques described in this book, in order to determine the function and regulation of mammalian genes. Completion of the sequencing of the human genome has revealed that there are between 20,000 and 25,000 protein-coding genes *(1)*; however, we understand the in vivo gene function of an only small proportion of these *(2)*. It is clear that a co-ordinated effort is required to functionally annotate the complete genome in order to enhance

Elizabeth J. Cartwright (ed.), *Transgenesis Techniques*, Methods in Molecular Biology, **vol. 561**
DOI 10.1007/978-1-60327-019-9_18, © Humana Press, a part of Springer Science+Business Media, LLC 2009

our understanding of gene function and provide insights into human health and disease.

For many years the large-scale generation of genetically modified *Drosophila*, worms, and zebrafish has been carried out; however, while these non-mammalian animal models have provided important insights into aspects of gene function it is the mouse that is considered to be the most desirable model to study mammalian gene function. Recently, a number of public funding bodies across the world have undertaken to fund the large-scale generation of genetically modified mice.

There are two main approaches that can be taken to identify mammalian gene function; these are the phenotype-driven approach and the genotype-driven approach. The genotype-driven approach includes classic transgenesis, in which a known candidate gene is analysed, as described in many chapters of this book. Gene trap mutagenesis is a gene-driven approach in which both known and unknown genes can be mutated. This is a method that is well suited to the large-scale approach and has been used successfully for a number of years. The phenotype-driven approach, on the other hand, involves the generation of mice with random genetic modifications which are then screened on the basis of their modified characteristics and traits.

2. Mouse ENU Mutagenesis

Some of the first major efforts to produce mouse mutants on a large scale involved ENU mutagenesis. ENU (ethylnitrosourea) is a powerful chemical mutagen which can efficiently introduce point mutations randomly into the genome of treated mice. Male mice are treated with ENU, mated to wild-type females, and their F1 progeny are screened for mutant phenotypes *(3, 4)*; this type of screening will reveal dominant traits. By further breeding and analysing the G3 generation, it is then possible to identify recessive phenotypes. The fundamental premise of this type of screen is that no a priori assumptions are made regarding the genes and their function within any particular pathway, thereby removing investigator bias towards the function of the gene. ENU mutagenesis is invaluable as it can induce many types of allele as a result of point mutations leading to complete or partial loss of gene function or even gain of function.

The key to the success of such a mutagenesis project is the ability to identify desired mutants using robust phenotypic screens. When screening such a large number of mice it is vital that the analyses carried out to identify the mutants are quick and technically simple to perform, inexpensive, reliable, and ideally non-invasive.

By using such a phenotype-driven approach mutants can be screened for an array of morphological, behavioural, biochemical, immunological, sensory etc abnormalities. Specialist screens can be developed to identify very specific desired characteristics. This type of approach is immensely powerful for identifying mutants with potentially, clinically relevant phenotypes. There are several large-scale ENU mutagenesis programmes, each of which tends to focus its efforts on the identification of a particular sub-set of phenotypes (*see* **Table 1**).

Although this technique is extremely efficient at introducing a large array of mutations into the genome it has its drawbacks: (1) the mutated gene must be identified by mapping and positional cloning, which are very costly and time-consuming processes (although with improved mutation detection technology this step will become quicker and inevitably cheaper); (2) the random nature of the mutation process is also a disadvantage when wishing to systematically mutate every gene in the genome.

Table 1
International mouse ENU mutagenesis programmes

Mutagenesis programme	URL	Primary goal
The Harwell ENU Mutagenesis Programme	www.mut.har.mrc.ac.uk/	To generate F1 progeny of mutagenised animals for genome-wide dominant phenotypic screens
Jax-HLBS: Mouse Heart, Lung, Blood and Sleep Disorder Centre	pga.jax.org/index.html	To determine factors underlying heart, lung, blood, and sleep diseases through phenotypic characterisation
Neuroscience Mutagenesis Facility	nmf.jax.org/index.html	To develop new neurological mouse models
Reproductive Genomics	reprogenomics.jax.org/index.html	To develop mouse models of male and female infertility
Munich ENU Project	www.helmholtzmuenchen.de/en/ieg/group-genome-project/enu-screen/index.html	To develop mouse models of inherited human disease
Mouse Functional Genomics Research Group, RIKEN GSC	www.gsc.riken.go.jp/Mouse/	To develop mouse models of common human diseases by screening for late-onset tumorigenesis and senescence-related phenotypes

Although not all programmes are actively producing new mutants, existing mutants are available as frozen embryos/sperm, see appropriate website for details

3. Large-Scale Gene Knockout Programmes

Over the years it has become clear that the gene knockout approach is a very powerful tool for providing information regarding gene function; however, only a relatively small proportion of genes have so far been knocked out. Many of the gene knockout models that have been produced have not been fully utilised for a number of reasons, for example, they may be covered by an intellectual property agreement and therefore their full details are not available to the research community, mice may be on a variety of backgrounds strains which may prohibit easy comparison with other knockouts, or they may have a phenotype which could not be fully analysed by the original researcher. *See* **Note 1** for a list of websites containing databases of gene knockout/mutant mice that have been generated. It is therefore the large-scale, co-ordinated production of gene knockout mice, which is currently the focus of several new mouse mutagenesis projects.

These projects have the same fundamental goal – high-throughput generation of targeted mutations in mouse ES cells covering every gene in the genome and the generation (on a smaller scale) of gene knockout mice. The generation of mutant ES cells through these systematic approaches means there will be more comparability between knockout mice; in addition, mutant ES cells, knockout mice, and phenotypic data will be readily accessible to researchers world-wide. The goal is that the cost of obtaining mutant ES cells or mice (likely as frozen embryos or sperm) should be minimal, thereby making gene knockout mice available to all researchers, not just those with specialist transgenesis skills and equipment.

The scale of such a co-ordinated programme is both its advantage and disadvantage; overall, large-scale mutagenesis projects will be both time and cost efficient. However, such projects will present a number of complex technical, organisational, and logistical problems, and the overall costs will be considerable.

Three consortia, EUCOMM, KOMP, and NorCOMM, are currently producing knockout ES cells and a number of other consortia are also preparing to join this international effort to knockout every gene in the genome.

- EUCOMM: European Conditional Mouse Mutagenesis Programme
  - www.eucomm.org
  - EUCOMM is a European Union funded consortium
- KOMP: Knockout Mouse Project
  - www.nih.gov/science/models/mouse/knockout
  - KOMP is funded by NIH in the USA

- NorCOMM: North American Conditional Mouse Mutagenesis Programme
  - www.norcomm.org
  - NorCOMM is a Canadian project funded by Genome Canada

3.1. Targeting Strategies

All three consortia use a combination of gene targeting and gene trapping to generate mutant alleles. Both of these approaches have their advantages and disadvantages: Gene targeting leads to specific gene deletion in a highly efficient manner; it allows for flexible construct design enabling not only null mutants to be generated but also knock-ins and conditional mutations. However, it is cumbersome, as it requires detailed knowledge of the structure of the gene of interest and for each target gene a different vector must be produced. Gene trapping on the other hand is much more suitable for high-throughput mutagenesis and uses random insertion of the vector into any gene to generate sequence, expression and functional information for that gene. Only a single, or small number of vectors, need to be designed and utilised to trap thousands of genes *(5)*. Trapping however is mainly suitable for genes that are highly expressed in ES cells and in some cases a true null allele will not be produced. It has therefore been considered prudent for these two techniques to be used in parallel in the large-scale mouse knockout programmes.

By analysing the phenotypes of GM mice generated by conventional gene knockout and gene trap strategies it is apparent that between 15 and 30% of genes are essential during development and their deletion leads to embryonic lethality, which clearly precludes functional analysis in the adult *(6–8)*. Conditional gene targeting is widely used to circumvent the problems of embryonic lethality by allowing the temporal and spatial deletion/mutation of the gene of interest. Both EUCOMM and NorCOMM have taken advantage of this approach by using conditional gene targeting and conditional gene trapping strategies by incorporating FRT and loxP recombination sites into the targeted allele, allowing the generation of both null mutant mice and conditional knockouts *(9)*. KOMP has opted to focus its efforts on gene targeting to generate deletion mutations from which null mutant mice can subsequently be generated, as well as using a conditional gene targeting strategy. *See* **Table 2** for further details regarding these programmes.

3.2. Origin of Embryonic Stem Cells: 129 vs C57BL/6

A major advantage to the efforts to understand gene function using the mouse as a model has been completion of the sequencing of the mouse genome *(10)*. However, of concern is that it is the C57BL/6 mouse strain that has been sequenced and the majority of ES cells currently in use have been derived from the 129 mouse strain. Historically the 129 mouse strain has been

Table 2
Large-scale mouse mutagenesis programmes

	EUCOMM	KOMP	NorCOMM
URL	www.eucomm.org	www.nih.gov/science/models/mouse/knock-out	www.norcomm.org
Funding	FP6, EU	NIH, USA	Genome Canada, Canada
Targeting strategies	Conditional gene trapping Conditional targeted trapping Conditional gene targeting	Gene targeting Conditional gene targeting	Conditional gene trapping Conditional gene targeting
Number of cells to be mutated	12,000 trapped 8,000 targeted	8,500 targeted	10,000 trapped 2,000 targeted
Research community involved in prioritisation of genes to be targeted	Yes	Yes	Yes
Origin of ES cells	129 and C57Bl/6	129 and C57Bl/6	129
Other services	Collect, distribute and archive existing Cre-expressing lines Generate 20 ligand inducible Cre transgenic lines Training in mutagenesis strategies	Acquire and archive 1,000 existing mouse knockouts and phenotyping data not already available in public domain www.informatics.jax.org/external/ko/ Generation of robust C57Bl/6 ES cell lines	Repository of Cre-recombinant promoters
Collaborative programmes	www.eumodic.org – a programme for comprehensive mouse phenotyping www.europhenome.org – database of phenotyping data www.emma.org – the European Mutant Mouse Archive which cryopreserves and distribute mutant strains	www.knockoutmouse.org – the data co-ordination centre for KOMP hosted at the Jackson Labs	

readily accessible to the generation of extremely good ES cells; however, the strain presents a number of problems in several fields of research including immunology, neuroscience, and physiology *(11, 12)* (also *see* **Note 2**). It has now been decided that although it may be necessary to begin the large-scale mutagenesis projects using 129 ES cells, every effort will be made to generate and subsequently use robust ES cells from C57BL/6 mice in order to take full advantage of the sequence information available.

3.3. Generation of Mice from Mutant ES Cells

All the mutated ES cells from these programmes will be archived as a frozen source which will be readily available for researchers from around the world to order. These will be made available at a minimal cost and will require the researcher to complete a material transfer agreement (MTA). In addition to the mutated ES cells it will also be possible to order targeting vectors and genetically modified mice (usually as frozen embryos or sperm – *see* **Note 3**).

The programmes are also dedicated to producing a small number of mouse mutants (350–500 for each programme) from the ES cell resources. These will be produced primarily to validate their systems in order to verify that successful null mutant mice can be generated from the targeted ES cells within their resource; all mice will then be made publicly available. EUCOMM also have a reciprocal agreement with another EU funded programme called EUMODIC (www.EUMODIC.org) to initially provide 650 mutant ES cell clones from which gene knockout mice will be generated and will subsequently undergo comprehensive phenotyping using the EMPReSSslim phenotyping protocols (*see* **Note 4** for details regarding the EUMODIC programme).

Individual researchers from around the world will also be able to access the projects websites and request the targeted ES cell line/s of their choice. The researcher will then use their animal facility of choice to generate the knockout mice. This will save a lot of time and financial resources for individual researchers and will open up the field of the analysis of mammalian gene function using gene knockout mice to many more researchers.

4. Conclusions

The large-scale generation of mouse mutants through co-ordinated programmes is very likely to meet our goal of producing a mutant for every gene in the genome, leading to the functional annotation of the genome. Access to gene knockout mouse models will become easier and less expensive to individual researchers, allowing more focus to be placed on determining gene function rather than generation of the mouse model itself. It is widely anticipated that the generation and analysis of mutant mouse models on such a large-scale will lead to major breakthroughs in our study of human gene function in health and disease in the coming years.

5. Notes

1. Archives and databases of gene knockout/mutant mice
   - International Mouse Strain resource: www.informatics.jax.org is a database containing details of available mouse mutants.
   - European Mutant Mouse Archive (EMMA): www.emma.org is an archive of cryopreserved mutant mouse lines.
   - Federation of International Mouse Resources (FIMRe): www.fimre.org is a collaborative organisation whose goal is to archive and provide strains of mice as cryopreserved embryos and gametes, ES cell lines, and live breeding stock to the research community.
   - Deltagen and Lexicon knockout mouse list: www.informatics.jax.org/external/ko/ is a resource of mouse lines available to academic and non-profit making organisations.
2. The Europhenome Mouse Phenotyping Resource website (www.europhenome.org) contains phenotyping data for a number of commonly used wild-type strains including both the 129SvPas and C57Bl6/J. To access this information from the home page click *Data Browser* under *Europhenome Tools.* Open the folder of interest under *Phenotyping Platforms,* then open the particular phenotypic analysis of interest – this will reveal pooled data for all the wild-type strains analysed for each parameter measured. Double clicking each parameter name will reveal strain and sex-specific information.
   The Mouse Phenome Database (http://www.jax.org/phenome) also contains phenotype information for many mouse strains. It provides a wealth of information regarding mouse strain characteristics and in many cases their response to inventions such as high-fat diet.
3. The over-riding remit of all large-scale mouse mutagenesis programmes is that the resources are made available to the international research community. This however will inevitably lead to the desire to ship large numbers of mice over long distances. It is desirable on several grounds that live mice are not shipped, but instead are transported as frozen embryos or sperm: (1) On economic grounds, it is very costly to maintain a repository of live mice and also to transport them. (2) For animal welfare, it is more acceptable to transport frozen stocks. (3) On logistic grounds, several airlines will no longer transport live mice. Shipment of frozen stocks does however have the disadvantage that the end-user needs to have the facilities to re-derive mice from these sources; such facilities are not routinely available in all animal units.

4. EUMODIC (www.EUMODIC.org) is an EU funded programme dedicated to the generation and comprehensive phenotypic analysis of gene knockout mice from large-scale ES cell resources. All the phenotyping protocols used under EUMODIC have been standardised and validated and are available from the EMPReSS resource (www.empress.har.mrc.ac.uk), which contains the SOPs (standard operating procedures). All of the phenotyping data from the knockout strains analysed will be available through the Europhenome Mouse Phenotyping Resource website (www.europhenome.org) described in **Note 2**.

References

1. International Human Genome Sequencing Consortium (2004) Finishing the euchromatic sequence of the human genome. *Nature*, **431**, 931–45.
2. Austin, C.P., Battey, J.F., Bradley, A., Bucan, M., Capecchi, M., Collins, F.S., Dove, W.F., Duyk, G., Dymecki, S., Eppig, J.T. et al. (2004) The knockout mouse project. *Nat Genet*, **36**, 921–4.
3. Hrabe de Angelis, M.H., Flaswinkel, H., Fuchs, H., Rathkolb, B., Soewarto, D., Marschall, S., Heffner, S., Pargent, W., Wuensch, K., Jung, M. et al. (2000) Genome-wide, large-scale production of mutant mice by ENU mutagenesis. *Nat Genet*, **25**, 444–7.
4. Nolan, P.M., Peters, J., Strivens, M., Rogers, D., Hagan, J., Spurr, N., Gray, I.C., Vizor, L., Brooker, D., Whitehill, E. et al. (2000) A systematic, genome-wide, phenotype-driven mutagenesis programme for gene function studies in the mouse. *Nat Genet*, **25**, 440–3.
5. Skarnes, W.C., von Melchner, H., Wurst, W., Hicks, G., Nord, A.S., Cox, T., Young, S.G., Ruiz, P., Soriano, P., Tessier-Lavigne, M. et al. (2004) A public gene trap resource for mouse functional genomics. *Nat Genet*, **36**, 543–4.
6. Zambrowicz, B.P., Abuin, A., Ramirez-Solis, R., Richter, L.J., Piggott, J., BeltrandelRio, H., Buxton, E.C., Edwards, J., Finch, R.A., Friddle, C.J. et al. (2003) Wnk1 kinase deficiency lowers blood pressure in mice: a gene-trap screen to identify potential targets for therapeutic intervention. *Proc Natl Acad Sci USA*, **100**, 14109–14.
7. Hansen, J., Floss, T., Van Sloun, P., Fuchtbauer, E.M., Vauti, F., Arnold, H.H., Schnutgen, F., Wurst, W., von Melchner, H. and Ruiz, P. (2003) A large-scale, gene-driven mutagenesis approach for the functional analysis of the mouse genome. *Proc Natl Acad Sci USA*, **100**, 9918–22.
8. Mitchell, K.J., Pinson, K.I., Kelly, O.G., Brennan, J., Zupicich, J., Scherz, P., Leighton, P.A., Goodrich, L.V., Lu, X., Avery, B.J. et al. (2001) Functional analysis of secreted and transmembrane proteins critical to mouse development. *Nat Genet*, **28**, 241–9.
9. Schnutgen, F., De-Zolt, S., Van Sloun, P., Hollatz, M., Floss, T., Hansen, J., Altschmied, J., Seisenberger, C., Ghyselinck, N.B., Ruiz, P. et al. (2005) Genomewide production of multipurpose alleles for the functional analysis of the mouse genome. *Proc Natl Acad Sci USA*, **102**, 7221–6.
10. Waterston, R.H., Lindblad-Toh, K., Birney, E., Rogers, J., Abril, J.F., Agarwal, P., Agarwala, R., Ainscough, R., Alexandersson, M., An, P. et al. (2002) Initial sequencing and comparative analysis of the mouse genome. *Nature*, **420**, 520–62.
11. Corcoran, L.M. and Metcalf, D. (1999) IL-5 and Rp105 signaling defects in B cells from commonly used 129 mouse substrains. *J Immunol*, **163**, 5836–42.
12. Crawley, J.N., Belknap, J.K., Collins, A., Crabbe, J.C., Frankel, W., Henderson, N., Hitzemann, R.J., Maxson, S.C., Miner, L.L., Silva, A.J. et al. (1997) Behavioral phenotypes of inbred mouse strains: implications and recommendations for molecular studies. *Psychopharmacology (Berl)*, **132**, 107–24.

Chapter 19

Dedicated Mouse Production and Husbandry

Lucie Vizor and Sara Wells

Summary

The complexity of mammalian biology, in combination with the ability to manipulate the embryonic genome in mice, has provided specific challenges in terms of mouse breeding, analysis, and husbandry. From the initial planning stage of a project involving the generation of novel, genetically altered mice, it is important to consider a number of potential issues involving the successful establishment and propagation of a transgenic line.

Key words: Background strain, Congenic, Inbred, Viability, Fertility, Phenotypic analysis, Welfare assessment, Breeding calculations, Husbandry

1. Introduction

It is now over a century since the mouse was first used as a laboratory model for studying mammalian biological processes *(1, 2)*. However, no scientist in the early twentieth century would have predicted the extent to which mice would be used throughout biological research today. The mouse has emerged as a key organism in the endeavour to study human disease because of its accessibility in practical terms, relatively short generation time, well-studied genetics (including a complete genome sequence), and a whole variety of applicable genetic manipulation technologies.

Even in the very early days of mouse research, it became obvious that not all mouse strains are the same and indeed, can be kept under the same conditions. Modifying mouse breeding and husbandry routines are therefore not new aspects of animal care. The conditions, methods, and techniques developed for wild type strains and those carrying spontaneous mutations, are

Elizabeth J. Cartwright (ed.), *Transgenesis Techniques*, Methods in Molecular Biology, vol. 561
DOI 10.1007/978-1-60327-019-9_19, © Humana Press, a part of Springer Science+Business Media, LLC 2009

just as applicable to the care of genetically altered (GA) transgenic mice.

With the arrival of transgenesis in the 1980s, there has been a dramatic increase in the number of different strains, especially over the last two decades where transgenic techniques have become more reliable, routine, and accessible. In the UK alone the use of GA mice has more than quadrupled, since 1995 *(3)*. Each different transgenic line can of course present its own challenges in terms of breeding, viability, welfare considerations, and phenotypic analysis; therefore, a note of caution on the instructions detailed within this chapter. We offer advice and suggestions based on our own experience and that of our colleagues within animal research. It is not that all these techniques will be successful with all transgenic lines. Indeed many variables including background strains, housing conditions, and even technical staff, may be responsible for mouse lines behaving differently in different institutions. It is important that any changes made, or techniques applied are suitable for the conditions and equipment available and their outcomes are analysed scientifically using the appropriate statistical tests. It is easy to alter the environment for individual mice and see improvements. However, conclusions should not be drawn without using statistically significant groups and relevant controls *(4, 5)*.

This chapter is divided into three sections with the intention of giving technical details for the establishment of a transgenic mouse colony, mouse production, and specialist mouse husbandry.

2. Establishment of a Transgenic Mouse Colony

2.1. Strain Background Selection

During the planning of a project using transgenic mice, it is important to consider the strain which is most appropriate for phenotyping and analysis (*see* **Note 1**). Many genes have different expression levels and patterns, depending on the mouse strain background. This is also applicable to the expression of both targeted and non-targeted transgenes (*see* **Note 2**). The phenotype observed from gene manipulations is therefore very much dependent on the strain of the mouse carrying it. Each inbred strain has a different set of genes which can differentially influence phenotypic traits, each strain having a unique set of characteristics *(6–8)*. Use of the correct breeding strategy in order to maintain isogenic or congenic stocks (*see* **Table 1**), can in some way negate the problems of background effects.

Table 1
Definition of genetic terms

Term	Definition
Inbred strain	A strain of mice that has been maintained for 20 generations or more, by only mating siblings
Isogenic	Genetically identical backgrounds (mice of an inbred strain)
Coisogenic	Nearly genetically identical, vary at a single locus (such as a targeted gene modification bred on to mice on the same background that the ES cells were derived from)
Congenic	Nearly genetically identical, vary at a single locus and the chromosomal region around that locus
Outcross	A mating where an individual is mated to a wild-type strain, different from its background strain
Backcross	A mating where an individual is crossed either to one of its parents or to a genetically identical mouse

Although developments and progress in ES cell culture and microinjection techniques have extended the range of strains available for transgenesis *(9, 10)*, there are still limitations to the background on which the initial embryo manipulation can be made.

2.1.1. Background Strains for Pronuclear Injection

Pronuclear microinjection can be performed on embryos from a number of pure inbred strains, including FVB and C57BL/6, or using F2 embryos generated by crossing strains such as CBA with C57BL/6. Some strain backgrounds show a lower efficiency of transgenic generation than others *(11)*. It is an important strategic decision to choose whether to use a strain which has been shown to be efficient in generating a high number of transgenics, but may not be ideal for analysis, or to microinject the most appropriate background strain for phenotyping, which may necessitate more time spent on pronuclear injection efforts. For example, FVB has been widely shown to be extremely effective for pronuclear injection, but has different behavioural and neuroanatomical traits than the well-characterized C57BL/6, and may therefore be unsuitable for classical neurological studies *(12, 13)*.

2.1.2. Embryonic Stem Cell Strains

Traditionally, ES cells used for targeted transgenesis have been derived from 129 substrains. The majority of knock out experiments have been completed on this background, because of the high efficiency and robustness of the cell lines. However, now there are ES cell lines available from C57BL/6 substrains *(14, 15)* which are being more widely used.

Whatever be the ES cell line used, a 129 or C57BL/6 substrain, coisogenicity (for definitions *see* **Table 1**) in the transgenic stock will only be obtained if chimaeras are bred with the exact substrain from which the cells are derived. With many different substrains of 129 and C57BL/6 mice, this is not a trivial task as it may involve importing the relevant strain into your institution.

Although there will always be an urgent desire to assess a chimaera for germ-line transmission, the initial mating schemes should not confound future analysis. Crossing a chimaera generated with 129 ES cells, carrying dominant alleles for agouti-white belly (A^w/A^w), with a C57BL/6 mouse carrying a recessive allele at the agouti locus (a/a), may give you an early indication of germ-line transmission by virtue of coat-colour (offspring carrying the ES genome will be agouti-white bellied). This generation are F1 hybrids, carrying 50% of genetic information from both 129 and C57BL/6. Analysis of this generation must, of course, involve using appropriate controls and the interpretation of the data gathered from F1 and subsequent generations must take into account the differences in backgrounds *(15)*.

2.1.3. Breeding to Congenicity

An alternative to breeding using isogenic lines (which in some cases may be difficult as the substrains may not always be available) is to breed the gene modification on to the background of your choice.

For a transgenic produced on a specific background such as 129P2Ola (the donor strain), the first cross to another strain, for example BALB/cByJ (the recipient strain), is called an outcross **(Fig. 1)**. The resultant F1 generation, contain 50% of the genome of each strain. Crossing the F1 individuals which carry the transgene with BALB/cByJ again (N_2- this is called the first backcross), will reduce the contribution of the 129 strain to 25%. Backcrossing to BALB/cByJ and genotyping for the transgene at every generation will, at N_{10}, result in a mouse line where the gene modification is classified as congenic on BALB/cByJ **(Table 2)**. At this point, the transgenic line is theoretically identical to the parental strain, apart from the area around the gene modification, the length of which will vary depending upon the random recombination events, which have occurred during the many generations of breeding.

At N_{10}, it is estimated that on average this region is approximately 20 cM and 99.9% of the total genome should be BALB/cByJ *(8)*. Elimination of the residual genome on which the manipulation was done, can be further enhanced by techniques such as marker-assisted selection, which allows breeding stock to be chosen, which contains the smallest amount of DNA from the donor strain *(16)*.

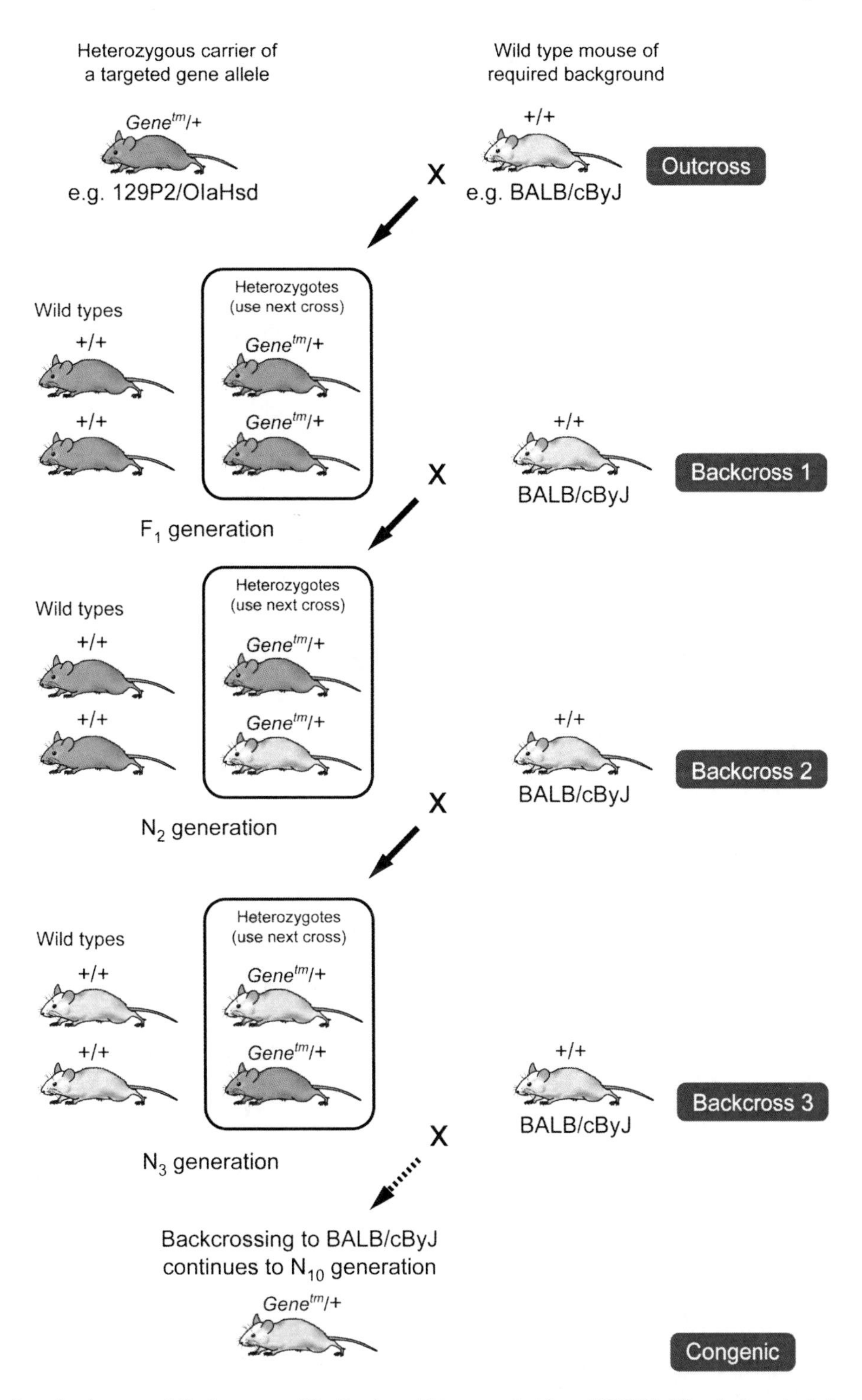

Fig. 1. Breeding to congenicity. A gene modification (gene^{tm}) is generated from 129P2Ola ES cells. In order to breed this modification onto another background (BALB/cByJ in this case), a carrier heterozygote is first crossed with the chosen strain (this is called an outcross). The resultant gene^{tm}/+ mice of this and subsequent generations are backcrossed to BALB/cByJ until the tenth generation. The mice are now considered as being congenic and therefore, genetically identical to BALB/cByJ, apart from the gene^{tm} region.

Table 2
Breeding to congenicity

Generation	% genetic contribution of donor:recipient
F1	50:50
N2	25:75
N3	12.5:87.5
N4	6.25:93.75
N5	3.13:96.87
N6	1.56:98.44
N7	0.78:99.22
N8	0.39:99.61
N9	0.20:99.8
N10	0.10:99.9

These breeding schemes involve a significant investment in time and genotyping resources. Of course, the importance of the background strain to the system being explored must warrant such an endeavour and obvious delay to the project.

2.2. Establishment of a New Colony

2.2.1. Chimaera and Founder Transgenic Mating

The first cross of a chimaera, or a founder transgenic, to a wild-type mouse can start to reveal any problems, which may be encountered whilst trying to propagate and expand a new transgenic line. Furthermore, during the production of the first generation, there may be additional problems that may need to be overcome.

Germ line transmission from chimaeras is very much dependent on the characteristics of the ES cell line *(17, 18)*. Even chimaeras with coat colours indicating a high contribution from the manipulated ES cell, can be unsuccessful in transmitting the transgene. The gene knockout may result in haploinsufficiency, where having only a single copy of the gene is not enough for survival (*see* **Note 3**). The result of this will be that in the first generations bred from a chimaera, there are pups containing the ES cell genome, but no carriers of the manipulated gene. In the example above, of a chimaera generated from 129 ES cells crossed with a C57BL/6 mouse, there would be pups with an agouti white-bellied coat, but none carrying the gene modification.

Lack of carriers in the first generation from founder's pronuclear transgenics, can indicate that the transgene integration occurred after cell division in the microinjected embryo and cells containing the transgene are not highly represented in the germ cells.

2.2.2. Breeding a New Transgenic Line

Once germline transmission of the transgene has been accomplished, the new transgenic line must be expanded to suit experimental needs. There are a number of common problems and concerns which any mouse breeder should be aware of at this stage:

1. *Viability.* If there are fewer carriers than predicted by usual Mendelian inheritance, there may be a problem with the viability of the transgene. The genetic manipulation may reduce the likelihood of the carrier mice to survive and thrive. Viability problems can occur from conception to adulthood. Some husbandry techniques, including supplementary feeding and different housing conditions may be applicable to overcome viability problems (*see* **Subheading 4**).
2. *Fertility.* The genetic modification may reduce the fecundity of a mouse strain. This can differentially affect males and/or females and may require the intervention of either different husbandry techniques and/or assisted reproductive technologies.
3. *Maternal behaviour.* Transgenic females may, by virtue of their phenotype, be poor mothers. This should always be considered as a possibility after losing whole or part litters of pups from transgenic mothers.
4. *Longevity.* It should never be assumed that a transgenic mouse will live as long or breed as prolifically as its wild-type parental strain. It is important to breed founders and chimaeras as soon as possible and to cryopreserve transgenic lines early in the breeding programmes.
5. *Phenotypic penetrance.* Penetrance is the indices of the proportion of mice carrying a transgene that shows the expected phenotype. Penetrance is not the property of the genetic alteration, but of the genetic background. Breeding the transgene onto different strain backgrounds can greatly alter the penetrance *(19)*.
6. *Sex-linkage.* If the transgene has integrated into the X chromosome, or this is the location of the targeted gene in ES cells, it will be inherited in a sex-linked manner, that is, only female carriers will be obtained from mating transgenic males.

2.3. Welfare Assessment

Whilst establishing a new GA colony, it may be necessary to assess the welfare of the transgenic mice. It is of no scientific or ethical advantage to breed mice with health problems, which can confound data collection or cause uncontrolled suffering to the animal. However, it is inevitable in efforts to use mice to model and recreate human and other mammalian disease conditions that GA mice will be generated, which are cause for concern. On the other hand, not all illness and welfare concerns will be attributable to a transgene. A small proportion of wild type mice also get sick and this will be reflected in a line carrying a transgene as well.

In order to control, alleviate, and reduce their impact on animal health, it is imperative to record the impact of the genetic modification throughout the life of the mouse. If no direct intervention can be taken to reduce the particular deficit (for example a physical abnormality or a progressive tumour growth), humane endpoints should be agreed with the veterinary and animal care staff. These endpoints set the limits of the suffering of the animal, beyond which the advantage in terms of scientific knowledge can not be justified by the increase in animal suffering.

2.3.1. Neonatal Welfare Assessment

Potentially a phenotype can manifest itself from conception onwards. Perinatal death and sickness have commonly been recorded as the consequence of a genetic modification *(20)*. When examining young pups (*see* **Table 3**), keep the disturbance to the mother and the nest to a minimum by always returning the pups to the home cage as quickly as possible. Young pups lose heat very quickly, so avoid placing them on cool surfaces.

2.3.2. Post-Weaning Welfare Assessment

Dysmorphology

An early indicator of the susceptibility to health problems may be an obvious or subtle dysmorphology. For example, an abnormal gait may be the first indication of a progressive motor neurone disease *(21)*. Observe the appearance and behaviour of mice in their home cage, with littermates, if available. Scruff the mice in

Table 3
Indicators of ill health which can be seen when examining neonatal mice

Characteristic	Normal	Abnormal
Colour	Pink	Purple/blue/white
Activity	Periodic wriggling	Reduced movement, hyperactivity, completely still
Size	The same as littermates	Small or large, left or right side
Milk Spot (*see* **Note 4**)	Present	Absent
Oedema (*see* **Note 5**)	Not present	Ranging from slight fluid retention under the skin to complete swelling of the pup
Position in the cage	In the nest	Strewn around the cage
Morphology	The same as littermates	Dysmorphic, anomalies in size or shape of physical characteristics
Respiration	Regular	Irregular or slow

order to inspect the physical characteristics of their coat, skin, genitalia, head shape, ears, eyes, nose, teeth, vibrissae, limbs, paws, digits, nails, tail, and anal area.

Weights

The weight of an animal is an important phenotypic parameter, when assessing welfare. GA mice can be large, or small, as a consequence of their genetic modification, but otherwise are healthy. Others fail to thrive, become overtly obese, and may require some special and specific husbandry intervention to ensure their wellbeing. Progressive weight loss in adult mice is a key indicator of ill-health and typically the loss of 10–15% body weight should warrant intervention.

General Welfare Assessment

When examining mice post weaning, they must be examined both in their home cage and whilst being handled, looking at their behaviour and movement. If the symptoms do not confound the continued collection of data, or is actually the phenotype being studied, their progression should be noted at regular intervals. In order to record the health and welfare of your mouse colony, use a comprehensive list, such as in **Table 4**, which you should regularly update with any different warning signs of ill health you may perceive.

Table 4
Indicators of welfare concerns in mice post-weaning

Characteristic	Abnormal observation
Activity	Lethargy, over activity, huddled, hyperactivity, repetitive movement
Breathing	Rapid, shallow, laboured, irregular
Skin colour	Purple, blue
Gait	Unco-ordinated, reluctant to move, limb dragging, limb paralysis
Vocalization	Struggles and squeaks loudly when handled, loud vocalization in the cage
Coat	Rough, unkempt, wounds, hair thinning, infected wounds, severe hair loss, bleeding, self-mutilation
Eyes	Wet, very dry, with a discharge, eyes partially or completely closed, dried blood
Nose	Very wet, runny, bloody or coagulated discharge
Faeces	Soft pellets, blood in faeces, runny discharge, soiled anus
Urination	Excessively wet bedding, wetness or staining of the fur around urethra
Body condition	Small, large, obese, loss of muscle fat, chronic weight loss
Dehydration	Skin less elastic, skin tenting, eyes sunken
Food intake	Lacking or increased consumption
Water intake	Frequent/constant drinking

3. Mouse Production

3.1. Breeding a GA Colony

It is good scientific and ethical practice to be able to accurately calculate the number of mice needed for statistically relevant analysis. Power analysis takes in to account the type of statistical analysis employed for an experiment, as well as what differences in data measurement constitute biological or clinical importance *(22)*. There are useful on-line resources to assist with calculating sample size and designing powerful equations, which minimize mouse use *(23, 24)*.

After establishing how many animals of what genotype you require, breeding calculations should be used to estimate the number of matings required in order to fulfil experimental needs without excessive breeding. Mice for replacement breeding stock must also be included in all breeding calculations.

The important parameters for calculating how many matings you will need from a breeding colony are listed below (e.g. *see* **Fig. 2**).

1. Genotypes and controls needed
2. Genotype of parents available
3. % non-productive matings
4. Average litter size
5. Pre-weaning mortality
6. Time between litters
7. Unexpected mortalities

If establishing a new GA colony, a good approximation of these will be to use the data from the wild type colony used to generate the transgenic. However, after several litters these figures should be revisited and adjusted appropriately.

3.2. Setting Up Matings

3.2.1. Selecting Productive Mating Stock

There are several factors to take into account when actually setting up the matings:

1. That age of sexual maturity has been reached (6–7 weeks of age for C57BL/6 females).
2. Fertile lifespan beyond which the mice will not be productive.
3. Economical breeding life. Mice can only sustain maximum breeding capacity for a limited amount of time and beyond which the litter size and frequency may decrease (typically this may be any time from 6 months onwards).
4. Impact of early or delayed matings. In females the viability of a first litter can depend on the age of a female at the time of first matings *(25)*. The ability to reproduce for the first time may also be affected in males if the first introduction to female mice is late in adulthood.

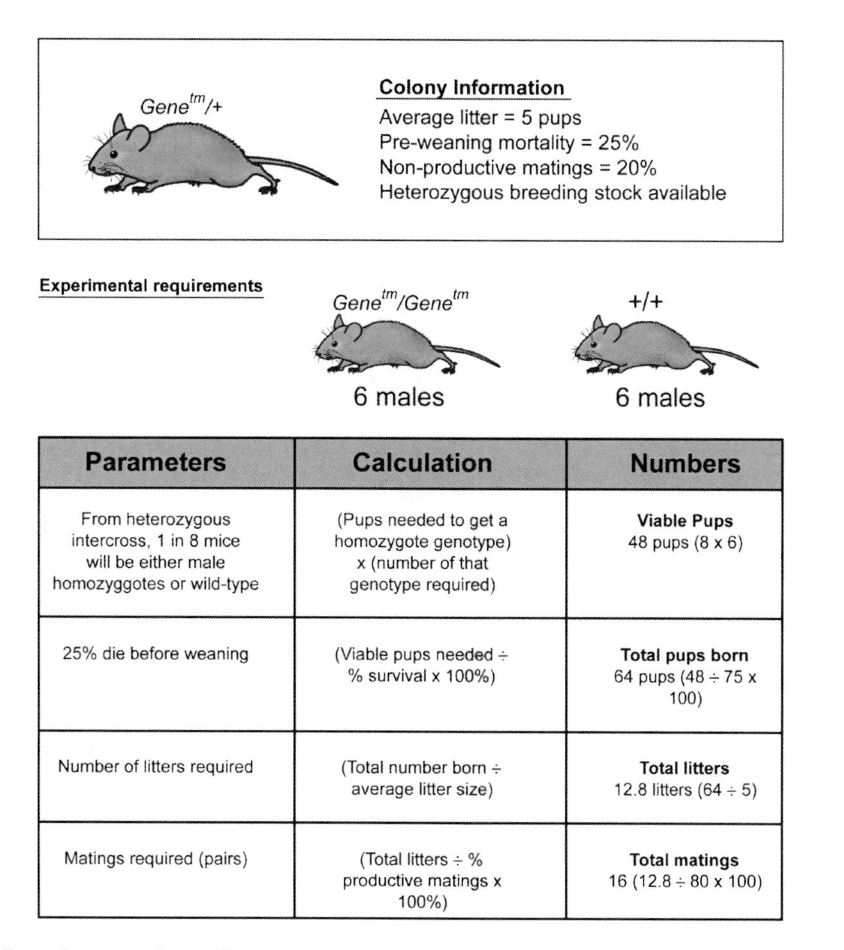

Parameters	Calculation	Numbers
From heterozygous intercross, 1 in 8 mice will be either male homozyggotes or wild-type	(Pups needed to get a homozygote genotype) x (number of that genotype required)	**Viable Pups** 48 pups (8 x 6)
25% die before weaning	(Viable pups needed ÷ % survival x 100%)	**Total pups born** 64 pups (48 ÷ 75 x 100)
Number of litters required	(Total number born ÷ average litter size)	**Total litters** 12.8 litters (64 ÷ 5)
Matings required (pairs)	(Total litters ÷ % productive matings x 100%)	**Total matings** 16 (12.8 ÷ 80 x 100)

Fig. 2. Breeding calculations. A specific number of homozygous male mice are required. The total number of matings which need to be set up from the breeding stock available is calculated using the parameters, which affect the number of pups available for experiments.

3.2.2. Trio or Harem Matings and Travelling Males

To minimize caging requirements and maximize the opportunity for a male transgenic to pass on the genetic modification, he can be housed with more than one female at any one time. However, a trio mating may not produce double the number of pups compared to a pair mating, as having two litters in the cage increases the chance of newborns being trampled by older pups. It is sometimes difficult to attribute a litter to a specific female and therefore not easy to remove a young litter to avoid this.

Travelling of the male between the cages of females every few days can increase the total number of pups sired by the transgenic over a specific time span. Of course, this does rely on a fecund and non-aggressive stock and uses more cages at any one time (although it may be for a shorter time overall). Moving the male from the females before they litter also precludes the postpartum mating occurring which is common in mice and routinely leads to litters every 20–21days.

3.2.3. Homozygous Colonies

It may be possible to breed a genetic modification to homozygosity and then maintain the colony by intercrossing homozygotes. This may reduce the number of mice being bred and eliminate unnecessary wild-type or heterozygous mice that are produced in matings involving heterozygotes. When first intercrossing heterozygotes to produce homozygotes, all pups must be assessed from birth and the genotypes of offspring studied to ensure that the homozygote phenotype does not affect viability. There are some disadvantages to maintaining a transgene with a homozygous stock. This selective breeding scheme may lead to genetic drift from original baseline controls (*see* **Note 6**) and the matings do not produce control groups for experimentation.

4. Specialist Mouse Husbandry

4.1. Variables of Mouse Husbandry

Many different aspects of animal care will affect the health, welfare, and productiveness of mice colonies *(26)*. Depending on the requirements of the analysis and phenotyping, it may be imperative that mice are kept in a consistent environment. There are many features of an animal facility that will be predetermined by the institution or governing authority *(27)*. However, specialist items such as enrichment, food supplements, and husbandry routines can all be introduced or altered in attempts to overcome problems of fertility, viability or aggressive behaviour.

4.1.1. Bedding and Nesting Materials

Bedding and nesting such as shredded paper, paper tissue, and wood chips are essential enrichment for mice *(28)*. They should be provided in adequate amounts to allow the mice to build nests *(26)*. Care should be taken in what type of material is provided, as some are not suitable for neonates or are incompatible with the phenotypic traits of some lines *(29)*.

4.1.2. Environmental Enrichment

An addition to cages such as shelters retreats and tubes promote the development of a range of behaviours and promote good health *(30)*. However, caution should be taken when choosing enrichment as some will differentially alter behavioural and biological traits (*see* **Note 6**). Mouse lines behave differently with different cage additions and when changing the enrichment, it is important that mice are carefully monitored for the first few days and any effects on baseline data are considered. Although a plethora of enrichments are commercially available and widely used, it should not be forgotten that the best enrichment for a mouse may well be another mouse. Therefore, housing mice individually, especially within closed-caging units such as individually ventilated cages (IVCs) should be avoided, if at all possible.

4.1.3. Food Supplements

In addition to the regular chow provided to mice, supplements can be added at times of additional need, such as weight loss due to ill health and overgrown teeth, or for pups which are not being given sufficient nutrition by their mothers. Almost every animal facility will have a range of extra foodstuffs, which at some point have been given to enhance mouse survival and growth. These include gels, wet food, sunflower kernels, and even peanut butter. Again the choice of what supplement you use may affect the final analysis (especially metabolic parameters) and should always have the same standard of sterility/cleanliness as the usual chow offered.

4.1.4. Husbandry Routines

Odours, noise, and physical disturbances are all stressful for laboratory animals *(31)*. It is therefore imperative that researchers working on GA mice are familiar with the husbandry routines and discuss changes with the animal care staff, if appropriate.

5. Summary

When establishing a new GA colony breeding strategy, strain background and appropriate controls for the planned analysis must be decided from the outset. Whilst establishing and expanding a new transgenic line, welfare and health problems may occur, but there are many well tried and tested husbandry techniques and routines which may help overcome these problems. A gene modification does not inherently imply that these mice will behave differently or experience any increased suffering compared to the wild type strains from which they were generated.

6. Notes

1. http://www.informatics.jax.org/external/festing/search_form.cgi has information on the biological characteristics of inbred strains and their use in research.
2. There are many examples of the phenotype of gene knockouts being dependent upon background strain. These include EGF receptor *(32, 33)*, heparin co-factor II *(34,)* and keratin 8 *(35)*.
3. Haploinsufficiency is a condition that arises, when the normal phenotype requires the protein product of both alleles and the loss of one of the alleles results in an abnormal phenotype.

4. Milk in the stomach of a neonatal pup is a sign that the pup is being fed. You can look for this by scruffing the young and examining their abdomens.
5. Oedema is the build up of fluid within the mouse and can usually be seen in a neonate, particularly around the head and neck area.
6. Genetic drift may occur if a sub-strain is separated from the parental strain for many generations. Within a colony there may be some residual heterozygosity from the parental line which, as a consequence of keeping a closed colony, may be bred to homozygosity. Any spontaneous mutations which occur may also be fixed in the colony.
7. Environmental enrichment can have significant effects on behavioural characteristics *(36)*, modify the aetiology of disease, *(37)* and increase aggressive behaviour *(38)*.

References

1. Hedrich, H. J. and Bullock, G. R. (2004) The Laboratory Mouse. Academic, New York.
2. Festing, M. F. and Fisher, E. M. (2000) Mighty mice. *Nature*, **404**, 815.
3. http://www.homeoffice.gov.uk/rds/scientific1.html.
4. Festing, M. F. (1994) Reduction of animal use: experimental design and quality of experiments. *Lab Anim*, **28**, 212–221.
5. van Wilgenburg, H., van Schaick Zillesen, P. G. and Krulichova, I. (2003) Sample Power and ExpDesign: tools for improving design of animal experiments. *Lab Anim (NY)*, **32**, 39–43.
6. Linder, C. C. (2006) Genetic variables that influence phenotype. *ILAR J*, **47**, 132–140.
7. Jacobson, L. H. and Cryan, J. F. (2007) Feeling strained? Influence of genetic background on depression-related behaviour in mice: a review. *Behav Genet*, **37**, 171–213.
8. Papaioannou, V. E. and Behringer, R. R. (2005) Mouse phenotypes: a handbook of mutation analysis. Cold Spring Harbour Laboratory Press, Cold Spring Harbour.
9. Taketo, M., Schroeder, A. C., Mobraaten, L. E., Gunning, K. B., Hanten, G., Fox, R. R., Roderick, T. H., Stewart, C. L., Lilly, F., Hansen, C. T., et al. (1991) FVB/N: an inbred mouse strain preferable for transgenic analyses. *Proc Natl Acad Sci USA*, **88**, 2065–2069.
10. Keskintepe, L., Norris, K., Pacholczyk, G., Dederscheck, S. M. and Eroglu, A. (2007) Derivation and comparison of C57BL/6 embryonic stem cells to a widely used 129 embryonic stem cell line. *Transgenic Res*, **16**, 751–758.
11. Auerbach, A. B., Norinsky, R., Ho, W., Losos, K., Guo, Q., Chatterjee, S. and Joyner, A. L. (2003) Strain-dependent differences in the efficiency of transgenic mouse production. *Transgenic Res*, **12**, 59–69.
12. Mineur, Y. S. and Crusio, W. E. (2002) Behavioral and neuroanatomical characterization of FVB/N inbred mice. *Brain Res Bull*, **57**, 41–47.
13. Bothe, G. W. M., Bolivar, V. J., Vedder, M. J. and Geistfeld, J. G. (2004) Genetic and behavioural differences among five inbred mouse strains commonly used in the production of transgenic and knockout mice. *Genes Brain Behav*, **3**, 149–157.
14. Auerbach, W., Dunmore, J. H., Fairchild-Huntress, V., Fang, Q.Auerbach, A. B., Huszar, D. and Joyner, A. L. (2000) Establishment and chimera analysis of 129/SvEv- and C57BL/6-derived mouse embryonic stem cell lines. *Biotechniques*, **29**, 1024–1028, 1030, 1032.
15. Estill, S. J., Fay, K. and Garcia, J. A. (2001) Statistical parameters in behavioral tasks and implications for sample size of C57BL/6J:129S6/SvEvTac mixed strain mice. *Transgenic Res*, **10**, 157–175.
16. Visscher, P. M., Haley, C. S. and Thompson, R. (1996) Marker-assisted introgression in backcross breeding programs. *Genetics*, 144, 1923–1932.
17. Brown, D. G., Willington, M. A., Findlay, I. and Muggleton-Harris, A. L. (1992) Criteria that optimize the potential of murine embryonic stem cells for in vitro and in vivo developmental studies. *In Vitro Cell Dev Biol*, **28A**, 773–778.

18. Lui, X., Wu, H., Loring, J., Hormuzdi, S., Disteche, C. M., Bornstein, P. and Jaenisch, R. (1997) Trisomy eight in ES cells is a common potential problem in gene targeting and interferes with germ line transmission. *Dev Dyn*, **209**, 85–91.
19. Nadeau, J. H. (2001) Modifier genes in mice and humans. *Nat Genet Rev*, **2**, 165–174.
20. Bridges, J. P. and Weaver, T. (2006) Use of transgenic mice to study lung morphologenesis and function. *ILAR J*, **47**, 22–31.
21. Nicholson, S. J., Witherden, A. S., Hafezparast, M., Martin, J. E. and Fisher, E. M. (2000) Mice, the motor system, and human motor neuron pathology. *Mamm Genome*, **11**, 1041–1052.
22. Festing, M. F. W. and Altman, D. G. (2002) Guidelines for the design and statistical analysis of experiments using laboratory animals. *ILAR J*, **43**, 244–258.
23. http://www.nc3rs.org.uk/.
24. http://www.StatPages.net.
25. Wang, M.-H. and Vom Saal, F. S. (2000) Maternal age and traits in offspring. *Nature*, **407**, 469–470.
26. Jennings, M., Batchelor, G. R., Brain, P. F., Dick, A., Elliot, H., Francis, R. J., Hubrecht, R. C., Hurst, J. L., Morton, D. B., Peters, A. G., Raymond, R., Sales, G. D., Sherwin, C. M. and West, C. (1998) Refining rodent husbandry: the mouse. *Lab Anim*, **32**, 233–259.
27. http://www.ncbi.nlm.nih.gov/entrez/query.fcgi?cmd=Retrieve&db=PubMed&list_uids=12391400&dopt=Abstract Guide for the care and use of laboratory animals (1996) Institute of Laboratory Animal Resources, Commission on Life Sciences National Research Council, National Academy, Washington, DC.
28. Van de Weerd, H. A., Van Loo, P. L. and Baumans, V. (2004) Environmental enrichment: room for reduction? *Altern Lab Anim*, **32**, 69–71.
29. Bazille, P. G., Walden, S. D., Koniar, B. L. and Gunther, R. (2001) Commercial cotton nesting material as a predisposing factor for conjunctivitis in athymic nude mice. *Lab Anim*, **30**, 40–42.
30. Olsson, I. A. and Dahlborn, K. (2002) Improving housing conditions for laboratory mice: a review of "environmental enrichment". *Lab Anim*, **36**, 243–270.
31. Balcombe, J. P., Barnard, N. D. and Sandusky, C. (2004) Laboratory routines cause animal stress. *Contemp Top Lab Anim Sci*, **43**, 42–51.
32. Threadgill, D. W., Dlugosz, A. A., Hansen, L. A., Tennenbaum, T., Lichti, U., Yee, D., LaMantia, C., Mourton, T., Herrup, K., Harris, R. C., et al. (1995) Targeted disruption of mouse EGF receptor: effect of genetic background on mutant phenotype. *Science*, **269**, 230–234.
33. Sibilia, M. and Wagner, E. F. (1995) Strain-dependent epithelial defects in mice lacking the EGF receptor. *Science,*, **269**, 909.
34. Aihara, K., Azuma, H., Akaike, M., Ikeda, Y., Sata, M., Takamori, N., Yagi, S., Iwase, T., Sumitomo, Y., Kawano, H., Yamada, T., Fukuda, T., Matsumoto, T., Sekine, K., Sato, T., Nakamichi, Y., Yamamoto, Y., Yoshimura, K., Watanabe, T., Nakamura, T., Oomizu, A., Tsukada, M., Hayashi, H., Sudo, T., Kato, S., and Matsumoto, T. (2007) Strain-dependent embryonic lethality and exaggerated vascular remodeling in heparin cofactor II-deficient mice. *J Clin Invest*, **117**, 1514–1526.
35. Baribault, H., Penner, J., Iozzo, R. V. and Wilson-Heiner, M. (1994) Colorectal hyperplasia and inflammation in keratin 8-deficient FVB/N mice. *Genes Dev*, **8**, 2964–2973.
36. Tucci, V., Lad, H. V., Parker, A., Polley, S., Brown, S. D. M. and Nolan, P. M. (2006) Gene-environmental interactions differentially affect mouse strain behavioural parameters. *Mamm Genome*, **17**, 1113–1120.
37. Spires, T. L. and Hannan, A. J. (2004) Nature, nurture and neurology: gene-environmental interactions in neurodegenerative disease. *FEBS J*, **272**, 2347–2361.
38. Haemisch, A., Voss, T. and Gärtner, K. (1994) Effects of environmental enrichment on aggressive behavior, dominance hierarchies, and endocrine states in male DBA/2J mice. *Physiol Behav*, **56**, 1041–1048.

Chapter 20

Biological Methods for Archiving and Maintaining Mutant Laboratory Mice. Part I: Conserving Mutant Strains

Martin D. Fray

Summary

The mouse is now firmly established as the model organism of choice for scientists studying mammalian biology and human disease. Consequently, a plethora of novel, genetically altered (GA) mouse lines have been created. In addition, the output from the large scale mutagenesis programmes currently under way around the world will increase the collection of GA mouse strains still further. Because of the implications for animal welfare and the constraints on resources, it would be unreasonable to expect anything other than those strains essential for ongoing research programmes to be maintained as breeding colonies. Unfortunately, unless the redundant strains are preserved using robust procedures, which guarantee their recovery, they will be lost to future generations of researchers.

This chapter describes some of the preservation methods currently used in laboratories around the world to archive novel mouse strains.

Key words: Cryopreservation, Archive, Mouse, Spermatozoa, Embryos, Ovary, Vitrification, Freeze-drying

1. Introduction: Cryopreserving the Mouse

In 1972, Whittingham et al. *(1)* and Wilmut *(2)* independently published the first methods for the successful cryopreservation of mouse embryos. Until that time the only way to reliably maintain a mouse line was through continuous breeding and the potential benefits of embryo freezing mouse stocks were immediately recognised *(3)*. Not only could non-essential stocks be economically preserved and removed from the animal rooms, but critical mouse models could also be archived and safeguarded against

Elizabeth J. Cartwright (ed.), *Transgenesis Techniques*, Methods in Molecular Biology, vol. 561
DOI 10.1007/978-1-60327-019-9_20, © Humana Press, a part of Springer Science+Business Media, LLC 2009

disease, fire, genetic contamination, and many other catastrophes. As a natural extension, embryo banks were established in the early 1970s at several major laboratories, notably at the MRC Mammalian Genetics Unit, Harwell, UK; the National Institutes of Health, Maryland, USA, the Centre for Animal Resources & Development (CARD), Kumamoto University, Kumamoto, Japan, and the Jackson Laboratory, Maine, USA.

The successful cryopreservation of murine ovarian tissue and the restoration of fertility following the orthotropic grafting of frozen/thawed tissue was first reported by Parrott in 1958 *(4)* and 1960 *(5)*. Although these investigations showed that, viable young could be produced, the number of surviving oocytes was low and the reproductive life of the host females was shortened. Further progress in this area remained limited until the 1990s, but the technique is now well documented *(6–9)*. During the intervening years a successful method for freezing of unfertilized mouse oocytes was published by Whittingham *(10)*.

Fowl sperm exposed to glycerol were the first germ cells to be frozen/thawed successfully *(11)*; a procedure that was soon repeated in man and in a wide variety of domestic species. However, for reasons that remain unclear *(12–15)*, mouse spermatozoa proved far more difficult to freeze/thaw successfully and cryobiologists had to wait until the 1990s, before reliable methods became available *(16–20)*. Although various alternative protocols have been published using differing cryoprotective agents, the combination of 18% raffinose and 3% skim milk *(18, 19)* has been adopted as the standard method by the majority of laboratories.

The following text describes the background and the methodologies commonly used for the cryo-preservation and long-term storage of germplasm collected from genetically altered mice. Readers wishing to consider this article in the general context of animals used in bio-medical research are referred to *(21)*.

1.1. Cryopreservation of Embryos

Embryo freezing protocols can be divided into two broad categories; slow equilibration methods and rapid non-equilibration methods.

Successful cryopreservation of cells using slow equilibration methods is achieved by the gradual dehydration of blastomeres as water is drawn from the intracellular compartments by extracellular ice crystallisation combined with ingress of a permeable cryoprotectant. However, the degree of dehydration achieved during freezing is determined by the duration of the controlled cooling phase, i.e. the temperature reached before the cells are plunged into liquid nitrogen. What is more, the degree of dehydration determines how the cells should be thawed. In essence, cells cooled to –70 to –80°C are highly dehydrated and need to be thawed slowly to allow adequate rehydration. On the other hand, embryos cooled to –30 to –35°C are only moderately dehydrated and can be thawed more rapidly.

Slow equilibration methods for freezing mouse oocytes and embryos are usually on the basis of the use of 1.0–2.0 M solutions of permeating cryoprotectants such as dimethylsulphoxide (DMSO), glycerol, propane-1,2-diol, or ethane-1,2-diol. In these protocols the embryos are equilibrated with the cryoprotectant before being cooled, and then frozen under carefully controlled conditions. During this procedure the cells undergo a characteristic series of volumetric fluctuations brought about by the changing osmotic environment to which they are exposed. For a more detailed explanation of these changes, see Glenister and Rall *(22)*.

In a similar vein, embryo and oocyte vitrification procedures allow cryobiologists to exploit the benefits afforded by rapid non-equilibration freezing methods. Vitrification refers to the physiochemical process by which solutions solidify into a glass state during rapid cooling, i.e. without crystallization of the solvent or solute *(22)*. Successful vitrification is achieved by exposing embryos to high concentrations of cryoprotectant, typically ~5.0 M, combined with very rapid cooling (>1000°C/min). In contrast to slow equilibration freezing, vitrification is a non-equilibration procedure, i.e. the embryos do not equilibrate with the cryoprotectant, but are partially dehydrated by it. Numerous vitrification protocols exist in the literature, e.g. *(23, 24)*.

1.1.1. Slow Equilibration Rate Freezing

The embryo freezing procedure described by Whittingham, Leibo and Mazur in 1972 *(1)* involved slow cooling (<1°C/min) to −80°C with the embryos suspended in a solution of 1.0 M DMSO. This was followed by slow warming (~10°C/min) and the stepwise dilution/removal of the DMSO. This protocol is essentially similar to the slow equilibration freezing methods followed by many laboratories. A second slow equilibrium method commonly used employs 1.5 M propane-1,2-diol as the cryoprotectant and a 1.0 M sucrose solution as the diluent combined with a cooling rate of 0.3-0.5°C/min to −30°C. The embryos are contained in 0.25 mL plastic insemination straws and require rapid warming (>1,000°C/min) for survival. Details of this method were originally published by Renard and Babinet *(25)*.

The major advantages of the slow equilibration freezing methods are their reliability coupled with the fact that the cryoprotectants they use are relatively non-toxic, which allows a certain amount of tolerance, when collecting large numbers of embryos for freezing. At the MRC-Harwell, we find that in most instances it is possible to recover sufficient live-born offspring to re-establish a breeding colony from one straw of approximately 20 frozen embryos. One disadvantage of this procedure is that precise control of the cooling rate generally requires costly programmable freezing machines, although cheaper alternatives are coming in the market.

1.1.2. Embryo Vitrification

The principal advantage of vitrification over slow equilibration freezing methods is that the embryos are not subject to chilling injury or blastomere damage resulting from intra- or extracellular ice crystal formation. However, to avoid toxicity effects of the vitrification solutions it is necessary to exercise precise control over the temperature and the time the embryos are exposed to the vitrification solutions. A more detailed overview of the biophysics of vitrification can be found in Glenister and Rall *(22)*.

Vitrification methods do benefit from not requiring any expensive equipment. In fact the only equipment needed is a vessel suitable for holding liquid nitrogen.

1.2. Cryopreservation of Spermatozoa

The principal advantage of sperm freezing is that the technique is very simple and requires no expensive equipment. In keeping with the vitrification and ultra-rapid cooling methods for embryos, all that is required for sperm freezing is a suitable container to provide a gradient of liquid nitrogen vapour such as a polystyrene box with a lid. An additional benefit is that spermatozoa from a large number of males (>50) can be frozen in a single day by a small laboratory team. These factors introduce one of the primary benefits of mouse sperm freezing, which is its potential to economically cryopreserve the huge numbers of mutants that are being generated from the large-scale mutagenesis programmes. Recovery of these mutations is then routinely achieved by in vitro fertilization *(20)*. It is now commonplace to set up parallel archives of matched sperm and tissue samples, for DNA extraction, during mutagenesis screens so that mutations identified retrospectively following DNA analysis can be recovered by in vitro fertilization *(17, 26)*.

One drawback to sperm cryopreservation is the marked differences in the fertility of frozen/thawed sperm, a phenomenon which is determined by the genetic background of the male donor *(12, 27)*. For example, C57BL/6 and 129 sperm are notoriously difficult to cryopreserve successfully. Nevertheless, the selected use of assisted fertilization techniques such as partial zona digestion (Doyle A, the Jackson Laboratory), sperm selection *(28)*, and partial zona dissection *(29)*, has yielded reasonable fertilization rates on these difficult backgrounds. However, recent publications indicate that the addition of the cholesterol acceptor, methyl- β cyclodextrin *(30)* to post-thawing equilibration media or the addition of monothioglycerol to the standard cryo-protective agents before freezing *(31)* improves the fertilization rates of frozen/thawed C57BL/6 sperm.

For a more detailed overview of cryopreservation of mouse spermatozoa, see Nakagata *(32)*.

1.2.1. Freeze Drying of Spermatozoa

Technological advances have made it possible to obtain live born pups from freeze-dried mouse sperm and sperm frozen in the absence of cryoprotectants *(1, 33–36)*. Live born pups have also even been resurrected from spermatozoa recovered from the whole testis frozen at –20°C for 15 years *(37)*.

Unfortunately, the sperm preserved under these conditions is immotile and fertilization can only be achieved with the use of micro insemination techniques. Nevertheless, in skilled hands over 50% of microinsemination derived embryos can develop into normal live offsprings *(34, 38)*.

One big advantage of storing freeze-dried spermatozoa over other methods is that the sperm remains viable after long-term storage at 4°C *(35)*. Further improvements in the freeze-drying procedure will increase its efficiency, which is likely to make the freeze-drying of spermatozoa more widespread in the future, not least because it eliminates the expense and inconvenience of liquid nitrogen storage and the shipment of frozen samples.

1.3. Cryopreservation of Oocytes

The successful cryopreservation of mouse metaphase II oocytes was first reported by Whittingham *(10)* who suspended oocytes in a solution of 1.5 M DMSO and then slowly cooled them to –80°C. It was subsequently discovered that the addition of 10% foetal calf serum to the cryoprotectant solution improved oocyte survival rates, possibly by preventing changes to the zona pellucida that occur during freezing and thawing *(39)*. During the intervening years, Nakagata *(40)* published a method for the successful vitrification of mouse oocytes.

1.4. Cryopreservation of Ovarian Tissue

Cryopreservation of ovarian tissue followed by orthotropic transplantation into recipient mice has been possible for many years *(4, 5, 41)*. Although numerous laboratories have used ovary freezing to good effect, the transplantation of frozen/thawed ovarian tissue has some technical drawbacks. Not the least is the fact that the number of recipients that conceive after transplantation is often reduced compared with control animals. Another complication is the need to transfer the explants into histocompatible or immunodeficient recipient females. Despite these drawbacks, ovary freezing can be used to cryoconserve female gametes collected at various stages of post-natal development, including prepubertal mice *(7–9, 42)*.

2. Materials

2.1. Cryopreservation of Embryos

2.1.1. Preparation for Slow Equilibration Rate Embryo Freezing

1. Dissecting microscope.
2. Surgical instruments for embryo collection.
3. 133 mm (0.25 mL) plastic semen straws (Planar UK; A201).
4. Programmable freezing machine, e.g. Kryo 360M (Planer UK) or Bio-cool IV (FTS systems).

5. Small dewar suitable for holding liquid nitrogen.
6. Liquid nitrogen.
7. Analar grade propane-1,2-diol (Sigma).
8. Analar grade sucrose (Merck).
9. 35 mm culture dishes (Falcon).
10. M2 medium (Sigma).
11. Capillary sealant (Fisher scientific) or heat sealer.
12. M2 medium containing 1.5 M propane-1,2-diol (PROH). To prepare this solution, pipette 4.4 mL of M2 medium into a culture tube and add 0.6 mL PROH. Filter sterilise into a 35 mm culture dish.
13. 1.0 M sucrose in M2 medium. To prepare this solution dissolve 1.71 g sucrose in 5 mL M2 medium and filter sterilise into a 35 mm culture dish.

2.1.2. Preparation for Embryo Vitrification

1. Dissecting microscope.
2. Small dewar suitable for holding liquid nitrogen.
3. Liquid nitrogen.
4. Cooling block (IWAKI Co. Ltd; CHT-100, or NALGENE).
5. Culture dishes (Nunc).
6. Freezing tubes (Nunc).
7. Cryoprotectant (DAP213): 2.0 M DMSO, 1.0 M Acetamide and 3.0 M propane-1,2-diol made up in PB1 medium.
8. PB1 medium: per 100 mL, 800 mg NaCl, 13.2 mg $CaCl_2$, 20 mg KH_2PO_4, 20 mg KCl, 10 mg $MgCl_2$, 289.9 mg Na_2HPO_4, 100 mg glucose, 100 U Penicillin G, 3.6 mg Na-pyruvate.

2.2. Cryopreservation of Spermatozoa

2.2.1. Preparation for Vapour Phase Method of Sperm Cryopreservation

1. Dissecting microscope.
2. Rack for supporting cryotubes.
3. 1.8 mL cryotubes (Nunc).
4. 1.5 mL microfuge tubes.
5. Raffinose (Sigma).
6. H_2O (Sigma).
7. Skim milk (Becton Dickinson).
8. 35 mm culture dishes (Falcon).
9. 15 mL tubes (Falcon).
10. 30 g needle/insulin syringe (Beckton Dickinson).
11. Fine watchmaker's forceps.

12. Deep polystyrene box with lid suitable for holding liquid nitrogen.
13. Small dewar suitable for holding liquid nitrogen.
14. Liquid nitrogen.
15. Hot block held a 37°C.
16. Mature male mice at least 8 weeks old which have not recently mated.
17. Cryoprotective agent (CPA):18% raffinose, 3% skim milk. To prepare this solution place 9 mL embryo tested H_2O into a 15 mL, screw top Falcon tube and equilibrate to 60°C in a water bath. Add 1.8 g raffinose and dissolve by gentle inversion. Add 0.3 g skim milk and dissolve by gentle inversion. Make up to 10 mL if necessary. Aliquot CPA solution into microfuge tubes and centrifuge at 14,000rpm for 10 min. Tip off supernatant and filter (0.45 µm syringe end filter) 1.1 mL aliquots into cryotubes. Store at –20°C for up to 3 months.

2.2.2. Preparation for Freeze Drying Spermatozoa

1. Dissecting microscope.
2. Long-necked 2 mL glass ampoules (Wheaton, Millville, NJ).
3. 1.5 mL polypropylene microcentrifuge tube (Fisher Scientific, Pittsburgh, PA).
4. Freeze-dry systems (e.g. Labconco Co., Kansas City, MO).
5. Small dewar suitable for holding liquid nitrogen.
6. Liquid nitrogen.
7. Freeze-drying solution: 50 mM NaCl, 50 mM EGTA, 10 mM Tris–HCl (pH 7.4) made up in sterile distilled water. Adjust to pH 8.0 by adding 1.0 M HCl. Filter sterilize freeze-drying solution before use.

2.3. Preparation for the Cryopreservation of Oocytes

1. Dissecting microscope.
2. 133 mm (0.25 mL) plastic semen straws (Planar UK; A201).
3. An automatic freezer e.g. Kryo 360M (Planer UK) or Biocool IV (FTS systems).
4. Small dewar suitable for holding liquid nitrogen.
5. Liquid nitrogen.
6. M2 medium (Sigma).
7. Analar grade DMSO (BDH; 103234L, MW 78.13, d = 1.099–1.101 g/mL).
8. Foetal calf serum (ICN Flow).

9. Hyaluronidase (Sigma).
10. Capillary sealant (Fisher scientific) or heat sealer.
11. M2 medium, plus 10% foetal calf serum. The pH of this solution needs to be adjusted to between pH 7.2–7.4 with 10 μL 2.0 M HCl/10 mL medium.
12. Cryoprotectant: 1.5 M dimethylsulphoxide (DMSO) made up in M2 medium, plus 10% foetal calf serum. Make up fresh each day and do not filter sterilize.
13. Hyaluronidase (300 μg/mL) made up in M2 medium.

2.4. Preparation for the Cryopreservation of Ovarian Tissue

1. Dissecting microscope.
2. Plastic 1.8 mL cryotubes (Nunc).
3. Programmable freezing machine e.g. Kryo 360 (Planer UK) or Bio-cool III (FTS systems).
4. Small dewar suitable for holding liquid nitrogen.
5. Liquid nitrogen.
6. M2 medium (Sigma).
7. Analar grade DMSO (BDH; 103234L, MW 78.13, d = 1.099–1.101 g/mL).
8. Foetal calf serum (ICN Flow).
9. M2 medium, plus 10% foetal calf serum. The pH of this solution needs to be adjusted to pH 7.2–7.4 with 10 μL 2 M HCl/10 mL medium.
10. Cryoprotectant: 1.5 M DMSO made up in M2 medium, plus 10% foetal calf serum. Add 1.07 mL DMSO to 8.93 mL M2 medium, plus 10% foetal calf serum. Make up fresh each day and do not filter sterilize.

3. Methods

The following methods should be read in conjunction with guidance (*see* **Notes 1–4**).

3.1. Methods for Cryopreservation of Embryos

3.1.1. Method for Slow Equilibration Method of Embryo Freezing

This procedure is based on the method of Renard and Babinet *(25)* as modified by Glenister and Rall *(22)*.

Freezing Procedure

1. Start the programmable freezer and equilibrate to the start temperature of –6°C.
2. Collect the embryos in M2 medium and screen carefully for abnormalities, working at room temperature.
3. After harvesting, embryos should be washed three times in M2 medium (2 mL of M2/wash). If more than one stock is being frozen on the same day ensure that each stock is handled with a unique pipette and has its own set of labelled wash dishes.
4. Prepare the straws as shown in **Fig. 1**. Take a 133-mm straw and using a metal rod push the plug from A to B, i.e. to within 75 mm of the open end of the straw. Attach the identifying label to location A.
5. With a fine pen draw three lines on each straw (*see* **Fig. 2**). The first mark should be 20 mm from the plug; the second mark should be 7 mm from mark 1, and the third mark 5 mm from mark 2. Use graph paper as a guide, or make the marks on a rack supporting the straws.
6. To fill the straws: aspirate 1.0 M sucrose to mark 3 using a 1-mL syringe, then aspirate air so that the sucrose meniscus reaches mark 2. Then aspirate 1.5 M PROH so that the sucrose meniscus reaches mark 1. Finally, aspirate air until the column of sucrose reaches half way up to the plug and forms a seal with the polyvinyl alcohol.
7. Pipette the embryos into a 150 μL drop of 1.5 M PROH and equilibrate for 15 min.

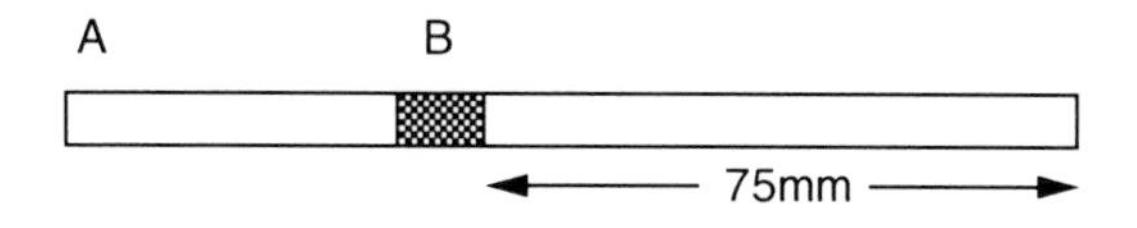

Fig. 1. Illustrates the appearance of a straw once the PVA/cotton plug has been relocated to within 75mm of the open end.

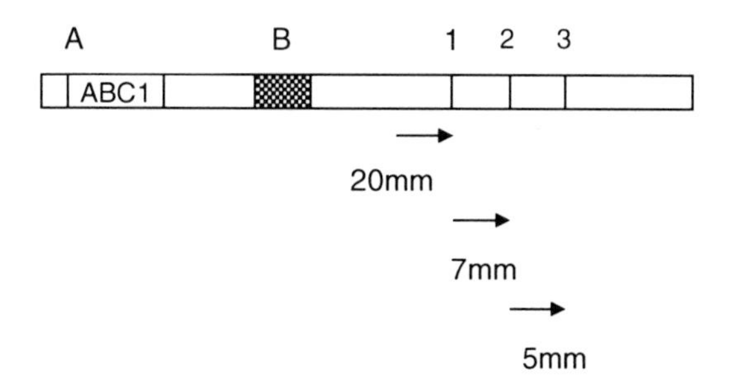

Fig. 2. Illustrates the location of the unique identity label and the position of the index marks used as a gauge when filling a straw with sucrose and ProH in preparation to receiving the embryos.

8. Using a finely drawn pipette, load the embryos into the 1.5 M PROH fraction of each straw and seal the straws using capillary sealant or a heat sealer (*see* **Fig. 3**).
9. Place the straws into the programmable freezing machine cooled to –6°C.
10. Wait 5 min to equilibrate.
11. Initiate ice crystal formation (referred to as seeding) in the sucrose fraction by touching it near the plug with a cotton wool bud or the tips of forceps pre-cooled in liquid nitrogen.
12. Wait 5 min, then check that the ice has migrated to the embryo fraction.
13. Cool at 0.3°C–0.5°C/min to –30°C.
14. Plunge the straws directly into liquid nitrogen.
15. Transfer the straws to the appropriate liquid nitrogen storage vessel, taking care at all times to keep the embryo fraction submerged in liquid nitrogen.

3.1.2. Method for Vitrification of Embryos

This simple vitrification method describes the procedure developed at the University of Kumamoto and is commonly used throughout Japan *(43)*.

Freezing Procedure

1. Place four separate drops (100 μL/drop) of PB1 supplemented with 1.0 M DMSO (1 M DMSO solution), into a dish.
2. Pipette up to 120 embryos in one of four drops and divide them into three groups. This will be sufficient to freeze up to 40 embryos in each of three freezing tubes.
3. Place each group of embryos into a separate drop and then transfer the embryos, in 5 μL of 1.0 M DMSO solution, into a freezing tube.
4. Load each freezing tube into the block cooler, set to 0°C and wait 5 min.
5. Add 45 μL of cryoprotective solution (DAP213), pre-cooled to 0°C into each freezing tube and wait 5 min.

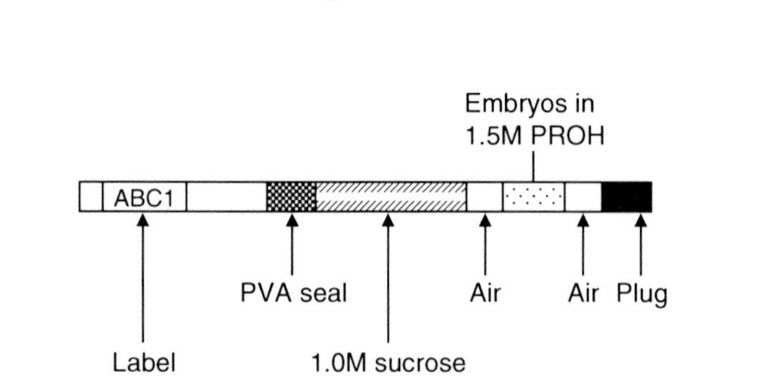

Fig. 3. Illustrates the appearance of a sealed straw after the embryos have been transferred into the ProH fraction.

6. Immediately fix the freezing tubes to a cane and plunge the samples directly into liquid nitrogen.

3.2. Methods for Cryopreservation of Spermatozoa

This method describes the procedure followed in numerous cryopreservation laboratories and is on the basis of the work originally published by Nakagata et al. *(19)* and modified by Glenister and Rall *(22)*.

3.2.1. Vapour Phase Method for the Cryopreservation of Sperm

1. Prepare the cooling apparatus by placing a platform (e.g. the insert from a Gilson tip box) into the polystyrene box suitable for holding liquid nitrogen. This platform is used to support the cryotube rack. Carefully pour liquid nitrogen into the polystyrene box so that it just covers the platform. Then place a rack suitable for holding cryotubes on top of the platform, ensuring that liquid nitrogen 'cannot' make direct contact with the cryotubes themselves.
2. Replace the lid on the polystyrene box and allow it to fill with vapour. Top up the liquid nitrogen reservoir when necessary during the freezing session, but don't allow the level to make contact with the cryotubes.
3. Thaw 1.0 mL of CPA per male and bring to 37°C, using a heated plate or CO_2 incubator. Mix by inversion, if there is any precipitation.
4. Pipette 1.0 mL CPA into a 35-mm culture dish and place on a heated plate at 37°C. Working under a dissecting microscope, dissect the vasa deferentia and cauda epididymides from each male (prepare one male at a time) and clean off any excess adipose or vascular tissue.
5. Transfer the tissues to the CPA. Release the sperm from the cauda epididymides by making six or seven cuts across the surface with a 30-G needle/insulin syringe. Then gently expel the sperm from the vasa deferentia by 'walking' two pairs of fine forceps along its surface.
6. Disperse the sperm by gently shaking the dish for 30 s and then incubate for 10 min at 37°C.
7. Aliquot 100 μL of the sperm suspension into each of eight cryotubes; use a wide bore pipette tip to reduce the risk of damaging the sperm. Replace the screw cap and tighten.
8. Transfer the cryotubes to the pre-cooled freezing apparatus and leave in the liquid nitrogen vapour phase for 10 min.
9. After 10 min have elapsed, plunge cryotubes into liquid nitrogen. Store in a liquid nitrogen refrigerator until required.

3.2.2. Freeze Drying Method for the Preservation of Mouse Sperm

This is the method described by Kaneko et al *(34)*.

Freezing Procedure

1. Aliquot 1 mL of the freeze-drying solution into a 1.5 mL polypropylene microfuge tube and warm to 37°C. Prepare 1 tube/male.
2. Dissect both caudae epididymides from each male using a pair of small scissors. Prepare one male at a time.
3. Lance the distal end of each epididymis with a pair of small forceps.
4. Collect the sperm mass and transfer it to the bottom of a microfuge tube containing freeze-drying solution.
5. Allow the spermatozoa to disperse into the freeze drying solution by incubating the sperm suspension for 10 min at 37°C.
6. After 10 min, transfer the upper 800 μL of the sperm suspension to another microfuge and discard the debris.
7. Pipette 100 μL aliquots of sperm suspension into long-necked glass ampoules.
8. Dip the glass ampoules, containing the sperm suspension, in liquid nitrogen.
9. Connect the ampoules to the freeze-drying systems and lyophilize for 4 h.
10. Using a gas burner, flame-seal each ampoule.
11. Store ampoules at 4°C until required.

3.3. Method for the Cryopreservation of Oocytes

This procedure is on the basis of the method by Carroll et al. *(39)* and can be used to freeze immature and mature mouse oocytes.

Freezing Procedure

1. Set the programmable freezer to 0°C.
2. Prepare the cooling apparatus by making an ice slurry in a low-sided container. Cover the ice slurry with aluminium foil and place two glass boiling tubes on top of the foil.
3. In addition, place a glass embryo dish on top of the foil. The embryo dish should contain 3.0 mL of 1.5 M DMSO in M2 medium, plus foetal calf serum. Leave the cooling apparatus in the refrigerator until needed.
4. Label sufficient straws for the freezing session and then prepare them as follows. Using a marker pen draw three lines on each straw; draw the first line 17 mm from the plug (mark 1), draw the second line 87 mm from the plug (mark 2), and draw third 95 mm from the plug (mark 3), *see* **Fig. 4**. Use graph paper as a guide, or make the marks on a rack supporting the straws.
5. Aspirate 1.5 M DMSO to mark 3. Aspirate air so that the meniscus reaches mark 2 and then aspirate 1.5 M DMSO

so that the meniscus reaches mark 1. Aspirate air until the column of cryoprotectant wets the PVA plug.

6. Place the straws horizontally in the pre-cooled boiling tubes.
7. Chill the remaining 1.5 M DMSO solution on ice.
8. Harvest the cumulus oocyte masses 15–16 h after the donor females were injected with hCG. Remove cumulus cells with hyaluronidase (300 μg/mL) and incubate at 37°C in M2 medium, plus foetal calf serum.
9. Don't allow the oocytes to cool before they have been placed in the DMSO.
10. Transfer the oocytes to the pre-cooled 1.5 M DMSO taking care to transfer only a small amount of M2 medium. Leave the oocytes to chill for 2 min, on ice, before loading them into the straws. Because the oocytes are not fully equilibrated at this stage they appear shrunken.
11. Using a finely drawn pipette transfer a cohort (~30) of oocytes into the large fraction of 1.5 M DMSO in each pre-cooled straw and seal the straw using a capillary sealant or heat sealer (*see* **Fig. 5**).
12. Return it to the pre-cooled glass boiling tube and leave them on ice.

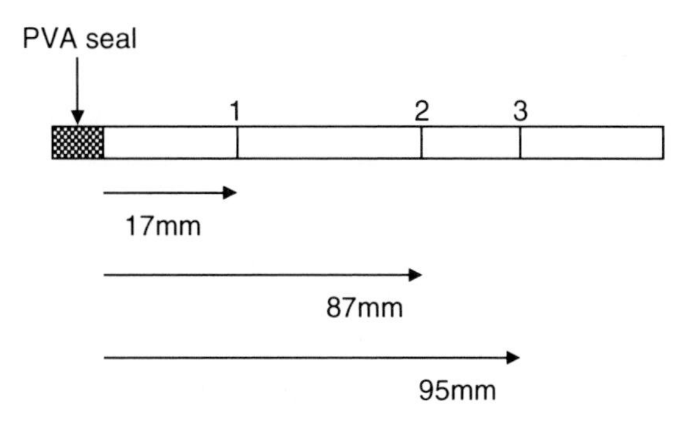

Fig. 4. Illustrates the position of the index marks used as a gauge when filling a straw with DMSO in preparation to receiving the oocytes.

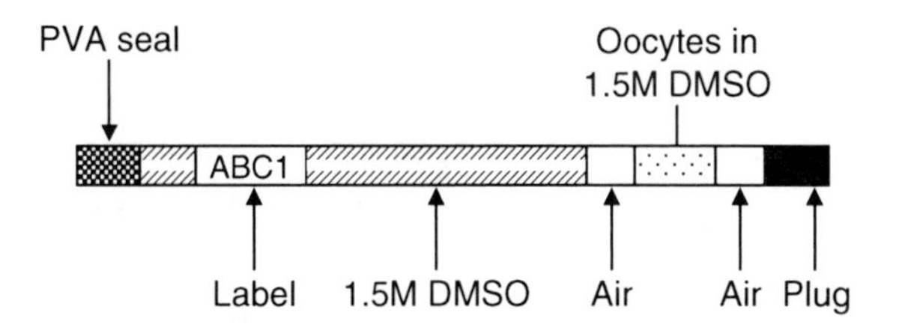

Fig. 5. Illustrates the appearance of a sealed straw after the oocytes have been transferred to the DMSO fraction.

13. Work quickly during the loading procedure to minimize the oocytes exposure to room temperature and keep the watch glass/oocytes cool.
14. Total exposure to 1.5 M DMSO at 0–4°C is 12 min.
15. Load the straws into the programmable freezer, again reducing the exposure to room temperature.
16. Cool the straws to −8°C, at 2°C/min.
17. Hold the straws for 5 min, at −8°C.
18. Initiate ice crystal formation (referred to as seeding) by touching the DMSO fraction near the meniscus with a pair forceps or cotton wool bud, previously cooled in liquid nitrogen.
19. Wait 5 min, then check that the ice has migrated to the fraction containing the oocytes, then cool at 0.3°C/min to −40°C.
20. Then plunge the straws into liquid nitrogen.
21. Transfer the samples to a liquid nitrogen storage vessel taking care to keep the straws submerged at all times.

3.4. Method for the Cryopreservation of Ovarian Tissue

Entire ovaries harvested from immature mice or segments of ovarian tissue (1–2 mm^3) taken from reproductively mature mice can be successfully frozen *(8, 9, 42)*. As a rule of thumb, ovaries taken from mice up to 10 days of age can be frozen whole but the ovaries of adult mice should be cut into quarters before being placed in the cryoprotectant. The method described below is on the basis of the method first described by Sztein et al *(9)*.

Freezing Procedure

1. Set the programmable freezer to cool and hold at 0°C.
2. Prepare the cooling apparatus by making an ice slurry in a low-sided container. Cover the ice slurry with aluminium foil and place a glass embryo dish on top of the foil. Leave the container in the refrigerator until needed.
3. Label sufficient cryotubes for the freezing session. Into each cryotube dispense 0.3 mL of 1.5 M DMSO. Embed the cryotubes into the ice slurry.
4. Pipette 3.0 mL of 1.5 M DMSO solution in the pre-chilled embryo dish and leave on ice.
5. Collect the ovaries in M2 medium, plus foetal calf serum. Dissect away any attached mesentery or fat.
6. Transfer the ovaries into the 1.5 M DMSO solution in the pre-chilled embryo dish. Take care to transfer only a small amount of M2 medium.
7. Leave the ovaries on ice for 2 min before placing them in the cryotubes. Between 1 and 4 ovaries can be loaded into each cryotube.

8. Secure the cryotube's screw cap and place the tube back on ice.
9. The total exposure to 1.5 M DMSO at 0-4°C should be no more than 20 min.
10. Load the cryotubes into the programmable freezing machine and cool to –8°C at 2°C/min.
11. Wait 5 min to allow the samples to equilibrate at –8°C.
12. Initiate ice crystal formation (referred to as seeding) by touching the meniscus of the DMSO fraction with a pair forceps or cotton wool bud, previously cooled in liquid nitrogen.
13. After 5 min, confirm that ice crystals are visible in all the cryotubes, then continue to cool the samples at 0.3°C/min down to –40°C.
14. Then increase the cooling rate to 10°C/min down to -150°C.
15. Plunge the cyrotubes into liquid nitrogen.
16. Transfer the samples to liquid nitrogen storage vessel.

3.5. When Is a Stock Safely Archived?

Irrespective of the procedure chosen to cryopreserve embryos and gametes, it is critically important to ensure that the material has been archived successfully and in sufficient quantities to service the likely future demand for the stock. This will help to avoid having to re-freeze the line too frequently.

When archiving spermatozoa it should be sufficient to freeze down 40–60 aliquots harvested from 4-6 males/strain. A test IVF needs to be performed to confirm the stock can be recovered. Similarly when considering embryo freezing, sufficient embryos should be placed in the bank to allow for repeated rederivation attempts. At the MRC-Harwell we apply rules that loosely allow for a stock to be rederived ten times or more before re-archiving becomes essential. For example, genetically altered lines are frozen in the homozygous state, 200 or more embryos would be frozen in order to safeguard the line. This rule can also be applied to inbred strains because all surviving progeny will be of the required genotype. Where a gene of interest is segregated in the cross set up to provide embryos for cryopreservation, the number of embryos to be frozen will be increased to ~500. However, it should be stressed that genetic background exerts a profound effect on the viability of frozen/thawed embryos *(44, 45)* and it is critical that a viability test is performed on each strain. The viability test should compare the number of embryos thawed to the number of pups born with the desired genotype. Using these data the archivist can calculate whether sufficient embryos have been banked, or whether additional material needs to be frozen.

In order to ensure the long-term survival of frozen stocks it is important to prepare for any eventuality that may befall the archived materials. To this extent it is advisable to split the frozen samples between two or more separate storage vessels. These vessels should then be housed in separate locations (local or geographical) where they will not be subjected to the same risk factors, e.g. fire, flood, earthquake, loss of liquid nitrogen supply etc.

3.6. Long-Term Survival of Cryopreserved Germplasm

Experience indicates that frozen mouse embryos will remain viable indefinitely and are unlikely to be affected by long-term exposure to low level/background radiation. For example, studies investigating the cumulative effects of background radiation demonstrated that low dose exposure equivalent to 2000 years of normal background radiation had no discernible effect on embryonic survival, *in vitro* or *in vivo* development, breeding performance, or the incidence of mutations *(46, 47)*. What is more, numerous stocks have been re-established from the MRC-Harwell, NIH, TJL and other laboratories following twenty or more years of storage in liquid nitrogen and none have shown any detrimental effects.

Live mice have also been recovered from spermatozoa thawed out after being cryopreserved for many years *(37, 48)* and there is no reason to believe that long-term storage will have an adverse effect on the fertility of frozen/thawed mouse spermatozoa. Further support for this statement is provided by the data on bovine spermatozoa which can retain its fertilising capacity after being frozen for 37 years *(49)*.

3.7. Concluding Remarks

It is now accepted that the routine cryopreservation of germplasm and ovarian tissue is pivotal to mouse genetics research. Even though banking mutant lines as sperm, oocytes and/or ovarian tissue is not commonly practised, embryo freezing is an established procedure exploited by numerous laboratories. Embryo freezing is also the most reliable means of cryopreserving mouse strains and offers the advantages that the diploid genotype is archived and the embryo transfer skills required to recover live mice are widely available. In addition, the ubiquitous access to embryo transfer skills makes disseminating stocks as embryos relatively straight forward. The principle disadvantage to embryo freezing is that it is a rather skilled process and frequently requires large numbers of time mated/superovulated female mice to provide a sufficient number of embryos to securely bank the stock.

Freezing sperm on the other hand is relatively simple, rapid, and cheap and has the advantage that large numbers of animals can be produced from one aliquot of frozen spermatozoa using IVF techniques. Fortunately, the dissemination of cryopreserved spermatozoa is becoming more common place, particularly now that more laboratories have developed the necessary skills to handle frozen spermatozoa and perform IVF and/or micro-insemination.

Although the fertility of frozen/thawed spermatozoa is currently dependent upon the genetic background of the donor mouse, a series of recent publications suggest these difficulties may soon be overcome. A more intractable drawback to sperm freezing is that only the haploid genotype is conserved. If the original genetic background is required in the future, the appropriate strain of oocyte donors would also have to be made available.

4. Notes

1. Use a standard superovulation protocol to enhance oocyte/embryo yields.
2. A permanent method of identifying the cryo-containers needs to be considered. The MRC-Harwell use self-adhesive printed cryolabels (Brady UK) to label the straws, but other possibilities exist including the use of an 'indelible' marker pen.
3. Samples are often lost through the inappropriate handling of vials/straws, when they are being removed from dry shippers or secure archives. In order to avoid premature thawing of your samples always place the vials/straws in liquid nitrogen holding vessels immediately after they have been removed from the main archive. The sample transfer should be done as quickly as possible.
4. Liquid nitrogen will enter poorly sealed vials/straws. If this occurs there is a risk that the vials/straws will explode as the liquid nitrogen rapidly converts to its gaseous state. In order to prevent the destruction of valuable samples and possible injury, it is important to ensure that the vials/straws are properly sealed before freezing. Plastic semen straws should be hermetically sealed at both ends and screw topped vials should be securely tightened.

   When screw topped cryotubes are tightened securely, before they are placed in the freezing apparatus, they do not present a problem, particularly if they are stored on the vapour phase. However, on occasions liquid nitrogen does enter cryotubes. A situation that is easily identified because fluid filled cryotubes sink when placed in liquid nitrogen. Fortunately, the contents of these vials can still be recovered, if the liquid nitrogen is allowed to escape by loosening the top of the vial and placing it on the bench top, before the designated thawing protocol is started.

References

1. Whittingham, D.G., Leibo, S.P. and Mazur, P. (1972) Survival of mouse embryos frozen to -196 degrees and -269 degrees C. *Science*, **178**, 411–414
2. Wilmut, I. (1972) The effect of cooling rate, warming rate, cryoprotective agent and stage of development on survival of mouse embryos during freezing and thawing. *Life Sci II*, **11**, 1071–1079
3. Whittingham, D.G. (1974) Embryo banks in the future of developmental genetics. *Genetics*, **78**, 395–402
4. Parrott, D.M.V. (1958) Fertility of orthotopic ovarian grafts. *Stud Fertil*, **9**, 137
5. Parrott, D.M.V. (1960) The fertility of mice with orthotopic ovarian grafts derived from frozen tissue. *J Reprod Fert*, **1**, 230
6. Carroll, J. and Gosden, R.G. (1993) Transplantation of frozen-thawed mouse primordial follicles. *Hum Reprod*, **8**, 1163–1167
7. Candy, C.J., Wood, M.J. and Whittingham, D.G. (2000) Restoration of a normal reproductive lifespan after grafting of cryopreserved mouse ovaries. *Hum Reprod*, **15**, 1300–1304
8. Gunasena, K.T., Villines, P.M., Critser, E.S. and Critser, J.K. (1997) Live births after autologous transplant of cryopreserved mouse ovaries. *Hum Reprod*, **12**, 101–106
9. Sztein, J., Sweet, H., Farley, J. and Mobraaten, L. (1998) Cryopreservation and orthotopic transplantation of mouse ovaries: new approach in gamete banking. *Biol Reprod*, **58**, 1071–1074
10. Whittingham, D.G. (1977) Fertilization in vitro and development to term of unfertilized mouse oocytes previously stored at -196 degrees C. *J Reprod Fertil*, **49**, 89–94
11. Polge, C., Smith, A.U. and Parkes, A.S. (1949) Revival of spermatozoa after vitrification and dehydration at low temperatures. *Nature*, **164**, 666
12. Songsasen, N. and Leibo, S.P. (1997) Cryopreservation of mouse spermatozoa. II. Relationship between survival after cryopreservation and osmotic tolerance of spermatozoa from three strains of mice. *Cryobiology*, **35**, 255–269
13. Nakagata, N. (2000) Cryopreservation of mouse spermatozoa. *Mamm Genome*, **11**, 572–576
14. Sztein, J.M., Farley, J.S., Young, A.F. and Mobraaten, L.E. (1997) Motility of cryopreserved mouse spermatozoa affected by temperature of collection and rate of thawing. *Cryobiology*, **35**, 46–52
15. Sztein, J.M., Farley, J.S. and Mobraaten, L.E. (2000) In vitro fertilization with cryopreserved inbred mouse sperm. *Biol Reprod*, **63**, 1774–1780
16. Tada, N., Sato, M., Yamanoi, J., Mizorogi, T., Kasai, K. and Ogawa, S. (1990) Cryopreservation of mouse spermatozoa in the presence of raffinose and glycerol. *J Reprod Fertil*, **89**, 511–516
17. Okuyama, M., Isogai, S., Saga, M., Hamada, H. and Ogawa S. (1990) In vitro fertilization (IVF) and artificial insemination (AI) by cryopreserved spermatozoa in mouse. *J Fertil Implant*, 7, 116
18. Takeshima, T., Nakagata, N. and Ogawa, S. (1991) Cryopreservation of mouse spermatozoa. *Exp Anim*, **40**, 493–497
19. Nakagata, N. (1994) Cryopreservation of embryos and gametes in mice. *Exp Anim*, 43, 11–18
20. Thornton, C.E., Brown, S.D. and Glenister, P.H. (1999) Large numbers of mice established by in vitro fertilization with cryopreserved spermatozoa: implications and applications for genetic resource banks, mutagenesis screens, and mouse backcrosses. *Mamm Genome*, **10**, 987–992
21. Mazur, P., Leibo, S.P. and Seidel, G.E., Jr. (2008) Cryopreservation of the germplasm of animals used in biological and medical research: importance, impact, status, and future directions. *Biol Reprod*, **78**, 2–12
22. Glenister, P.H. and Rall, W.F. (2000) In: Abbott, C., JAckson, I.J (eds.), *Mouse genetics and transgenics, a practical approach*. Oxford University Press, Oxford, pp. Chap. 2
23. Rall, W.F. and Wood, M.J. (1994) High in vitro and in vivo survival of day 3 mouse embryos vitrified or frozen in a non-toxic solution of glycerol and albumin. *J Reprod Fertil*, **101**, 681–688
24. Kasai, M. (1997) Cryopreservation of mammalian embryos. *Mol Biotechnol*, 7, 173–179
25. Renard, J.P. and Babinet, C. (1984) High survival of mouse embryos after rapid freezing and thawing inside plastic straws with 1-2 propanediol as cryoprotectant. *J Exp Zool*, **230**, 443–448
26. Quwailid, M.M., Hugill, A., Dear, N., Vizor, L., Wells, S., Horner, E., Fuller, S., Weedon, J., McMath, H., Woodman, P. et al. (2004) A gene-driven ENU-based approach to generating an allelic series in any gene. *Mamm Genome*, **15**, 585–591
27. Nakagata, N. and Takeshima, T. (1993) Cryopreservation of mouse spermatozoa from inbred and F_1 hybrid strains. *Exp Anim*, **42**, 317

28. Bath, M.L. (2003) Simple and efficient in vitro fertilization with cryopreserved C57BL/6J mouse sperm. *Biol Reprod*, **68**, 19–23
29. Nakagata, N., Okamoto, M., Ueda, O. and Suzuki, H. (1997) Positive effect of partial zona-pellucida dissection on the in vitro fertilizing capacity of cryopreserved C57BL/6J transgenic mouse spermatozoa of low motility. *Biol Reprod*, **57**, 1050–1055
30. Takeo, T., Hoshii, T., Kondo, Y., Toyodome, H., Arima, H., Yamamura, K., Irie, T. and Nakagata, N. (2008) Methyl-beta-cyclodextrin improves fertilizing ability of C57BL/6 mouse sperm after freezing and thawing by facilitating cholesterol efflux from the cells. *Biol Reprod*, **78**, 546–551
31. Ostermeier CG, W.M., Farley, JS, Taft RA (2008) Conserving, distributing and managing genetically modified mouse lines by sperm cryopreservation. *Plos ONE*, **3**, e2792
32. Nakagata, N. (2000) Cryopreservation of mouse spermatozoa. *Mamm Genome*, **11**, 572
33. Wakayama, T. and Yanagimachi, R. (1998) Development of normal mice from oocytes injected with freeze-dried spermatozoa. *Nat Biotechnol*, **16**, 639–641
34. Kaneko, T., Whittingham, D.G., Overstreet, J.W. and Yanagimachi, R. (2003) Tolerance of the mouse sperm nuclei to freeze-drying depends on their disulfide status. *Biol Reprod*, **69**, 1859–1862
35. Ward, M.A., Kaneko, T., Kusakabe, H., Biggers, J.D., Whittingham, D.G. and Yanagimachi, R. (2003) Long-term preservation of mouse spermatozoa after freeze-drying and freezing without cryoprotection. *Biol Reprod*, **69**, 2100–2108
36. Kusakabe, H., Szczygiel, M.A., Whittingham, D.G. and Yanagimachi, R. (2001) Maintenance of genetic integrity in frozen and freeze-dried mouse spermatozoa. *Proc Natl Acad Sci U S A*, **98**, 13501–13506
37. Ogonuki, N., Mochida, K., Miki, H., Inoue, K., Fray, M., Iwaki, T., Moriwaki, K., Obata, Y., Morozumi, K., Yanagimachi, R. et al. (2006) Spermatozoa and spermatids retrieved from frozen reproductive organs or frozen whole bodies of male mice can produce normal offspring. *Proc Natl Acad Sci U S A*, **103**, 13098–13103
38. Kaneko, T. and Nakagata, N. (2006) Improvement in the long-term stability of freeze-dried mouse spermatozoa by adding of a chelating agent. *Cryobiology*, **53**, 279–282
39. Carroll, J., Wood, M.J. and Whittingham, D.G. (1993) Normal fertilization and development of frozen-thawed mouse oocytes: protective action of certain macromolecules. *Biol Reprod*, **48**, 606–612
40. Nakagata, N. (1989) High survival rate of unfertilized mouse oocytes after vitrification. *J Reprod Fertil*, **87**, 479–483
41. Parkes, A.S. (1957) Viability of ovarian tissue after freezing. *Proc R Soc Lond B Biol Sci*, **147**, 520–528
42. Cox, S.L., Shaw, J. and Jenkin, G. (1996) Transplantation of cryopreserved fetal ovarian tissue to adult recipients in mice. *J Reprod Fertil*, **107**, 315–322
43. Nakao, K., Nakagata, N. and Katsuki, M. (1997) Simple and efficient vitrification procedure for cryopreservation of mouse embryos. *Exp Anim*, **46**, 231–234
44. Dinnyes, A., Wallace, G.A. and Rall, W.F. (1995) Effect of genotype on the efficiency of mouse embryo cryopreservation by vitrification or slow freezing methods. *Mol Reprod Dev*, **40**, 429–435
45. Byers, S.L., Payson, S.J. and Taft, R.A. (2006) Performance of ten inbred mouse strains following assisted reproductive technologies (ARTs). *Theriogenology*, **65**, 1716–1726
46. Whittingham, D.G., Lyon, M.F. and Glenister, P.H. (1977) Long-term storage of mouse embryos at–196 degrees C: the effect of background radiation. *Genet Res*, **29**, 171–181
47. Glenister, P.H., Whittingham, D.G. and Lyon, M.F. (1984) Further studies on the effect of radiation during the storage of frozen 8-cell mouse embryos at -196 degrees C. *J Reprod Fertil*, **70**, 229–234
48. Kaneko, T., Yamamura, A., Ide, Y., Ogi, M., Yanagita, T. and Nakagata, N. (2006) Long-term cryopreservation of mouse sperm. *Theriogenology*, **66**, 1098–1101
49. Liebo, S.P., Semple, M.E. and Kroetsch, T.G. (1994) In vitro fertilisation of oocytes by 37-year-old cryopreserved bovine spermatozoa. *Theriogenology*, **42**, 429

Chapter 21

Biological Methods for Archiving and Maintaining Mutant Laboratory Mice. Part II: Recovery and Distribution of Conserved Mutant Strains

Martin D. Fray

Summary

The mouse is now firmly established as the model organism of choice for scientists studying mammalian biology and human disease. Consequently, large collections of novel genetically altered mouse lines have been deposited in secure archives around the world. If these resources are to be of value to the scientific community, they must be easily accessible to all researchers regardless of their embryological skills or geographical location.

This chapter describes how the archiving centres attempt to make the strains they hold visible and accessible to all interested parties, and also outlines the methods currently used in laboratories around the world to recover mouse strains previously archived using the methods highlighted in this manual (*see* Chapter 20).

Key words: Cryopreservation, Archive, Mouse, Spermatozoa, Embryos, IVF, Ovary, Vitrification, Freeze-drying

1. Introduction

1.1. Dissemination of Mouse Models as Frozen Materials

Bio-medical research now functions in an international setting and in keeping with this collaborative environment, there is an ever increasing need to move mouse strains around the world. However, shipping live mice around the world is expensive, inconvenient, and fraught with problems such as incorrect handling, complex customs regulations, and attendant welfare issues.

Elizabeth J. Cartwright (ed.), *Transgenesis Techniques*, Methods in Molecular Biology, vol. 561
DOI 10.1007/978-1-60327-019-9_21, © Humana Press, a part of Springer Science+Business Media, LLC 2009

Animal health status is also of great importance, as most facilities are extremely reluctant to accept live mice that may carry micro-organisms new to their mouse units. Historically, lengthy quarantine procedures coupled with caesarean rederivation have been used to circumvent these problems. However, some pathogens such as mouse hepatitis virus (MHV) can cross the placenta and may not be removed by caesarean rederivation *(1)*. In contrast, embryo transfer has been shown to eliminate most viruses including Sendai virus, mouse rotavirus, MHV, reovirus3, Theiler's encephalomyelitis virus, mouse norovirus, and mouse adenovirus, provided careful washing of the embryos in a sterile medium is carried out before embryo transfer *(2–5)*. In recognition of these difficulties, and the imperative to attain the highest welfare standards, every effort should be made to transport mouse strains as germplasm, when it is practical to do so.

The introduction of liquid nitrogen dry-shipping containers has made the dissemination of frozen embryos and gametes a simple and safe procedure (*see* **Note 1**). Dry shippers absorb liquid nitrogen into a heavily insulated lining so that the contents can remain at liquid nitrogen temperature for approximately 2 weeks. This should provide sufficient cooling capacity to protect the samples during transit even if the journey time is unexpectedly extended. Because dry-shippers do not actually contain free flowing liquid nitrogen they are not classed as "dangerous goods" by the International Air Transport Association (IATA), which minimises the cost and bureaucracy associated with the shipping procedure.

1.2. Mouse Mutant Resources

Researchers who do not wish to archive/distribute their own lines may submit them to one of the public archiving centres around the world. These archiving centres offer many advantages over do-it-yourself archiving and their services are often free. The paper by Rockwood et al. *(6)* introduces many of the benefits of using these public resource centres.

In 2006, the principle mouse repository and resource centres established a formal collaborative initiative called the Federation of International Mouse Resources (FIMRe). This federation was set up in recognition of the inescapable fact that no single archiving centre could cryopreserve/distribute the huge number of new mouse models generated by the systematic mutagenesis programmes, which have been established around the world. FIMRe's collective goal is to archive and provide strains of mice as cryopreserved embryos and gametes, as well as ES cell lines and live breeding stock to the research community.

The FIMRe website (http://www.fimbre.org) has links to all the major archiving centres around the world. Through these links the scientific community can review the publicly accessible databases maintained by each archiving centre. These databases list all the strains that can be made available to the scientific community. In addition, many laboratories also submit their strains to the International Mutant Strain Resource (IMSR) website (http://www.informatics.jax.org/imsr/index.jsp). The IMSR is building a "one stop genetic resource" for scientists looking for a particular mouse line.

On a less formal basis several countries have established mouse locator networks, such as the Mouse Locator in the UK *(7)* and Kia7souris in France *(8)*. These networks aim to establish a dialogue between scientists wishing to locate a strain of interest and those institutions that may hold the material, but as yet have not made it freely available to the scientific community.

To encourage the deposition of mouse strains into public archives, the receiving centres usually offer a free cryopreservation service providing the strain can be made available to the scientific community. The deposition of new mutant lines may also be accompanied by a period of restricted public access, if details of the mutation have not been published. This enables researchers to archive their mouse lines without prematurely releasing them to the public. Binding material transfer agreement can also be drawn up to protect the originators beneficial rights.

In the UK, "free" public archiving services are offered by the Frozen Embryo and Sperm Archive (FESA) which is sited at the MRC-Harwell (http://www.har.mrc.ac.uk)

1.3. In Vitro Fertilisation

In vitro fertilisation is an extremely versatile technique and the potential to produce large numbers of embryos by IVF is being exploited in many ways. For example IVF techniques can be used for the rapid production of substantial numbers of backcross progeny, to rescue lines that fail to breed naturally and to produce large age matched cohorts for studying age related phenotypes, i.e. late onset deafness. The generation of large cohorts of IVF derived embryos is also being exploited in order to accelerate the cryopreservation of mouse strains. So called cryo-IVF, where IVF derived 2-cell embryos are frozen, can be used successfully to bank novel mouse mutations on most common backgrounds *(9, 10)*. Cryo-IVF not only accelerates the archiving process but also reduces the number of superovulated females required to freeze down a strain and represents a significant refinement in the archiving procedure. Although the simultaneous generation of the large batches of embryos offers many benefits *(11, 12)* it is important not to lose sight of the technical effort required to set up these large IVF and embryo transfer sessions.

2. Materials

2.1. Thawing Embryos Frozen Using the Slow Equilibration Method

1. 35-mm culture dishes (Falcon).
2. 60-mm culture dishes (Falcon).
3. Metal rod (16 g) and scissors.
4. Small dewar suitable for holding liquid nitrogen.
5. Liquid nitrogen.
6. M2 medium (Sigma).
7. Timer.

2.2. Thawing Vitrified Embryos

1. Small dewar suitable for holding liquid nitrogen.
2. Liquid nitrogen.
3. Analar grade sucrose (Merck).
4. HFT medium *(13)*.
5. Timer.
6. PB1 medium containing 0.25 M sucrose.
7. PB1 medium: per 100 mL – 800 mg NaCl, 13.2 mg $CaCl_2$, 20 mg KH_2PO_4, 20 mg KCl; 10 mg $MgCl_2$, 289.8 mg Na_2HPO_4, 100 mg glucose, 100 U Penicillin G, 3.6 mg Na–pyruvate.

2.3. Thawing Frozen Sperm Using the Vapour Phase Method

1. Water bath set to 37°C.
2. Small dewar suitable for holding liquid nitrogen.
3. Liquid nitrogen.
4. In vitro fertilisation dishes (*see* **Subheading 2.7**).
5. Timer.
6. IVF culture media e.g. Cook's Media (Cook Australia; K-RVFE-50).

2.4. Recovering Freeze Dried Sperm and Microinsemination

1. Distilled water.
2. Hepes-CZB medium.
3. Polyvinyl pyrrolidone (PVP-360).
4. Hepes-CZB medium, plus 12% Polyvinyl pyrrolidone.

2.5. Thawing Cryopreserved Oocytes

1. Water bath set to 30°C.
2. Heated block set to 37°C.
3. Small dewar suitable for holding liquid nitrogen.
4. Liquid nitrogen.
5. Metal rod (16 g) and scissors.
6. M2 medium (Sigma).

7. Foetal calf serum (ICN flow).
8. Timer.
9. M2 medium plus 10% foetal calf serum. The pH of this solution needs to be adjusted to pH 7.2–7.4 with 10 μL 2 M HCl/10 mL medium.

2.6. Thawing Cryopreserved Ovarian Tissue

1. Water bath set to 30°C.
2. Heated block set to 37°C.
3. Small dewar suitable for holding liquid nitrogen.
4. Liquid nitrogen.
5. M2 medium (Sigma).
6. Foetal calf serum (ICN flow).
7. Timer.
8. M2 medium, plus 10% foetal calf serum. The pH of this solution needs to be adjusted to pH 7.2–7.4 with 10 μL 2 M HCl/10 mL medium.

2.7. In Vitro Fertilisation

1. Dissecting microscope and instruments.
2. 37°C incubator gassed with 5% CO_2 in air.
3. 35-mm dishes for oviduct collection (Falcon).
4. 60-mm dishes for IVF and embryo culture (Falcon).
5. Embryo tested mineral oil (Sigma).
6. IVF culture media, e.g. Cook's Media (Cook Australia; K-RVFE-50).

3. Methods

The thaw/reconstitution protocols in this chapter compliment procedures described in Part I – Conserving mouse strains, of this manual and should be read in conjunction with guidance (*see* **Notes 1–4**).

3.1. Procedure for Thawing Embryos Frozen Using the Slow Equilibration Method

This thawing procedure is on the basis of the freezing method first described by Renard and Babinet *(14)* and modified by Glenister and Rall *(15)*.

1. Transfer the selected straws to a bench top liquid nitrogen dewar.
2. Using forceps, hold the straw near the label (opposite end to the embryo fraction) for 30 s in air and then place in a water bath at room temperature until the contents of the straw thaw completely (*see* **Note 2**).

3. Wipe the straw with a clean paper tissue.
4. Cut off the capillary sealant and also cut through the cotton/PVA seal leaving about half the seal in place to act as a plunger.
5. Using a metal rod, expel the entire liquid contents of the straw into a 35-mm Falcon dish. Don't touch the drop of fluid with the end of the straw, as embryos may adhere to it.
6. Wait 5 min, during which the embryos will shrink considerably.
7. Transfer the embryos to a drop of M2 medium. They will rapidly take up water and assume normal appearance.
8. After 5 min, wash the embryos through a second drop of M2 medium.
9. Embryos from the 2-cell, to the blastocyst stage of development can be transferred into the oviducts of 0.5 day pseudopregnant recipient. However, blastocysts may be transferred into the uterus of 2.5 day pseudopregnant recipient. If necessary culture the embryos in M16 or KSOM before conducting the embryo transfers.
10. Follow the surgical embryo transfer procedures described by Nagy et al. *(16)*.

3.2. Procedure for Thawing Embryos Frozen by Vitrification

This thawing procedure is on the basis of the simple vitrification method described by Nakao et al. *(17)*.

1. Transfer the selected vial to a bench top liquid nitrogen dewar.
2. After removing the sample from the liquid nitrogen, open the cap of the cryo-vial (*see* **Note 3**) and let it stand at room temperature for 30 s.
3. Put 0.9 mL of PB1 containing 0.25 M sucrose into the cryo-vial and thaw the sample quickly by gently pipetting the media up and down.
4. Pour the contents of the freezing tube into a culture dish.
5. Recover the embryos, and transfer them in a drop of HTF medium.
6. After 5 min, take the embryos through two washes of fresh HTF medium. In the case of blastocysts, wash the embryos after 60 min.
7. Follow the surgical embryo transfer procedures described by Nagy et al. *(16)*.

3.3. Procedure for Thawing Sperm Frozen Using the Vapour Phase Method

This method describes the procedure followed in numerous cryopreservation laboratories and is on the basis of the work originally published by Nakagata et al. *(18)* and modified by Glenister and Rall *(15)*.

1. Using forceps, hold the cryotube in the air for 30 s, and then thaw rapidly by placing in a 37°C water bath.
2. Take special care that the tube has not filled with liquid nitrogen before plunging into the water bath (such tubes may explode –*see* **Note 3**). If liquid nitrogen is present in the tube, loosen the lid and wait for it to evaporate.
3. Gently agitate the cryotube to mix the sperm. Using a wide bore pipette tip, take 5–10 μL aliquots and pipette gently into 500 μL fertilisation drops (*see* **Subheading 3.7** – IVF protocol). There should be enough sperm for 9× fertilisation drops.

3.4. Procedure for the Re-hydration of Freeze Dried Sperm Spermatozoa and Micro-insemination

This re-hydration procedure is on the basis of the method described by Kaneko et al. *(19)*. In experienced hands more than 70% of the oocytes will survive sperm injection and more than 90% of surviving oocytes should fertilise normally.

Although freezing sperm in the absence of conventional cryoprotective agents, combined with micro-insemination appears to be an efficient means of archiving mouse lines, there is some evidence that the recipient ooctyes may only partially repair DNA fragmentation in the spermatozoa and that these effects may only manifest themselves as the mice age *(20)*.

1. Open an ampoule and rehydrate spermatozoa by adding 100 μL of sterile distilled water.
2. Collect a small volume of the sperm suspension and mix 40–50 μL of Hepes-CZB medium containing 12% PVP.
3. Collect intact-looking spermatozoa with heads still attached to the tails.
4. Draw a single spermatozoon, tail first, into the injection pipette in such a way that its neck (the junction between the head and tail) is at the opening of the pipette.
5. Separate the head from the tail by applying a few piezo-pluses to the neck region.
6. Inject sperm head only into each oocyte.
7. Culture injected oocytes in CZB medium in 5% CO_2 and 95% air at 37°C and allow them to develop to the 2-cell stage.
8. Follow the surgical embryo transfer procedures described by *(16)*.

3.5. Procedure for Thawing Cryopreserved Ooctyes

This thawing procedure is on the basis of the method by Carroll et al. *(21)*.

1. Transfer the selected straw from the storage vessel to a small dewar filled with liquid nitrogen.
2. Using a pair of forceps pick the straw up at the labelled end, i.e. furthest away from the oocytes and hold it in air, for 40 s.

Then place the straw in a water bath set at 30°C until the straw's contents thaw completely.

3. Dry the outside of the straw with a clean paper tissue.
4. Cut off the capillary sealant and also cut through the PVA seal leaving about half of the seal in place to act as a plunger. Then carefully expel its contents by pushing the seal down the straw with a metal rod into a 35-mm dish.
5. Incubate the oocytes at room temperature for 5 min, then add 0.8 mL M2 medium, plus FCS to the dish and gently mix.
6. Incubate the oocytes for a further 10 min at room temperature.
7. Transfer the oocytes to a fresh dish of M2 medium, plus FCS and incubate at room temperature for a further 10 min.
8. Wash the oocytes in a second dish of M2 medium, plus FCS and incubate in this medium at 37°C until ready to carry out the IVF. This incubation period should not last longer than 30 min.
9. The oocytes should be washed through the appropriate IVF medium before being transferred to the fertilisation drops.

3.6. Procedure for Thawing Cryopreserved Ovarian Tissue

This thawing procedure is on the basis of the method published by Sztein et al. *(22)*.

1. Transfer the frozen cryotubes to a small dewar containing liquid nitrogen. Make sure no liquid nitrogen is trapped in the cryotube (such tubes may explode–*see* **Note 3**). If liquid nitrogen is present in the tube, loosen the lid and wait for it to evaporate.
2. Using pre-cooled forceps, hold the cryotube in air at room temperature for 40 s and then place the vial in water at 30°C until the contents completely thaw.
3. Using fine forceps remove the ovaries from the cryotube. Place into 1–2 mL of M2 medium, plus FCS, at room temperature.
4. Gently swirl the dish to mix the solutions. Wait 5 min at room temperature.
5. Transfer the ovaries to a second dish of M2 medium, plus FCS at room temperature, and leave for 5 min.
6. Transplant to a suitable recipient female as soon as possible after thawing. Follow the surgical ovarian transfer procedures described by Nagy et al. *(16)*.

3.7. Procedure for In Vitro Fertilisation

Several robust protocols have been published for performing IVF with mouse gametes. Although the type of culture medium (MEM, HTF, T6) used may vary between these different methods, the basic principles remain constant. The IVF system employed at the MRC-Harwell is on the basis of the method published by Sztein et al. *(23)* and is as follows:

3.7.1. Preparation of IVF Dishes

1. On the day before the oocytes are to be harvested set up the IVF dishes, as described below.
2. Prepare 1 × 35 mm oviduct collection dish for every three females of the oocyte donor stock. Each dish should contain 2 ml of Cook's medium. Incubate the dishes overnight at 37°C, in 5% CO_2 in air.
3. Prepare 1 × 60 mm dish/three females of the oocyte donor stock. Each dish should contain 1 × 500 μL drop of Cook's medium for the fertilisation step, plus 4 × 150 ul drops of Cook's medium for the culturing/washing steps (*see* **Fig. 1**). Overlay the drops with mineral oil and incubate overnight at 37°C, in 5% CO_2 in air.
4. If freshly harvested sperm is to be used, prepare a sperm dispersal dish by placing a 500 μL drop of Cook's medium into the centre of a 60-mm culture dish. Overlay with mineral oil and equilibrate overnight at 37°C, in 5% CO2 in air (this step can be omitted when using frozen/thawed sperm).

3.7.2. Sperm Sample Preparation: Freshly Harvested Sperm

1. The male should be at least 8 weeks old, and not have been mated within three days of sperm collection.
2. Sacrifice the male and dissect out the cauda epididymides.
3. Place the cauda epididymides in the oil surrounding the sperm dispersal drop.
4. Nick the apex of each cauda epididymus with a pair of fine (iris) scissors and tease out the bleb of sperm that emerges from the incision. Using a pair of fine forceps, drag the bleb of sperm into the dispersal drop.

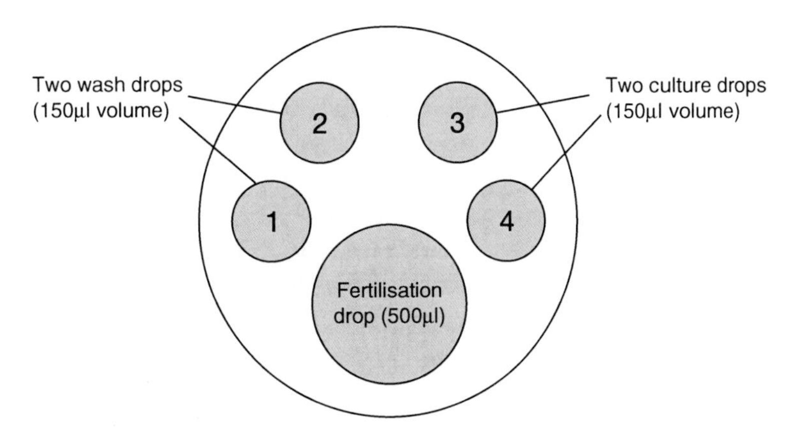

Fig. 1. Illustrates the organisation of an IVF dish after the culture media has been dispensed.

5. Place the dispersal drop inside a CO_2 incubator at 37°C for approximately 60 min to allow the sperm to disperse.
6. After the sperm have been dispersed, pipette 5–10 μL sperm into each fertilisation drop using a wide-bore tip and return the dishes to the incubator.

3.7.3. Sperm Sample Preparation: Cryopreserved Sperm

Thaw the sperm sample in accordance with the relevant freezing/thawing protocol (*see* **Note 4**) and add the recommended volume of sperm to each fertilisation drop. However, if the sperm has been frozen down using the method outlined in this manual, proceed as described in **Subheading 3.3**, of this chapter.

3.7.4. Harvesting Oocytes

1. Dissect the oviducts from superovulated females ~14 h after their hCG injection and place them into a dish of Cook's medium (no more than six oviducts/dish).
2. Under a dissecting microscope and using fine forceps, gently release the cumulus masses into the Cook's medium.
3. Gently draw all the cumulus masses up into a wide bore pipette tip in the minimum amount of Cook's medium (<30 μL in total) and transfer the cumulus masses to one of the fertilisation drops containing spermatozoa.
4. Repeat **Steps 1-3** for each fertilisation dish in succession so that the minimum time elapses between killing the oocyte donors and returning the fertilisation dishes to the incubator.
5. Incubate the dish at 37°C, in 5% CO_2 in air for approximately 5 h to allow fertilisation to occur.

3.7.5. Washing and Culturing the Fertilised Oocytes

1. After incubating the fertilisation drop for ~5 h remove all the presumptive zygotes from the fertilisation drop and place them in wash drop 1 (*see* **Fig. 1**) Then move the oocytes to wash drop 2. During this process transfer as little cell debris as possible.
2. Divide the washed oocytes approximately equally between the culture drops 3 and 4.
3. Incubate overnight at 37°C, in 5% CO_2 in air and then harvest the 2-cell embryos.
4. Either transfer the 2-cell embryos to the oviducts of 0.5 days pseudopregnant foster mothers, or prepare the 2-cell embryos for cryopreservation according to a standard protocol, or transfer the embryos into KSOM for prolonged in vitro culture.
5. Follow the surgical embryo transfer procedures described by Nagy et al. *(16)*.

4. Notes

1. Samples are often irrevocably damaged through the inappropriate handling of vials/straws when they are being removed from dry shippers or secure archives. Always ensure frozen samples stay below the glass transition temperature of −150°C. Below this temperature water molecules are retained in a solid state and all biological reactions effectively cease. It is vitally important that frozen materials are exposed to ambient temperature for the absolute minimum amount of time. Particularly when samples are being transferred between the storage dewars, whether vapour or liquid phase, to avoid premature warming of the sample. When it is time to thaw a sample, the best way to safe guard the material is to transfer the samples from their storage location to a temporary/portable holding vessel where they can be immersed in liquid nitrogen until they are thawed.
2. Liquid nitrogen will enter poorly sealed vials/straws. If this occurs there is a risk that the vials/straws will explode as the liquid nitrogen rapidly converts to its gaseous state. In order to prevent the destruction of valuable samples and possible injury, it is important to ensure that the vials/straws are properly sealed before freezing. Plastic semen straws should be hermetically sealed at both ends.
3. When screw topped cryotubes are tightened up properly before they are placed in the freezing apparatus, they do not present a problem, particularly if they are stored on the vapour phase. However, on occasions liquid nitrogen does enter cryotubes, a situation that is easily identified, because fluid filled cryotubes sink, when placed in a liquid nitrogen holding vessel. Fortunately, the contents of these vials can still be recovered, if the liquid nitrogen is allowed to escape by loosening the top of the vial and placing it on the bench top before the designated thawing protocol is started.
4. It is important to realise that a huge number of cryopreservation protocols have been developed over the years and subtle differences in the freeze protocol need to be complimented by specific thawing procedures. Do not rely on standard in-house procedures and always obtain the specific details relevant to the material you are handing.

References

1. Harkness, J.E. and Wagner, J.E. (1983) *The Biology and Medicine of Rabbits and Rodents*, 3rd edition. Lea and Febiger, London
2. Morrell, J.M. (1999) Techniques of embryo transfer and facility decontamination used to improve the health and welfare of transgenic mice. *Lab Anim*, **33**, 201
3. Carthew, P., Wood, M.J., and Kirby, C. (1983) Elimination of Sendai (parainfluenza type 1) virus infection from mice by embryo transfer. *J Reprod Fertil*, **69**, 253–257
4. Carthew, P., Wood, M.J., and Kirby, C. (1985) Pathogenicity of mouse hepatitis virus for preimplantation mouse embryos. *J Reprod Fertil*, **73**, 207–213
5. Henderson, K.S. (2008) Murine norovirus, a recently discovered and highly prevalent viral agent of mice. *Lab Anim*, **37**, 314–320
6. Rockwood, S.F., Fray, M.D., and Nakagata, N. (2006) Managing success: mutant mouse repositories. In: J.P. Sundberg and T. Ichiki (Eds.), *Genetically Engineered Mice Handbook*, CRC Press, Boca Raton, FL, pp. 27–37
7. Bugeon, L. and Rosewell, I. (2003) "Mouse locator-UK": a networking tool for academic transgenic research in the UK. *Transgenic Res*, **12**, 637
8. Holzenberger, M., Tronche, F., Bugeon, L., and Larue, L. (2005) A French academic network for sharing transgenic materials and knowledge. *Transgenic Res*, **14**, 801–802
9. Suzuki, O., Asano, T., Yamamoto, Y., Takano, K., and Koura, M. (1996) Development in vitro of preimplantation embryos from 55 mouse strains. *Reprod Fertil Dev*, **8**, 975–980
10. Byers, S.L., Payson, S.J., and Taft, R.A. (2006) Performance of ten inbred mouse strains following assisted reproductive technologies (ARTs). *Theriogenology*, **65**, 1716–1726
11. Thornton, C.E., Brown, S.D., and Glenister, P.H. (1999) Large numbers of mice established by in vitro fertilization with cryopreserved spermatozoa: implications and applications for genetic resource banks, mutagenesis screens, and mouse backcrosses. *Mamm Genome*, **10**, 987–992
12. Marschall, S. and Hrabe de Angelis, M. (1999) Cryopreservation of mouse spermatozoa: double your mouse space. *Trends Genet*, **15**, 128–131
13. Quinn, P., Kerin, J.F., and Warnes, G.M. (1985) Improved pregnancy rate in human in vitro fertilization with the use of a medium based on the composition of human tubal fluid. *Fertil Steril*, **44**, 493–498
14. Renard, J.P. and Babinet, C. (1984) High survival of mouse embryos after rapid freezing and thawing inside plastic straws with 1-2 propanediol as cryoprotectant. *J Exp Zool*, **230**, 443–448
15. Glenister, P.H. and Rall, W.F. (2000) In: Abbott, C, Jackson, I.J. (ed.), Mouse Genetics and Transgenics: A Practical Approach. Oxford University Press, Oxford, pp. Chap. 2
16. Nagy, A.G.M., Vintersten, K., and Behringer, R. (2003) *Manipulating the Mouse Embryo: A Laboratory Manual*, 3rd edition. Cold Spring Harbor Laboratory Press, Cold Spring Harbor, NY
17. Nakao, K., Nakagata, N., and Katsuki, M. (1997) Simple and efficient vitrification procedure for cryopreservation of mouse embryos. *Exp Anim*, **46**, 231–234
18. Nakagata, N. (1994) Cryopreservation of embryos and gametes in mice. *Exp Anim*, **43**, 11–18
19. Kaneko, T., Whittingham, D.G., Overstreet, J.W., and Yanagimachi, R. (2003) Tolerance of the mouse sperm nuclei to freeze-drying depends on their disulfide status. *Biol Reprod*, **69**, 1859–1862
20. Fernandez-Gonzalez, R., Moreira, P.N., Perez-Crespo, M., Sanchez-Martin, M., Ramirez, M.A., Pericuesta, E., Bilbao, A., Bermejo-Alvarez, P., de Dios Hourcade, J., de Fonseca, F.R. et al. (2008) Long-term effects of mouse intracytoplasmic sperm injection with DNA-fragmented sperm on health and behavior of adult offspring. *Biol Reprod*, **78**, 761–772
21. Carroll, J., Wood, M.J., and Whittingham, D.G. (1993) Normal fertilization and development of frozen-thawed mouse oocytes: protective action of certain macromolecules. *Biol Reprod*, **48**, 606–612
22. Sztein, J., Sweet, H., Farley, J., and Mobraaten, L. (1998) Cryopreservation and orthotopic transplantation of mouse ovaries: new approach in gamete banking. *Biol Reprod*, **58**, 1071–1074
23. Sztein, J.M., Farley, J.S., and Mobraaten, L.E. (2000) In vitro fertilization with cryopreserved inbred mouse sperm. *Biol Reprod*, **63**, 1774–1780

Index